Energy-Efficient Communication Processors

Robert Fasthuber · Francky Catthoor
Praveen Raghavan · Frederik Naessens

Energy-Efficient Communication Processors

Design and Implementation for Emerging Wireless Systems

Springer

Robert Fasthuber
IMEC
Leuven
Belgium

Praveen Raghavan
IMEC
Leuven
Belgium

Francky Catthoor
IMEC
Heverlee
Belgium

Frederik Naessens
IMEC
Leuven
Belgium

ISBN 978-1-4614-4991-1 ISBN 978-1-4614-4992-8 (eBook)
DOI 10.1007/978-1-4614-4992-8
Springer New York Heidelberg Dordrecht London

Library of Congress Control Number: 2013938037

Printed on acid-free paper

Springer is part of Springer Science+Business Media (www.springer.com)

Preface

With great advances in technology over the last decades, the computational performance and energy efficiency of mobile devices has been significantly improved. This improvement has led to modern mobile devices, such as smartphones or tablets, which offer a multitude of interesting applications. For this reason, they have become an essential part of our lifestyle and can be found everywhere around us. It is expected that in the future even hundreds of mobile devices will be in personal use. All of these devices will have to communicate with each other, mostly by using a huge variety of different wireless communication standards.

To support this huge variety of standards in a cost-effective way, i.e. to avoid the use of many underutilized hardware resources, (1) high programming flexibility is required. Because of the joint demand for (2) high performance and long battery lifetime, these devices will have to be (3) extremely energy efficient. To achieve high performance and high energy efficiency, future designs need to be able to exploit the full potential offered by the latest process technologies, i.e. they need to be (4) technology scaling-friendly. Since the Non-Recurring Engineering (NRE) costs for products that leverage on future process technologies will become dramatically high (5) high reusability of designs and design effort across a large number of applications will be essential. Considering these joint requirements, it will become very challenging to continue the trend towards ever-more computationally powerful mobile devices. The purpose of this book is to contribute to solving this challenge.

A review of the state-of-the-art literature clearly shows the drawbacks of existing solutions and motivates the need for a new holistic design approach, which can cope with the joint set of requirements in a more optimal way. This book proposes such a design approach and demonstrates the feasibility on three case studies.

The core of this book is the energy-efficient technology-friendly Domain-Specific Instruction set Processor (DSIP) architecture template, which enables high reusability. This architecture template targets specifically the baseband functionality of emerging high-performance wireless communication systems, which is very difficult to implement under the given design constraints. To achieve high energy efficiency, innovative architecture concepts, such as software Single Instruction Multiple Data (SIMD), Distributed Loop Buffer (DLB) and Very-Wide Register

(VWR), have been combined together with consistent co-design flows. Technology friendliness, i.e. the capability of a design to significantly profit from technology scaling, is ensured by proactively coping with negative side effects of future Deep–Deep Sub-Micron (DDSM) technologies. In this book we focus mainly on the increasing influence of wires over transistors. We handle this issue by keeping explicitly the most important wires (most active, critical path) in the architecture template short and by proposing a compatible back-end semi-custom design flow which will lead to the desired layout.

This book includes three relevant case studies which demonstrate the application and feasibility of the proposed design approach. All three DSIP architecture template instances, i.e. the advanced Multiple-Input Multiple-Output (MIMO) detector for future LTE/WLAN standards, the high-speed Finite-Impulse Response (FIR) filter for emerging 60 GHz systems and the high-throughput Fast-Fourier Transformation (FFT) for 60 GHz and WLAN systems, have been designed and implemented in TSMC 40 nm technology. Although sufficiently fulfilling also of all other requirements, these designs are at least a factor of 2–3 more energy/area efficient than state-of-the-art programmable solutions. Thus, this result motivates the content of this book.

We expect this book to be of interest for academia in both ways, for describing the overall design approach of efficient architecture implementations, and for describing the proposed innovative architecture concepts in more detail. The goal of all projects, which have driven this research, was to obtain results that are relevant for industry. Since this book is reflecting this, we believe that the content is also of interest for senior architecture design engineers and for their managers in industry. This is specifically for those who want to make use of the proposed concepts in their own research and development or for those who wish to anticipate the evolution of commercially available design concepts over the next few years.

The material of this book is based on research that has been carried out at IMEC in the period of 2005–2012, partly in the context of European and national research projects. It has been a pleasure for us to work in this research domain and to co-operate with our project partners and with our colleagues from the analog and digital SSET-CSI group.

We would like to use this opportunity to thank all the people who have provided contributions and feedback in the direct focus of this book, both at IMEC and at other locations. In particular we want to mention David Novo, Min Li, Wim Van Thillo, Sofie Pollin, Ubaid Ahmad, Prashant Agrawal, Halil Kukner, Matthias Hartmann, André Bourdoux, Claude Desset, Hans Cappelle, Peter Debacker, Raf Appeltans, Tom Vander Aa, Veerle Derudder, Liesbet Van der Perre, Antoine Dejonghe, Bruno Bougard, Jos Huisken, Wim Dehaene and Paolo Ienne. Furthermore, we want to thank all the Master's students who have helped us in this research, in particular Alaa Medra, Imran Ali, Kostas Samaras and Vagelis Bebelis.

Finally, we hope that the reader will find this book useful and enjoyable and that the proposed concepts and results will contribute to the continued progress in this field.

Leuven, Belgium

Robert Fasthuber
Francky Catthoor
Praveen Raghavan
Frederik Naessens

Contents

1 Introduction and Motivation 1
1.1 The Smartphone of the Future. 1
1.1.1 Trends and Consequences 1
1.1.2 The Solution: A Highly Energy-Efficient SDR Platform . 6
1.2 Research Challenges to Enable Highly Energy-Efficient SDR Platforms 6
1.2.1 The Energy-Efficiency Gap 6
1.2.2 The Architecture Gap for Ultimately Scaled Technologies 7
1.2.3 The Productivity Gap 9
1.2.4 The Culture Gap for Design Paradigms 10
1.3 Key Concepts to Tackle Research Challenges and Related Gaps in the State of the Art 11
1.3.1 Employ an Architecture With a Well-Chosen Flexibility 11
1.3.2 Leverage on an Energy and Cost-Effective Architecture Template 12
1.3.3 Consider DDSM Technology Constraints During the Design of the Architecture 13
1.3.4 Employ Consistent, Predictable and Systematic Design Flows Around the Architecture Template 13
1.3.5 Ensure that Algorithm and Architecture are Well-Matched 14
1.3.6 Effectively Adapt to the Actual Requirements at Run-Time. 14
1.4 Proposed DSIP Architecture Template Design Approach 16
1.5 Main Focus and Contributions 17
1.5.1 Main Focus of this Book 17
1.5.2 Overview of the Main and Side Contributions 17
1.6 Structure of this Book 18
References 20

2 Background and Related Work 25
2.1 Background on Wireless Communication Systems. 25
2.1.1 The General Digital Wireless Communication System. . . 25
2.1.2 The Increasing Complexity of Wireless Communication Systems 26
2.1.3 Physical Layer Signal Processing in an Advanced Receiver 27
2.2 Background on Architecture Styles for Wireless Communication Systems 28
2.2.1 ASIC/rASIC. 29
2.2.2 ASIP (Application Processor). 30
2.2.3 ASIP (Baseband Processor) 31
2.2.4 DSIP. 32
2.2.5 DSP 32
2.2.6 Other Styles 33
2.3 Background on the Physical Layer System Design 33
2.3.1 Functionality Design 33
2.3.2 Algorithm Design 34
2.3.3 Partitioning 35
2.3.4 Architecture Design. 35
2.3.5 Software Mapping and Compilation 36
2.3.6 Hardware Implementation 36
2.4 Related Work on Functionality/Algorithm Design and Partitioning. 37
2.4.1 Functionality Design 37
2.4.2 Algorithm Design and Optimization 37
2.4.3 Partitioning 39
2.5 Related Work on Architecture Design 40
2.5.1 Design of the Platform 40
2.5.2 Design of an ASIC 40
2.5.3 Design of an Overall Processor Platform 41
2.5.4 Design of the Processing Elements of a Processor 43
2.5.5 Design of the Data Storage Hierarchy of a Processor . . . 44
2.5.6 Design of the Instruction C./S. Hierarchy of a Processor. 45
2.6 Related Work on Software Mapping/Compilation 46
2.6.1 Software Mapping. 46
2.6.2 Compilation 46
2.7 Related Work on Hardware Implementation 47
2.7.1 Implementation of the Platform 47
2.7.2 Implementation of a Block. 47
2.8 Related Work on Wireless Architectures and Templates. 51
2.8.1 ASIC/rASIC. 51
2.8.2 ASIP (Application Processor). 52

2.8.3 ASIP (Baseband Processor) 53
2.8.4 DSIP 54
2.8.5 DSP 56
2.9 Summary and Conclusions 56
References 57

3 The Proposed DSIP Architecture Template for the Wireless Communication Domain 69
3.1 An Effective Architecture Template for the Wireless Communication Domain 69
3.1.1 Considered Domain 69
3.1.2 Proposed Design Approach 70
3.2 Applied Design Approach to Define Architecture Template 73
3.2.1 Analysis and Definition of System and Algorithm Requirements 73
3.2.2 Analysis and Definition of Technology Constraints 75
3.2.3 Evaluation and Selection of Architectural Concepts 78
3.2.4 Definition of the Architecture Template 79
3.3 Requirements from Algorithm Perspective 79
3.3.1 Data Representation 80
3.3.2 Arithmetic and Logic Operations 80
3.3.3 Parallelization 82
3.3.4 Data Transfer Operations 83
3.3.5 Data Storage 84
3.3.6 Instruction Control 85
3.3.7 Summary 86
3.4 Employed Architectural Concepts 86
3.4.1 Data Representation 87
3.4.2 Arithmetic and Logic Operations 89
3.4.3 Parallelization 92
3.4.4 Data Transfer Operations 93
3.4.5 Data Storage 94
3.4.6 Instruction Control 95
3.5 Proposed Architecture Template 97
3.5.1 Top Level 97
3.5.2 Cluster Level 98
3.5.3 Engine Level 101
3.5.4 Slice Level 102
3.6 Scalability of the Architecture Template 107
3.6.1 Technology Scalability 107
3.6.2 Hardware/Instance Scalability: Design Space and Architecture Instantiation Design Flow 109
3.6.3 Hardware/Instance Scalability: Model to Define Flexibility in a Quantitative Manner 111

3.6.4 Hardware/Instance Scalability: Flexibility Evaluation of the Proposed Architecture Template . . . 112
3.6.5 Software/Run-Time Scalability . . . 121
3.7 Summary of Combined Innovative Concepts. . . . 121
3.7.1 Main Template-Specific Concepts. . . . 122
3.8 Conclusions . . . 123
References . . . 132

4 Case Study 1: DSIP Architecture Instance for MIMO Detection . . . 137
4.1 Motivation, Related Work and Contributions . . . 137
4.1.1 Context and Motivation . . . 137
4.1.2 Summary of Related Work. . . . 137
4.1.3 Main Contributions . . . 138
4.2 Background on Driver . . . 139
4.2.1 MIMO Detection . . . 139
4.2.2 Motivation for a Flexible Implementation . . . 140
4.3 Algorithm Optimizations and Characteristics . . . 141
4.3.1 Algorithm Choice and Applied Optimizations . . . 141
4.3.2 Algorithm Characteristics. . . . 144
4.4 Proposed DSIP Architecture Instance. . . . 145
4.4.1 Top Level . . . 145
4.4.2 Cluster Level . . . 146
4.4.3 Engine Level . . . 149
4.4.4 Slice Level. . . . 149
4.5 Software Mapping and Hardware Implementation . . . 155
4.5.1 Software Mapping and Scheduling . . . 155
4.5.2 Hardware Implementation and Results. . . . 156
4.5.3 Instance Scaling . . . 157
4.5.4 Run-Time Scaling . . . 157
4.6 Comparison and Discussion . . . 159
4.6.1 Implemented ASIC References . . . 159
4.6.2 Comparison to ASIC References. . . . 159
4.6.3 Flexible Implementations from Literature. . . . 163
4.6.4 Comparison to Flexible Implementations . . . 165
4.7 Conclusions . . . 166
References . . . 167

5 Case Study 2: DSIP Architecture Instances for FIR Filtering . . . 171
5.1 Motivation, Related Work and Contributions . . . 171
5.1.1 Context and Motivation . . . 171
5.1.2 Summary of Related Work. . . . 172
5.1.3 Main Contributions . . . 173
5.2 Background on Driver . . . 173
5.2.1 Matched Filter for the 60 GHz System . . . 174
5.2.2 Motivation for a Flexible Implementation . . . 174

5.3 Algorithm Optimizations and Characteristics 175
5.3.1 Algorithm Choice and Applied Optimizations 175
5.3.2 Algorithm Characteristics 179
5.4 Proposed DSIP Architecture Instances 180
5.4.1 Top Level 181
5.4.2 Cluster Level 183
5.4.3 Engine Level 184
5.4.4 Slice Level of the HW-SIMDi 185
5.4.5 Slice Level of the SW-SIMDi 187
5.5 Software Mapping and Hardware Implementation 194
5.5.1 Software Mapping and Scheduling 194
5.5.2 Hardware Implementation and Results 197
5.5.3 Throughput and Scalability 200
5.6 Comparison and Discussion 202
5.6.1 Implemented ASIC References 202
5.6.2 Processor References from Literature 203
5.6.3 Comparison (Normalized, Pessimistic) 203
5.6.4 Discussion 205
5.7 Conclusions 206
References 215

6 Case Study 3: DSIP Architecture Instance for FFT Computation 219
6.1 Motivation, Related Work and Contributions 219
6.1.1 Context and Motivation 219
6.1.2 Summary of Related Work 220
6.1.3 Main Contributions 220
6.2 Background on Driver 220
6.2.1 (I)FFT for High-Data Rate Standards 221
6.2.2 Motivation for a Flexible Implementation 222
6.3 Algorithm Optimizations and Characteristics 223
6.3.1 Algorithm Choice and Applied Optimizations 224
6.3.2 Algorithm Characteristics 226
6.4 Proposed DSIP Architecture Instance 228
6.4.1 Top Level 228
6.4.2 Cluster Level 230
6.4.3 Engine Level 231
6.4.4 Slice Level 232
6.5 Software Mapping and Hardware Implementation 236
6.5.1 Utilized Processor Design Tool Suite 236
6.5.2 Software Mapping and Scheduling 237
6.5.3 Hardware Implementation and Results 238
6.5.4 Throughput and Scalability 240

6.6 Comparison and Discussion . . . 241
6.6.1 ASIC References . . . 242
6.6.2 Flexible References from Literature . . . 242
6.6.3 Comparison (Normalized, Pessimistic). . . 242
6.6.4 Discussion . . . 245
6.7 Conclusions . . . 248
References . . . 249

7 Front-End Design Flow: Bridging the Algorithm-Architecture Gap . . . 251
7.1 Motivation and Issues, Overview of Proposal and Contributions . . . 251
7.1.1 Context and Motivation . . . 251
7.1.2 Algorithm-Architecture Co-Design for Traditional Architecture Styles . . . 252
7.1.3 Proposed Measures to Enable an Effective Algorithm-Architecture Co-Design Flow . . . 254
7.1.4 Main Contributions . . . 256
7.2 Proposed Architecture Template Instantiation Design Flow . . . 256
7.2.1 Algorithm Transformations . . . 258
7.2.2 Decisions on Data Representation . . . 259
7.2.3 Support of Arithmetic and Logic Operations . . . 260
7.2.4 Decisions on Parallelization . . . 261
7.2.5 Support of Data Transfers . . . 264
7.2.6 Data Storage Dimensioning . . . 264
7.2.7 Instruction Control Dimensioning . . . 265
7.3 Application on Case Study 1: MIMO Detector . . . 265
7.3.1 Algorithm Transformations . . . 265
7.3.2 Decisions on Data Representation . . . 266
7.3.3 Support of Arithmetic and Logic Operations . . . 266
7.3.4 Decisions on Parallelization . . . 266
7.3.5 Support of Data Transfers . . . 269
7.3.6 Data Storage Dimensioning . . . 270
7.3.7 Instruction Control Dimensioning . . . 270
7.4 Application on Case Study 2: FIR Filter. . . . 271
7.4.1 Algorithm Transformations . . . 271
7.4.2 Decisions on Data Representation . . . 272
7.4.3 Support of Arithmetic and Logic Operations . . . 272
7.4.4 Decisions on Parallelization . . . 272
7.4.5 Support of Data Transfers . . . 274
7.4.6 Data Storage Dimensioning . . . 275
7.4.7 Instruction Control Dimensioning . . . 275
7.5 Conclusions . . . 275
References . . . 276

8 Conclusions and Future Work 279
8.1 Conclusions 279
8.1.1 A Clear Need for New Design Approaches for Wireless Application Platforms 279
8.1.2 Proposed Design Approach for Wireless Application Platforms 279
8.1.3 Results are Very Promising 280
8.1.4 Main Messages 280
8.1.5 Main Contributions 281
8.2 Future Work 281
8.2.1 Remaining Challenges and Tasks 281
8.2.2 Planned Continuation of Work 284
References 284

Index 287

Acronyms

Wireless Communication Standards

2/3/4G	Communication standards of the 2nd/3rd/4th generation
3GPP	3rd Generation Partnership Project
ATSC	Advanced Television Systems Committee
DMB	Digital Multimedia Broadcasting
DVB	Digital Video Broadcasting
ECMA	European Computer Manufacturers Association
EV-DO	Evolution-Data Optimized
GPRS	General Packet Radio Service
GSM	Global System for Mobile communications
HSDPA	High Speed Download Packet Access
HSUPA	High Speed Upload Packet Access
IrDA	Infrared Data Association
LTE(-A)	Long-Term Evolution (-Advanced)
NFC	Near-Field Communication
P2P	Peer-to-Peer
UMTS	Universal Mobile Telecommunications System
WiMAX	Worldwide interoperability for Microwave Access
WLAN	Wireless Local Area Network
WMAN	Wireless Metropolitan Area Network
WPAN	Wireless Personal Area Network
WWAN	Wireless Wide Area Network

System/Functionality/Algorithm/Software

AFE	Analog Front-End
BAO	Basic Arithmetic Operation
BER	Bit-Error-Rate
BLCO	Boolean Logic/Comparison Operation
CDFG	Control and Data Flow Graph
CDMA	Code Division Multiple Access

CFO	Carrier-Frequency Offset
CORDIC	COordinate Rotation DIgital Computer algorithm
CP	ComParison
CR	Cognitive Radio
CSD	Canonical Signed Digit
DCD	Dichotomous Coordinate Descent algorithm
DFE	Digital Front-End
DFG	Data Flow Graph
DIF	Decimation in Frequency
DIT	Decimation in Time
DLP	Data Level Parallelism
DTSE	Data Transfer and Storage Exploration
EQ	EQualizer
FDM	Frequency-Division Multiplexing
FEC	Forward Error Correction
FFT	Fast Fourier Transformation
FIR	Finite Impulse Response
HO	Hard-Output
ILP	Instruction Level Parallelism
ITSE	Instruction Transfer and Storage Exploration
LDPC	Low Density Parity Check
LLR	Log-Likelihood-Ratio
LORD	Layered ORthogonal Lattice Detection
LOS	Line Of Sight
LR	Lattice Reduction
LRA-MMSE	Lattice-Reduction-Aided MMSE
LSB	Least Significant Bit
MAC	Multiply-and-ACcumulate operation
MCO	Multiplication with a Constant Operator
MCM	Multiple Constant Multiplication
MDF	Multi-path Delay Feedback
MFCSO	Modified Fixed-Complexity Soft-Output
MIMO	Multiple Input Multiple Output
ML	Maximum Likelihood
MMSE	Minimum Mean-Squared Error
MPSoC	Multi-Processor SoC
MSB	Most Significant Bit
MTT	Multi-pass Trellis Traversal
MVON	Multiplication with a Variable Operator and a multiplier which can adopt only a Narrow value range
MVOW	Multiplication with a Variable Operator and a multiplier which can adopt a Wide value range
NOP	No Operation
OFDM	Orthogonal Frequency-Division Multiplexing
Op	Operation

OSI	Open Systems Interconnection layers
PER	Packet Error Rate
PHY	PHYsical layer (lowest OSI layer)
QAM	Quadrature Amplitude Modulation
QPSK	Quadrature Phase-Shift Keying
QRD	QR orthogonal-triangular Decomposition
RD	ReaD
RF	Radio Frequency
RRDML	Reconfigurable Reduced Dimension Maximum Likelihood
SBP-LR	Scalable Block-based Parallel Lattice Reduction
SC	Single Carrier
SCO	Sampling Clock Offset
SDR	Software Defined Radio
SHF	SHift Factor
SIC	Successive Interference Cancellation
SIFS	Short InterFrame Space latency/timing
SNR	Signal-to-Noise Ratio
SO	Soft-Output
SoC	System on a Chip
SRRC	Square-Root-Raised-Cosine filter
SSFE	Selective Spanning with Fast Enumeration
SWL	Subword Length/Size
TDM	Time-Division Multiplexing
TLP	Task Level Parallelism
TSD	Tuple Search Detector
TSO	Trigonometric/Special Operation
WR	WRite
ZF	Zero-Forcing

Architectures/Architecture Styles

ADRES	Architecture for Dynamically Reconfigurable Embedded System
ASIC	Application Specific Integrated Circuit
ASIP	Application Specific Instruction set Processor
ASIP-AP	ASIP-Application Processor
ASIP-BP	ASIP-Baseband Processor
CG(R)A	Coarse Grained Reconfigurable Array (processor)
DSP	Digital Signal Processor
DSIP	Domain Specific Instruction set Processor
EVP	Embedded Vector Processor
FG(R)A	Fine Grained Reconfigurable Array (processor)
FPGA	Field-Programmable Gate Array
GPP	General Purpose Processor
GPU	Graphics Processing Unit

rASIC	reconfigurable ASIC
RISC	Reduced Instruction Set Computer
SDR	Software Defined Radio (processor)
SIMD	Single Instruction Multiple Data (processor)
SIMT	Single Instruction Multiple Threads (processor)
SODA	Signal-processing On-Demand Architecture
STA	Synchronous Transfer Architecture
TTA	Transport Triggered Architecture
VLIW	Very Long Instruction Word (processor)
VDSP	Vector DSP

Architecture Components/Elements

ADC	Analog-to-Digital Converter
ALU	Arithmetic and Logic Unit
AMBA	Advanced Microcontroller Bus Architecture
AOI	And-Or-Inverter
ASI	Application/Algorithm Specific Instruction
BAU	Basic Arithmetic Unit
BPE	Block Processing Engine
CLB	Centralized Loop Buffer
CLC	Cluster Level Control
RCU	Reusable Custom Unit
DEMUX	DE-MultipleXer
DMA	Direct Memory Access
DMEM	Data Memory
DSH	Data Storage Hierarchy
DSI	Domain-Specific Instruction
FF	Flip-Flop
FIFO	First-In First-Out
FU	Functional Unit
HA	Hardware Accelerator
HardSIMD	Hardware SIMD
HC	Hold Counter
ICache	Instruction Cache
ICH	Instruction Control Hierarchy
IMEM	Instruction Memory
ISH	Instruction Storage Hierarchy
LB	Loop Buffer
LC	Loop Controller
LU	Logic Unit
LUT(U)	Look-Up Table (Unit)
MRAM	Magnetoresistive Random Access Memory
MU	Multiplier Unit

MUX	Multiplexer
NoC	Network on Chip
NVM	Non-Volatile Memory
OICU	Online Instruction Computation Unit
PC	Program Counter
PE	Processing Element
RF	Register File
RRAM	Resistive Random Access Memory
SELC	Shared Engine Level Control
SHU	Shuffler
SIMD	Single Instruction Multiple Data
SoftSIMD	Software SIMD
SRAM	Static Random Access Memory
SSU	Subword Shuffler Unit
STLC	Shared Top Level Control
STR	STRide
STT-RAM	Sing Torque Transfer Random Access Memory
TW	TWiddle
VWR	Very Wide Register
VWR_SI	SI VWR Slice Interface
WSU	Word Shuffler Unit

Design/Tools/Automation/Technology

ADL	Architecture Description Language
BC	Best Case
CTS	Clock Tree Synthesis
DDSM	Deep Deep Sub-Micron technologies (65 nm and below)
DPG	Data-Path Generator
DRC	Design Rule Check
DVFS	Dynamic-Voltage and Frequency Scaling
EDI	Cadence Encounter Digital Implementation
EPS	Cadence Encounter Power System
ETS	Cadence Encounter Timing System
GP	General Purpose
HLE	High-Level Estimation
HDL	Hardware Description Language, e.g. VHDL and Verilog
HLS	High-Level Synthesis
HT	High-Throughput
HW	HardWare
IP	Intellectual Property
LA	Low-Area
LP	Low Power
NRE	Non-Recurring Engineering cost

PCB	Printed Circuit Board
RC	Resistance–Capacitance
RTL	Register Transfer Level
SDF	Standard Delay Format
SDP	Structured DataPath
SPEF	Standard Parasitic Exchange Format
SW	SoftWare
TC	Typical Case
TCL	Tool Command Language
TSMC	Taiwan Semiconductor Manufacturing Company, Limited
VCD	Value Change Dump
VHDL	Very high-speed integrated circuit Hardware Description Language
WC	Worst Case

Institutions/Divisions/Departments

ESAT	Department of Electrical Engineering, KUL
IMEC	Interuniversity MicroElectronics Centre
KUL	University of Leuven
NTNU	Norwegian University of Science and Technology
SSET	Smart Systems and Energy Technology, Division in IMEC

Metrics

bits	(Mega/kilo) bits
bytes	(Mega/kilo) bytes
bps	(Giga/Mega/kilo) bits per second
OPS	(Giga/Million) Operations Per Second
GE	Gate Equivalent
IPS	(Giga/Million) Instructions Per Second
J	(pico/femto) Joule
W	(milli) Watt

Chapter 1
Introduction and Motivation

1.1 The Smartphone of the Future

1.1.1 Trends and Consequences

Over the last decades, mobile communication devices have rapidly evolved [76]. Simple voice-only cellular phones are being replaced with modern smartphones [75]. To continue this evolution, the smartphone of the future has to offer higher computational performance, higher flexibility to program a broader range of application codes, higher energy efficiency, more technology scaling-friendliness and higher design reusability across several application domains, than current devices [24, 76]. Importantly, all these 5 requirements need to be jointly fulfilled. These requirements, which are summarized in Fig. 1.1, are driven by the following facts/trends.

1.1.1.1 The Need for Higher Computational Performance for the Increased Data Rates

In the 1990s, during the second generation of cellular phones (2G), the downlink data rate was in the order of several kilo bits per second (kbps) [75]. However, emerging 4G and 60 GHz communication standards specify data rates with several Giga bits per second (Gbps), which is 6 order of magnitudes higher [7, 40, 52, 77]. The need for higher data rates is driven by new applications, such as high-definition video streaming, augmented reality and online 3D games. In addition to the higher data rates, new applications, such as 3D gaming, demand shorter latencies. Thus, to cope with these requirements, mobile phones have to offer an increasing amount of computational performance and operate under strict timing guarantees (especially for latency) [4, 67, 75].

R. Fasthuber et al., *Energy-Efficient Communication Processors*,
DOI: 10.1007/978-1-4614-4992-8_1,

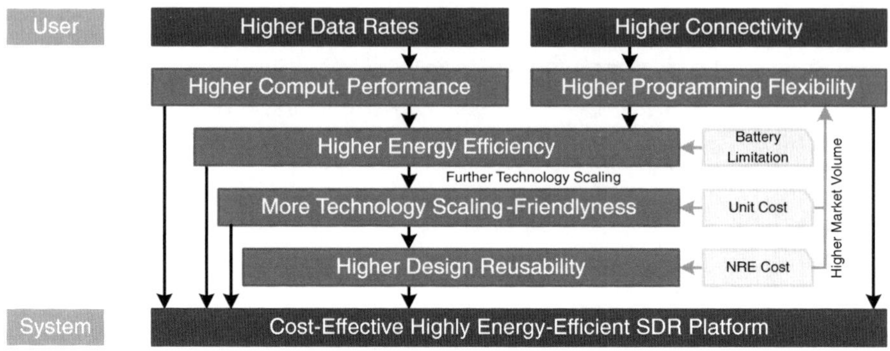

Fig. 1.1 Trends and main requirements for the implementation of future smartphones

1.1.1.2 The Need for Higher Programming Flexibility to Reduce Production Cost

With the increasing number of users and the demand for higher data rates, a shortage in available frequency spectrum has occurred [14]. To utilize the spectrum more efficiently, new techniques, such as Orthogonal Frequency-Division Multiplexing (OFDM) and Multiple Input Multiple Output (MIMO), and a more flexible spectrum use, e.g. different channel bandwidth and channel bonding, have been proposed [70, 72]. In addition, the utilization of new frequency spectrum, e.g. around 60 GHz, is ongoing [63]. The flexible spectrum use results in an increased number of modes [70]. To leverage on the techniques that enable a more efficient spectrum use, the new techniques have to be standardized. Because of a diversified market demand and because of fragmented industrial segments, not only one institution, but several competing institutions propose new standards [63]. For this reason, as shown in Fig. 1.2, the number of new wireless communication standards is rapidly increasing.

The mobile phone of the future has to provide seamless connectivity anywhere and anytime [24]. To enable this, the support of a wide variety of short-range to long-range communication standards is essential. The device has to provide connectivity to Wireless Personal Area Networks (WPAN)s (e.g. to connect to a headset), to Wireless Local Area Networks (WLAN)s (e.g. to connect to a desktop PC or TV), to Wireless Metropolitan Area Networks (WMAN)s (e.g. to connect a to a city hub), to cellular networks (e.g. for voice and video transmissions) and to broadcasting stations (e.g. to receive TV content) [36].

Especially for cellular networks, the migration to a new standard requires an upgrade of base stations which is typically very costly for the provider. Therefore the upgrade of all country-wide base stations may take several years. In addition, providers from other countries may not even consider a migration. Thus, the mobile phone has to support standards from different generations.

Considering the number of standards and the number of modes per standard, the device has to support an enormous and a further increasing number of standards/

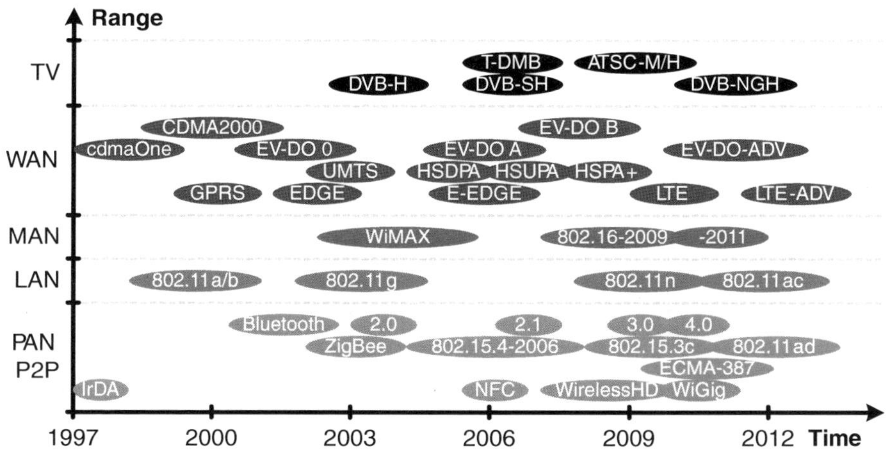

Fig. 1.2 The number of wireless communication standards is rapidly increasing. A modern smartphone has to enable short-range to long-rang data communication and has to offer backward compatibility. For this reason, it needs to support a wide variety of standards/modes and thus, flexibility/programmability (for cost effectiveness)

modes. In a traditional discrete solution, each of the standards/modes is implemented as a dedicated hardware chip. All the chips are then combined on the Printed Circuit Board (PCB). However, with the increasing number of standards/modes, this solution is becoming very costly (chip area, design/integration cost) and also the PCB size becomes too large for the available space [24, 68]. Because anyway only a limited number of connections can occur in parallel, i.e. only a limited number of standards/modes can be active at the same time, such a high amount of parallel hardware, which is never fully utilized, is clearly not needed. Thus, to reduce the production cost, which is becoming unacceptable for multiple custom implementations, the much more cost-effective approach is to reuse the same hardware resource for multiple standards/modes over time (time-sharing). However, to enable this kind of reuse, the hardware needs to be flexible, i.e. reconfigurable or re-programmable for different standards/modes. This need for flexibility, which leads to a significant cost reduction (because of less required hardware resources), has driven the development of mobile multi-standard Software Defined Radio (SDR) solutions [56].

1.1.1.3 The Need for Higher Energy Efficiency for Portability and Autonomy

Portable devices, such as mobile phones, are battery powered. The battery life time, i.e. the time until the battery needs to be recharged, depends on the capacity of the battery and the energy consumption of the mobile device. The development of new battery technologies results in a capacity improvement of about 10 % per

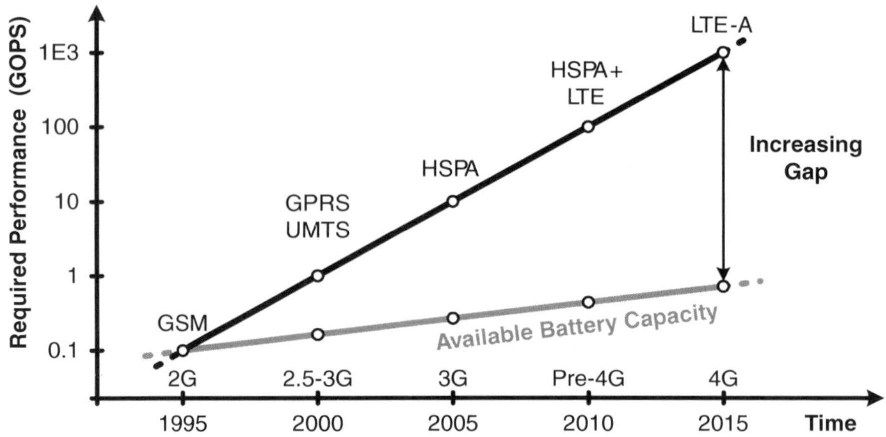

Fig. 1.3 Increasing energy efficiency gap: logarithmic increase of required computational performance versus linear increase of available battery capacity [67, 75]

year (estimated). Nevertheless, to enable higher portability/wearability, the size and weight of mobile phones has significantly decreased over the last years. This trend has consequences on the battery dimensions. Thus, the available battery capacity in mobile phones does almost not increase. In contrast, the increasing requirements on computational performance translate to an increase in energy consumption [75]. Because of this conflict, portable devices suffer from energy shortage. For instance, under average to moderately heavy use, the batteries of today's modern smartphones, mainly due to the display and the wireless connectivity, last not even one day [16, 22]. This energy shortage clearly limits the usability of such devices for the end users.

Figure 1.3 depicts this gap, i.e. referred as energy efficiency gap, between required computational performance and available battery capacity [75]. To counteract this problem, future mobile devices have to become much more energy efficient to maintain at least the current autonomy.

1.1.1.4 The Need for More Technology Scaling-Friendliness for Following Moore's Law

Technology scaling reduces the silicon area and therefore the unit cost, decreases the energy consumption and improves the performance. Therefore, technology scaling has been and is an essential measure to deal with the increasing complexity of designs. The mobile industry has become a major driver for technology scaling [6]. For instance, the first 14 nm tape-out of Intel will be a LTE chip and not as usual, a x86 processor [57]. Importantly, to fully leverage on the gain of technology scaling, the design must be optimized for exploiting the changing characteristics of the new technologies, i.e. it needs to be technology scaling-friendly, otherwise the potential improvement in delay and energy will not be achieved. Conventional design methods and styles are rapidly leveling off in terms of such improvements [48]. This is especially important for Deep Deep Sub-Micron (DDSM) technologies, i.e. 65 nm and below.

1.1.1.5 The Need for Higher Reusability at the Design Level for Reducing NRE Cost

With the scaling towards DDSM technologies, the technology-related design complexity and the chip development costs, i.e. the Non-Recurring Engineering (NRE) costs, have significantly been increased [6, 43, 67]. The technology-related design complexity increases mainly because of the stricter design rules (e.g. due to the smaller sizes and 3D integration) [15], the increasing number of design corners (e.g. due to voltage scaling and multiple voltage islands) and the increasing number of intermixed technology deviates (e.g. due to analog/digital-mix, different threshold voltages). The rising design, test and mask-set costs are major contributors to the rising chip development cost [15, 45]. For instance, according to INVOMEC [42], a 90 nm mask-set costs 700 k $, a 40 nm mask-set costs 1.35 M $ (1.9× more for 2 generations) and a 28 nm mask-set costs already 3.2 M $ (2.4× for only 1 generation). Because of the higher complexity, the re-spin rate approximately doubles when evolving from a given technology node to the next technology node [74]. Clearly, the increasing re-spin rate has a strong negative impact on the chip development cost. To ensure that the total cost, i.e. the sum of unit costs and NRE cost, in a new technology node is not rising, the NRE cost needs to be amortized over increasingly larger market volumes. For instance, based on a design example with 2 M logic gates and 8 Mbytes of memory, to maintain the same profit, the market volumes need to increase by a factor 9 when changing from 90 to 20 nm technology node. The increasing design complexity and the increasing re-spin rate have not only a strong negative influence on the chip development cost, but also on the time to market. However, in the fast evolving consumer industry, the importance of time-to-market is even increasing [3].

The need for multi-standard multi-mode systems, which offer high performance and which are highly energy efficient, has a significant impact on design complexity. Thus, the design complexity and the design time is not only increasing because of technology scaling, but also because of the increasing complexity of systems.

Considering these trends, the only viable solution is the reuse of design effort and the reuse of designs across a larger number of applications and even among application sub-domains with similar compatible characteristics [4]. To enable large market volumes, platform designs must be reusable across product generations and across products of a wide variety of market segments, e.g. U.S., Japan and Europe, low-end and high-end products.

1.1.2 The Solution: A Highly Energy-Efficient SDR Platform

To cope with the increasing need for flexibility and reusability, the Software Defined Radio (SDR) concept has been proposed [4, 36]. A SDR is a radio in which the system functionality is implemented on a reconfigurable and/or programmable platform. Thus, the migration to new standards/modes can generally be done in software and does not require the design of a new chip. As a consequence, SDR can lead to short time-to-markets, high market volumes and long product life cycles. SDR is therefore the ultimate solution from a business aspect. For this reason, academia and industry has strongly driven the development of SDR in the last years (for the consumer market) [4, 5, 36].

However, mainly because of the reduced energy efficiency, the wide employment of SDR in mobile devices is still a big challenge. Typical programmable architectures, such as General Purpose Processors (GPP)s and Digital Signal Processors (DSP)s, provide energy efficiencies that are several orders of magnitude lower than of Application Specific Integrated Circuit (ASIC) solutions [67]. Although Application Specific Instruction set Processors (ASIP)s are reducing the energy efficiency gap [51, 58], state-of-the-art ASIPs are still far away from the energy efficiencies of ASIC implementations. The fundamental reason is that state-of-the-art processors spend a significant amount of energy in non-computational hardware parts, such as in flexible interconnect and in instruction memory hierarchy/decoder [20]. Thus, new design solutions, which combine the benefit of SDR with close-to-ASIC energy efficiencies, are urgently needed.

1.2 Research Challenges to Enable Highly Energy-Efficient SDR Platforms

The practical implementation of a mobile SDR, which combines high performance, high energy efficiency, high flexibility, high reusability and technology scaling-friendliness is a huge challenge. Especially the followings four gaps are becoming very challenging:

1.2.1 The Energy-Efficiency Gap

As shown in Fig. 1.3, with the evolution of communication standards, the required computational performance is dramatically increasing [75]. Since the battery capacity enlargement is comparatively low, an increasing energy efficiency gap exists [67]. With the trend from ASICs towards SDR solutions, the energy efficiency gap is even further increasing. The fundamental reason behind is that, in a practical realization, flexibility and energy efficiency are conflicting design goals. As shown in Fig. 1.4, state-of-the-art architecture styles provide different flexibility/energy-efficiency trade-offs [50]. DSPs and Domain Specific Instruction set

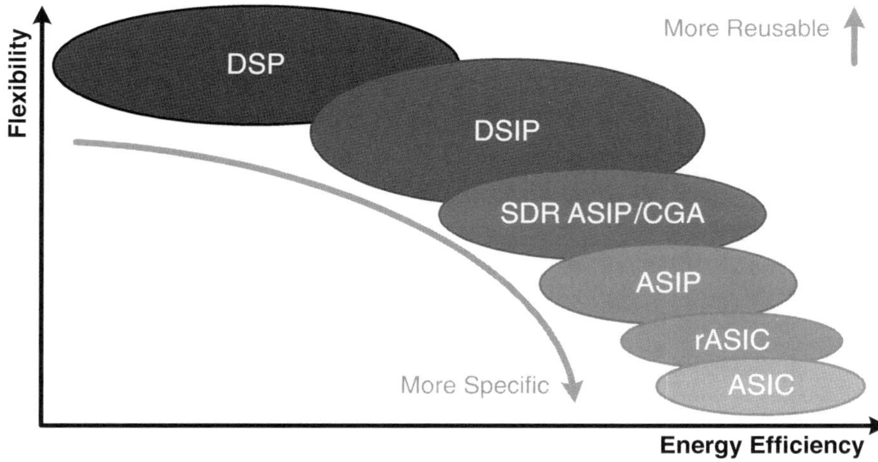

Fig. 1.4 State-of-the-art architecture styles offer different trade-offs in terms of flexibility and energy-efficiency. Based on [50]

Processors (DSIP)s offer a high degree of flexibility while ASICs and reconfigurable ASICs (rASIC)s offer maximal energy efficiency. The energy efficieny gap between both extremes is about a factor 10–100 [67], which is very significant. In between, ASIPs and Coarse Grain Array (CGA) processors for the complete wireless receiver, i.e. denoted as SDR ASIPs, SDR CGAs or SDR baseband processors, as well as ASIPs for sub-functionalities are positioned. Although ASIPs seem to offer a reasonable trade-off, the drawbacks of state-of-the-art solutions are still considerable. For instance, the difference in energy efficiency between ASICs and ASIPs is typically still an order of magnitude [20, 27, 67], which is unacceptably high. Besides, ASIPs are customized with Application/Algorithm Specific Instructions (ASI)s and/or ASIC Hardware Accelerators (HA)s, which limit the reusability. Thus, for the successful employment of SDR, the energy efficiency of programmable architectures needs to significantly be improved. To achieve this, further research is indispensable.

In this book, we focus on a fully programmable very light-weight DSIP solution that exploits many recent architectural innovations to achieve close-to-ASIC energy efficiency.

1.2.2 The Architecture Gap for Ultimately Scaled Technologies

Historically, because even basic technology-models were accurate enough, the technology aspects could easily be abstracted away from the architecture designer. For instance, in the 1990s, the transistor count was a sufficient metric for making

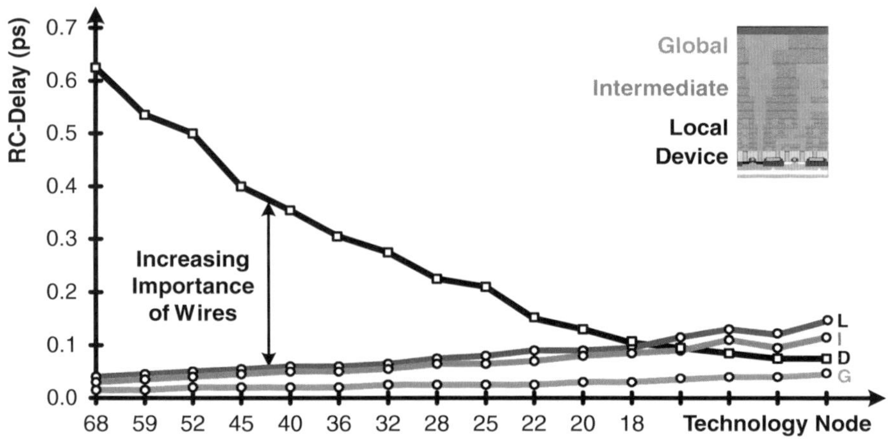

Fig. 1.5 With the scaling towards DDSM technologies, the delays of wires are becoming more important. Global wires scale better than local (M1) and intermediate wires. Figure based on [2, 43, 44]

representative architectural choices. However, with the scaling towards DDSM technologies and the increasing hardware complexity, technology aspects cannot longer be neglected at higher design levels [44, 73]. In general, the effects of DDSM technologies are bringing a new set of challenges, which require more and more considerations and countermeasures at architectural level [1, 13, 21, 23]. The decreasing reliability [62], the increasing variability [15, 39], the increasing leakage (especially in memories) [69] and the increasing wire influence [44] are amongst these effects. In this book, we consider especially the latter aspect. The other effects are tackled by concurrent efforts in the design community and they have less direct impact on the platform/architecture style selection.

As shown in Fig. 1.5, the technology scaling in the DDSM era further decreases the delay of transistors. However, in contrast to older technologies, the Resistance-Capacitance (RC)-delay of wires is increasing [2, 43, 44]. The increase is mainly caused by the resistance, which is a consequence of the decreased wire width and wire height. Because of this trend, the relative importance of wires compared to devices is rising. Thus, the architecture designer can no longer neglect the influence of wires on the performance metrics, i.e. clock frequency, area and energy consumption, of the design. The increased area and energy consumption is a consequence of buffering, which is a countermeasure to reduce the delay. As shown in Fig. 1.5, in technologies below 18 nm, wires will dominate these performance metrics even more.

However, state-of-the-art processor architectures and design approaches do not consider the wire influence sufficiently. For instance, in the ASIP design style, new ASIs are added to the basic instruction set. The ASIs, which generally consist of a mix of arithmetic, logic and multiplexing operators, are typically chosen based on code profiling [50]. Because the complexity of the ASIs is typically higher than the complexity of the basic instruction set [25, 38], the complexity and area of the Functional Units (FU)s increases significantly [25]. Thus, in this conventional approach, the amount of multiplexing and the wire lengths are very high [9, 20]. For this reason, when scaling between DDSM technology nodes, in which the wires are dominating, the performance metrics basically do not improve. DSIPs and DSPs are based on more basic and less complex instructions, therefore they are generally more suited for advanced technologies. Nevertheless, because of the amount of employed instructions and the employed datapath operators (e.g. complex multipliers in combination with complex adders), the amount of multiplexing as well as the number and lengths of wires is also very high. For example, the DSP TI TMS320C62x supports 87 instructions, which results in 20–29 sub-instructions per FU [41]. Thus, new architectures, which reduce the number of wires to enable a better routability and which reduce especially the wire lengths of heavily used connections, are needed.

Traditionally, a technology/layout-independent architecture design approach is applied. Thereby, the layout is either completely neglected or only a coarse floorplan, e.g. placement of memory macros, is considered. The datapath and control logic is automatically mapped onto standard-cell logic. Afterwards, the standard cells are automatically placed and routed. Because the architecture design neglects the fine-grain placement of datapath and control components and the associated routing, wire congestions and long wires may be present [17]. Thus, in DDSM technologies, the technology/layout-independent architecture design approach will lead more and more to efficiency, scalability and predictability issues. For this reason, new processor design approaches, which consider architecture and technology/layout together, i.e. which ensure that the full benefit of technology scaling is obtainable, are urgently needed.

That is another important focus of this book.

1.2.3 The Productivity Gap

A further issue is the increasing productivity gap, which is shown in Fig. 1.6 [43]. The productivity gap is the gap between the available technology/silicon capability, i.e. predicted by Moore's law and obtained by technology scaling [60], and the ability to exploit it. The reason for this increasing gap in the enormous increase in design complexity and the lack of increasing the design productivity in the same extent. The design complexity mainly increases because of (1) the increasing wireless system complexity, (2) the integration of an increasing number of mixed-signal designs into a single chip and (3) the increasing technology complexity. To bridge this gap, the design productivity needs to rise dramatically. Thus, future design approaches need to leverage on much more reuse and much more design automation than what is the case today [49].

In this book, we propose a design approach which enables high reusability across applications and standards and a high-degree of potential automation.

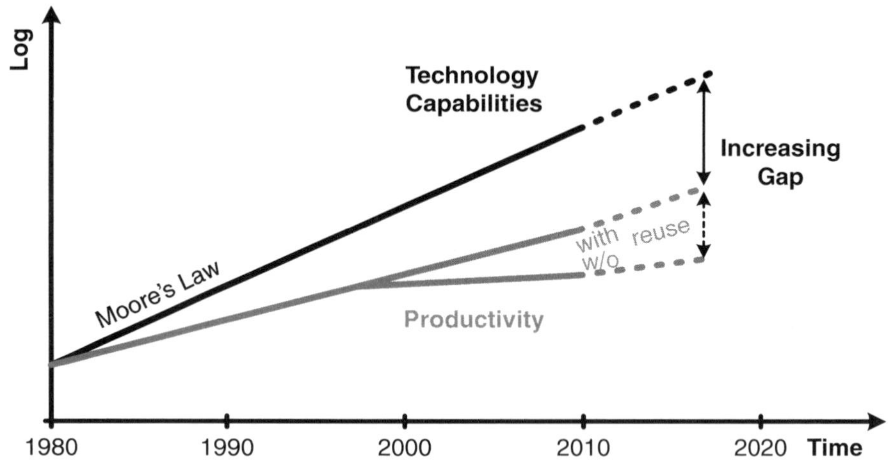

Fig. 1.6 Because of the increasing hardware design productivity gap [43], higher reusability is becoming essential

1.2.4 The Culture Gap for Design Paradigms

For the design of a mobile SDR system, researchers and engineers from different design communities are required. In general, the communities can be separated into (1) system, functionality and algorithm designers, into (2) architecture and platform designers and into (3) back-end designers, i.e. people who perform the tasks related to chip layout and integration. Each of the communities is specialized in its field. Because the complexity even within a field is very high, one community has typically little knowledge of the other fields. This causes a culture gap between these different design communities, which limits the exchange of information and the propagation of constraints.

Due to the following main consequences, this cultural gap is increasingly becoming an issue:

- Systems and algorithms are designed and optimized without considering the actual implementation cost. This results in algorithms which are either not suited for practical implementation at all or which have a very high implementation cost. Thus, algorithm research effort is wasted (productivity gap) and the required energy efficiency (energy efficiency gap) cannot be achieved.
- Hardware platforms are over-designed and under-utilized. Because the hardware platform designers do typically not have the full system knowledge, local decisions are made and therefore potential optimization opportunities are not exploited. The resulting platforms are expensive in area, i.e. chip cost, and in energy (energy efficiency gap).
- Design of architectures, that cannot efficiently be implemented in silicon. For instance, the design of ASIs reduces the clock cycle count. However, the

increased multiplexing, the increased wire lengths and the increased number of pipeline registers can result in an overall low energy efficiency (energy efficiency gap). Besides, as previously mentioned, the benefit of technology scaling may disappear (architecture gap).
- Technology-related issues, such as reliability, are solved at circuit level, although countermeasures at architecture or system level may be more efficient (energy efficiency gap).

In general, the culture gap is the most fundamental and challenging gap. Because of the culture gap, predictions are difficult to make, specifications cannot be met and re-spins are necessary. Thus, long time-to-markets and high costs are the consequences. For this reason, it is very essential to propose new innovations that can efficiently bridge the culture gap, but which are also compatible with predominant business models.

In this book, we propose a design approach that significantly contributes in bridging this cultural gap.

1.3 Key Concepts to Tackle Research Challenges and Related Gaps in the State of the Art

To tackle the fore-mentioned research challenges, i.e. to enable highly energy-efficient SDR platforms, we propose a solution that is based on 6 key concepts. Although these 6 key concepts are already mostly known to the research community, the combined exploitation of them is still largely missing in literature. In the following we will explain these 6 key concepts and highlight the main related research gaps in the state of the art.

1.3.1 Employ an Architecture With a Well-Chosen Flexibility

By employing more multiplexers, more control hardware and/or more storage, an architecture can be made more flexible. However, this additional hardware, which does not perform computations on data, causes an overhead in energy [20]. Therefore, flexibility and energy efficiency are in general two conflicting design objectives (e.g. compare DSP with ASIC). For this reason, it is very important to choose the right trade-off between them. To guide this decision, the actual required flexibility needs to be determined. The employed architecture should be customized so that it provides only the required amount of flexibility and not more than that. For example, today, a single DSIP is employed for the computation of several heterogeneous sub-functionalities [9]. Because it needs to offer the flexibility to support the sum of all sub-requirements, the energy efficiency is low. Thus, choosing the right trade-off, i.e exploiting this concept, is essential for bridging the energy efficiency gap. A design approach, which combines a reusable template with customizations can ensure low NRE costs while enabling a high energy efficiency.

For SDR, a wide variety of architecture styles and design approaches, which offer different trade-offs, have been proposed [19, 37, 47, 64, 71]. State-of-the-art DSIP and ASIP solutions are flexible and partly reusable but the energy efficiency compared to ASIC solutions is still very low, i.e. much more than a factor 3. Thus, it is essential to better understand the reasons for this energy efficiency gap and to propose a solution which can jointly fulfill the requirements of high performance, high energy efficiency, high flexibility, high reusability and technology scaling-friendliness.

1.3.2 Leverage on an Energy and Cost-Effective Architecture Template

In the conventional ASIC design approach, for every new implementation, a new architecture is designed and verified. Thereby the reuse from the previous implementations is generally very low. However, to bridge the productivity gap, design approaches, which leverage on a high amount of reuse, are essential. A high amount of reuse can be enabled by employing an architecture template [12, 37]. Basically, the architecture template is designed and verified once. For every new implementation, only the pre-defined template parameters, such as memory sizes or SIMD factors, are chosen. Because the general architecture structure is pre-defined and reused, it is feasible to develop tools for it, e.g. compilers. Thus, with an architecture-template based design approach, the design time and the design cost, i.e. the NRE cost, is significantly reduced.

Several architecture templates for the DSIP and ASIP architecture styles have been proposed [12, 35, 37, 46, 64, 71]. Nevertheless, the existing templates have either one or multiple of the following drawbacks: (1) Rather general processor architecture, which is energy inefficient. (2) The architecture template enables only a coarse configuration (e.g. number of FUs), which prevents to achieve a good match between algorithm and architecture. (3) The customization is mainly done by adding new instructions, which leads to low reusability at the back-end side. Therefore the need for an architecture template, which enables a medium-grain customization within a pre-defined design space, and which leads to truly energy-efficient implementations, still remains.

1.3.3 Consider DDSM Technology Constraints During the Design of the Architecture

As mentioned in Sect. 1.2.2, one of the technology-related constraints is the increasing influence of wires [44]. The amount and lengths of wires, especially when considering the application of a semi-custom back-end design flow , is mainly decided by the architecture designer. Thus, to ensure that the architecture can efficiently be implemented in advanced DDSM technologies, the architecture designer needs to take the wire influence into account. In general, it is essential to consider the wires that are most important, e.g. wires in the critical path, highly active wires and wires in areas where congestions are likely. Many different architectural options for reducing the amount and/or lengths of certain wires exist. For instance, employing an iterative instead of a parallel hardware unit (time multiplexing). Another example is the separation of a big memory into several smaller memories, which can then be placed closer to the hardware components, which read/write frequently. Because of the large exploration space, it is important to focus on pareto-optimal combinations of architectural options. The concept of considering technology-constraints at architectural level can significantly contribute in bridging the architecture and the energy-efficiency gap.

So far, most processor architectures for SDR are basically designed without considering the wire influence from the beginning on. Place and route results may show issues related to wires. However, this issues are then typically considered at a second phase. The insertion of pipeline registers, which is a typical measure during the second phase, can help to increase the clock frequency. However, this approach is not sustainable, because for a slightly different floorplan and/or technology, the inserted pipeline registers may not have been necessary. Nevertheless, without establishing a close link between architecture and layout, such issues are difficult to understand. For future DDSM technologies the wire influence is becoming more and more important [44]. Therefore, a need for new design approaches, which strongly couple architecture and technology/layout, still exists. In addition, a need for new architectures, which strongly consider the influence of wires (1) to enable an efficient implementation in DDSM technologies and (2) to enable a scaling to higher performances without becoming inefficient, is prevailing.

1.3.4 Employ Consistent, Predictable and Systematic Design Flows Around the Architecture Template

The architecture-template based design approach enables to employ consistent, predictable and systematic design flows around the architecture template. A front-end design flow can bridge the culture gap between the algorithm designers and the architecture designers. Cost models, e.g. energy consumption of certain architecture configurations, can guide algorithm optimizations. In addition to the front-end design flow, also a consistent back-end design flow can be established. For instance, by defining a clear link between architecture and layout, the back-end design flow can bridge the culture gap between architecture designers and layout designers. Since the design flows can be largely automated, the productivity gap can be drastically reduced.

For state-of-the-art DSIP and ASIP architecture templates, such as [12, 35, 37, 46, 64, 71], consistent predictable back-end design flows are still missing. Because of this, accurate cost models for the front-end design can hardly be established. Thus, a need for a comprehensive design approach, which employs consistent, predictable and systematic design flows around the architecture template, still exists.

1.3.5 Ensure that Algorithm and Architecture are Well-Matched

An algorithm can be well-suited for a certain processor architecture, but very unsuited for another processor architecture [8]. For instance, a completely indeterministic algorithm, which includes many conditions, cannot efficiently be executed on a deeply-pipelined parallel data processor. Therefore, to obtain high energy efficiency, which is essential to bridge the energy efficiency gap, the algorithm and the architecture need to be well-matched [53, 54]. By applying an algorithm-architecture co-design approach, i.e. the algorithm and the architecture are designed for each other, this can be ensured [55]. When considering a template-based design approach, in order to enable high reuse, it is important that the architecture template is well-suited for the targeted algorithms and that the targeted algorithms are optimized for the architecture template.

State-of-the-art algorithm-architecture co-design approaches focus mainly on ASIC and ASIP solutions [18, 50, 53]. For instance, with a High-Level Synthesis (HLS) tool, different algorithm-ASIC combinations can be evaluated [18]. Co-design approaches for ASIPs focus mainly on the design of ASIs and/or HAs [50]. Besides, the architecture cost models, which are used for guiding algorithm optimizations, are often very inaccurate, e.g. because the link between architecture and technology is not taken into account. Thus, algorithm-architecture co-design approaches, which are based on more accurate cost models and which also focus on DSIP architectures, are still needed.

1.3.6 Effectively Adapt to the Actual Requirements at Run-Time

In general, a wireless communication system is very dynamic [10]. For instance, the quality of the communication channel and the required data rates are varying over time. These dynamically varying environmental conditions translate into

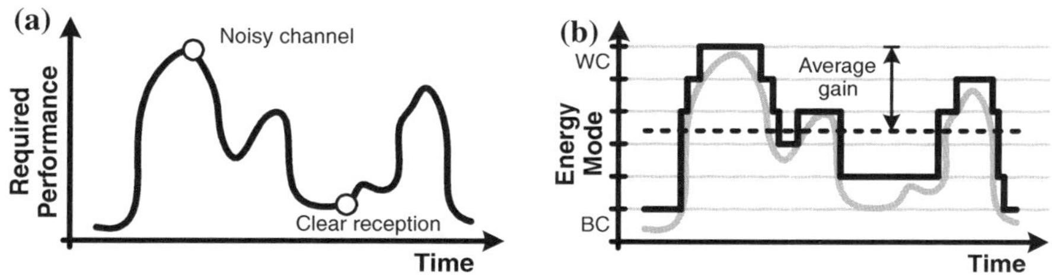

Fig. 1.7 Concept of dynamic run-time scaling: The system, which is controlled by a run-time manager, is dynamically adapting to the actual requirements so that a certain performance, e.g. bit error rate, can "just" be met. Thereby, since, e.g. during good channel conditions, not more computations are performed than needed, the average energy consumption is reduced. For the exploitation of this concept, a flexible hardware is needed

varying performance requirements for the communication device. Figure 1.7 illustrates this on an example. In case the communication device is far away from a base station, the communication channel will be strongly disturbed, i.e. the channel adds a lot of noise to the transmitted signal. Therefore, the communication device needs to use "strong" algorithms to recover the transmitted data. However, in case the communication device is close to the base station, the received data is almost completely error free. Therefore, the communication device can use algorithms with a very low complexity, since they are sufficient to correct the remaining few errors. The energy consumption of the communication device in the first case and in the second case will significantly be different. Thus, by implementing an adaptive system, which switches dynamically between different performance-energy modes, the average energy consumption can drastically be reduced [66].

To exploit this concept, which is based on system scenarios [11, 53, 59], the system needs to include a run-time manager and architectures that enable the flexibility for supporting multiple modes. The run-time manager, which is outside the scope of this book, monitors the environment and decides on the required performance-energy mode. Since the SDR paradigm is already demanding a flexible solution, the required overhead for supporting these performance-energy modes will be small. For this reason, this concept can considerably contribute in bridging the energy-efficiency gap. Besides, because traditional ASICs operate permanently in worst-case mode, this concept can further reduce the energy-efficiency gap between ASICs and processors [28].

Many systematic approaches for the design of dynamically adapting systems have been proposed [11, 59, 61, 65, 66]. The concept of dynamic run-time scaling has been demonstrated for the multimedia domain [59, 65, 66]. Although the concept has also been motivated for the wireless domain, practical implementations have basically not appeared in literature. For multiple deterministic modes with different performance requirements, mainly Dynamic-Voltage and Frequency Scaling (DVFS) is applied today. However, considering the high cost for power isolation and synchronization circuits for state-of-the-art processor architectures, only a scaling on a coarse architectural level is feasible. Besides, the supply voltage ranges, which can effectively be exploited with voltage scaling, are decreasing in DDSM technologies. In addition, DVFS may also not be applicable for latency sensitive applications. Thus, with state-of-the-art solutions, the full gain potential of dynamic run-time scaling can clearly not yet be sufficiently well exploited. Therefore, new solutions, which can efficiently exploit dynamic run-time scaling on a fine-grain granularity, are still needed. Besides, the effective design of the run-time monitor and the run-time manager are still not sufficiently addressed.

1.4 Proposed DSIP Architecture Template Design Approach

This book proposes the following overall design approach, which incorporates the fore-mentioned 6 key concepts to enable highly energy-efficient SDR platforms. An overview is provided in Fig. 1.8. Contrary to most related work, we select the DSIP instead of the ASIP architecture style. We combine the DSIP architecture style with the architecture template concept. Since we leverage, for every template architecture instance, on the same set of instructions, high reusability is ensured. Together with the front-end and the back-end design flow, a comprehensive solution for bridging the productivity and the culture gap is offered. Because of the DSIP architecture style, the energy efficiency is generally lower compared to more customized solutions. However, we include several measures to increase the energy efficiency significantly. First, during the design of the architecture template, we ensure, that the requirements from the domain-specific algorithm side and the technology side are well incorporated. The latter enables to bridge the architecture gap in terms of technology scaling and performance increase. Second, for the computations of the complete baseband functionality, we do not employ one rather general processor, but multiple more customized processors. Importantly, all the employed processors are based on the same DSIP architecture template. The customization of a DSIP architecture instance ensures that the algorithms and the architecture are well matched. In the course of the customization, values for the pre-defined template parameters are chosen. Thereby, for each instance, the desired flexibility/energy-efficiency trade-off is selected. Thus, this approach enables to fully bridge the energy-efficiency gap and it enables both programming flexibility and design reuse across application domains. Third, to further improve the energy efficiency , the concept of dynamic run-time scaling with system scenarios is effectively applied. Instead of DVFS on a coarse granularity, we apply mainly power gating on a fine-medium granularity. Since the architecture is programmable and tightly coupled to technology, the overhead for supporting multiple performance-energy modes is much smaller compared to the achievable gain. Thus, this decreases the energy gap again.

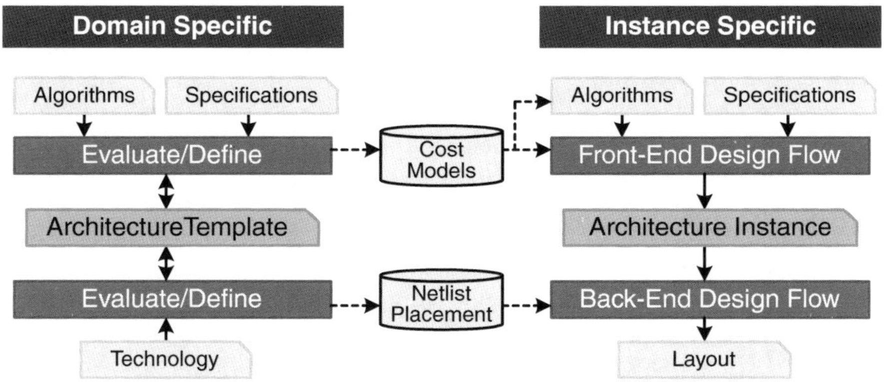

Fig. 1.8 Proposed DSIP architecture template and design flows

1.5 Main Focus and Contributions

1.5.1 Main Focus of this Book

This book focuses mainly on the architecture aspect of the proposed design approach. The core of this work is (1) the evaluation and implementation of architectural concepts and (2) the proposal of a scalable DSIP architecture template. The proposed template targets the implementation of the digital baseband functionality, i.e. ranging from the digital front-end to the outer modem. The proposed design flows and the results on the run-time scaling concept complete the core of this book.

1.5.2 Overview of the Main and Side Contributions

The main contributions of this book can be summarized as follows:

- Comparison of different architecture styles in terms of flexibility, energy efficiency and area efficiency [26–28, 32, 34]. Thereby the fundamental reasons for the energy-efficiency gap between processors and ASICs are analyzed and identified.
- Implementation and evaluation of existing architecture concepts for the DSIP architecture template [31, 34]. Thereby, the first implementation of an instruction memory hierarchy with distributed loop buffers and the first implementation of software Single Instruction Multiple Data (SIMD) is shown.
- Proposal, implementation and evaluation of new architecture concepts for the DSIP architecture template [31, 33]. For instance, the concept of employing multiple units, slices, engines, clusters and the way in which these components

are interconnected and in which they share hardware resources. Another example is the scaling with multiple Very-Wide Registers (VWR)s of different types and the scaling of the Subword-Shuffler Unit (SSU) with different stages. Beside the concepts which are mainly related to scalability, new approaches for the efficient computation of multiplications are proposed. This has also resulted in a patent application.
- The design of a scalable wire-friendly DSIP architecture template specifically for the digital wireless baseband domain [29–31]. Thereby the domain-specific requirements are analyzed and well-suited architecture components are combined accordingly. Because the architecture template is designed while keeping the wire influence and the layout in mind, it fulfills the requirements for advanced DDSM technologies.
- Application and evaluation of the proposed DSIP architecture template on three important case studies [32, 34]. The first case study is on an advanced high-performance MIMO detector for future LTE/WLAN standards, the second case study is on a high-speed FIR filter for emerging 60 GHz systems and the third case study is on a high-speed FFT for 60 GHz and WLAN systems. The comparison results of all case studies demonstrate the high energy efficiency, and the effective scalability to high performance, of the proposal.

In addition, this book offers the following side contributions:

- Proposal of a front-end design flow which complements the proposed DSIP architecture template. The application of the front-end design flow is demonstrated on case study 1 and 2.
- Proposal of a back-end design flow which complements the proposed DSIP architecture template. The application of the back-end design flow is demonstrated on the case study 2 (FIR Filter) and on a standard-cell memory.

  Since the proposed front-end design flow and the proposed back-end design flow are fully matched, i.e. they are fully consistent, the culture and the productivity gap can be bridged.
- For the case studies, the exploitation of the dynamic run-time scaling concept has been motivated [28, 32, 34]. The energy consumption for different performance-energy modes varies significantly. This can be exploited effectively to further reduce the overall energy consumption per task.

1.6 Structure of this Book

As illustrated in Fig. 1.9, the remaining book is structured as follows:

Chapter 2 provides background information and reviews related work. First, the wireless communication system, the proposed architecture styles for wireless systems and the general system design flow are described. Then, related work on design methodologies, on architectures and on templates is reviewed.

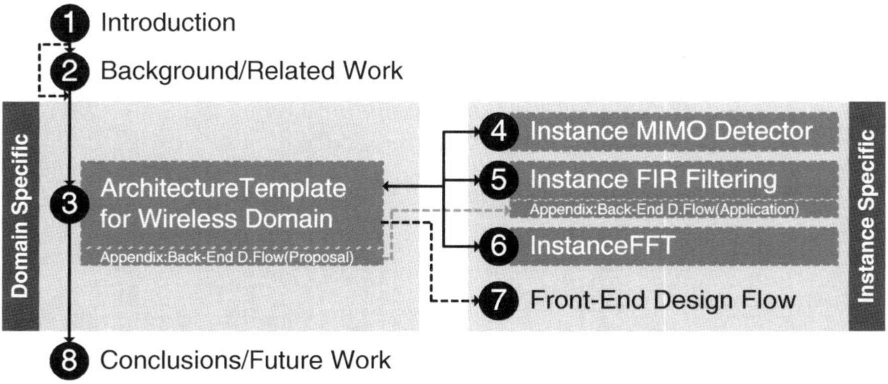

Fig. 1.9 Overview of the chapters/appendices of this book

Chapter 3 presents the proposed DSIP architecture template. After analyzing the domain-specific algorithm requirements, different architectural components are evaluated. Then, the selected components are combined to the proposed scalable architecture template.

Chapter 4 shows the application of the DSIP architecture template for the implementation of an advanced MIMO detector. The proposed MIMO detector processor targets 4G standards, such as LTE and IEEE 802.11ac.

Chapter 5 shows the application of the DSIP architecture template for the implementation of a high-throughput FIR filter. The proposed FIR filter processors targets 60 GHz standards, such as IEEE 802.15.3c and IEEE 802.11ad. In the course of this case study, an implementation for Software SIMD is proposed.

Chapter 6 shows the application of the DSIP architecture template for the implementation of a high-throughput FFT.

Chapter 7 proposes a systematic front-end design flow that enables a fast and efficient design of template instances. The application of the design flow is demonstrated on two case studies, the MIMO detector and the FIR filter.

Chapter 8 concludes this book, provides information on ongoing work and addresses future work.

The appendix of Chap. 3 presents initial ideas for a back-end design flow that is compatible with the proposed design concept, i.e. which ensures that the obtained layout represents the layout that is targeted by the architecture designer. This is a necessity to reduce the lengths of important wires and to enable scalability. In the appendix of Chap. 5 experimental results are shown.

References

1. Agarwal, M., Paul, B.C., Zhang, M., Mitra, S.: Circuit failure prediction and its application to transistor aging. In: IEEE VLSI Test Symposium (VTS), pp. 277–286 (2007). doi:10.1109/VTS.2007.22
2. Albelo, J.: Laser applications in advanced IC packaging. ELECTROIQ (2007). http://www.electroiq.com/articles/ap/print/volume-16/issue-5/features/laser-applications-in-advanced-ic-packaging.html
3. Amdocs Survey: Time to market grows in importance. Amdocs (2011). http://www.amdocs.com/News/Pages/TimetoMarketSurvey-2011.aspx
4. Anjum, O., Ahonen, T., Garzia, F., Nurmi, J., Brunelli, C., Berg, H.: State of the art baseband DSP platforms for software defined radio: a survey. EURASIP J. Wireless Commun. Netw. **2011**(1), 5 (2011). doi:110.1186/1687-1499-2011-5
5. Bagheri, R., Mirzaei, A., Heidari, M., Chehrazi, S., Mikhemar, M., Tang, W., Abidi, A.: Software-defined radio receiver: dream to reality. IEEE Commun. Mag. **44**(8), 111–118 (2006). doi:10.1109/MCOM.2006.1678118
6. Ballhaus, W., Pagella, A., Vogel, C.: A change of pace for semiconductor industry? PricewaterhouseCoopers. In: Technology, Media and Telecommunications (2009)
7. Baykas, T., Sum, C.S., Lan, Z., Wang, J., Rahman, M., Harada, H., Kato, S.: IEEE 802.15.3c: the first IEEE wireless standard for data rates over 1 Gb/s. IEEE Commun. Mag. **49**(7), 114–121 (2011). doi:10.1109/MCOM.2011.5936164
8. Berset, T., Catthoor, F.: Technical note TN-11-WATS-TIP2: high level estimation for independent component analysis. Technical report, Holst Centre, IMEC-NL (2012)
9. Bougard, B., De Sutter, B., Rabou, S., Novo, D., Allam, O., Dupont, S., Van der Perre, L.: A coarse-grained array based baseband processor for 100Mbps+ Software defined radio. In: Design, Automation and Test in Europe (DATE), pp. 716–721, IMEC (2008). doi:10.1109/DATE.2008.4484763
10. Bougard, B., Novo, D., Van der Perre, L., Catthoor, F.: Bridging the energy gap in size, weight and power constrained software defined radio: Agile baseband processing as a key enabler. In: IEEE International Conference on Acoustics, Speech and Signal Processing (ICASSP) pp. 5384–5387 (2008). doi:10.1109/ICASSP.2008.4518877
11. Carvalho, E.L.d.S., Calazans, N.L., Moraes, F.G.: Dynamic task mapping for MPSoCs. In: IEEE Des. Test Comput. 27(5), 26–35 (2010). doi:10.1109/MDT.2010.106
12. CEVA Inc.: CEVA-XC321 product note (2009). http://www.ceva-dsp.com/
13. Chandra, V.: Dependable design in nanoscale CMOS technologies: challenges and solutions. In: Workshop on Dependable and Secure Nanocomputing (WDSN), ARM (2009)
14. Chandra, A.: Spectrum management for future generation wireless based technology. In: European Wireless Technology Conference (EuWIT), pp. 201–205 (2009)
15. Cheng, Y.: A glance of technology efforts for design-for-manufacturing in nano-scale CMOS processes. Sci. China Series F: Inf. Sci. **51**(6), 807–818 (2008). doi:10.1007/s11432-008-0054-9
16. Cheng, R.: CES 2012 highlights: hallelujah! samsung promises all-day battery life on phones, CNET (2012). http://ces.cnet.com/8301-33363_1-57358219/
17. Chinnery, D., Keutzer, K.: Closing the Power Gap between ASIC & Custom: Tools and Techniques for Low Power Design. Springer, New York (2007)
18. Coussy, P., Gajski, D.D., Meredith, M., Takach, A.: An introduction to high-level synthesis. In: IEEE Des. Test Comput. **26**(4), 8–17 (2009). doi:10.1109/MDT.2009.69
19. Cupaiuolo, T., Siti, M., Tomasoni, A.: Low-complexity high throughput VLSI architecture of soft-output ML MIMO detector. In: Design, Automation and Test in Europe (DATE), pp. 1396–1401 (2010)
20. Dally, W.J., Balfour, J., Black-Shaffer, J.C., Harting, R.C., Parikh, V., Park, J., Sheffield, D.: Efficient embedded computing. Computer **41**(7), 27–32 (2008)

21. De Micheli, G.: Logic synthesis and physical design: Quo vadis? In: Design, Automation and Test in Europe (DATE), pp. 1–1. EPFL (2011)
22. Duan, R., Bi, M., Gniady, C.: Exploring memory energy optimizations in smartphones. In: International Green Computing Conference and Workshops (IGCC), pp. 1–8 (2011). doi:10.1109/IGCC.2011.6008591
23. Ernst, D., Das, S., Lee, S., Blaauw, D., Austin, T., Mudge, T., Flautner, K.: Razor: circuit-level correction of timing errors for low-power operation. IEEE Micro **24**(6), 10–20 (2004). doi:10.1109/MM.2004.85
24. Faintuch, A., Machness, R.: Trends and Challenges for Wireless Mobile Connectivity. Texas Instruments. EE Times (2008). http://www.eetimes.com/design/microwave-rf-design/4018989/
25. Fasthuber, R., Li, M., Novo, D., Raghavan, P., Van der Perre, L., Catthoor, F.: A soft-output near-ML MIMO baseband receiver For 75Mbps+ SDR. ICT Mobilesummit. In: IMEC (2009)
26. Fasthuber, R., Li, M., Novo, D., Van Der Perre, L., Catthoor, F.: Novel energy-efficient scalable soft-output SSFE MIMO detector architectures. In: International Conference on Embedded Computer Systems (IC-SAMOS). IMEC (2009). doi:10.1109/ICSAMOS.2009.5289228
27. Fasthuber, R., Li, M., Novo, D., Van Der Perre, L., Catthoor, F.: Exploration of soft-output MIMO detector implementations on massive parallel processors. J. Sig. Proc. Syst. (JSPS) **64**(1), 75–92 (2010). doi:10.1007/s11265-010-0499-0
28. Fasthuber, R., Li, M., Novo, D., Raghavan, P., Van der Perre, L., Catthoor, F.: Energy-efficient run-time scalable soft-output SSFE MIMO detector architectures. Trans. High-Perform Embed Architect Compilers (HiPEAC) (Special Issue SAMOS 2009) **5**(3), 1–20 (2011)
29. Fasthuber, R., Raghavan, P., Catthoor, F.: A programmable low energy massive-parallel architecture for wireless communication systems. In: Workshop in DATE: Designing for Embedded Parallel Computing Platforms (Architecture Session) (2011)
30. Fasthuber, R., Raghavan, P., Catthoor, F.: An highly-efficient processor template for wireless communication systems. In: DATE PhD Forum (2012)
31. Fasthuber, R., Agrawal, P., Raghavan, P., Catthoor, F., Van der Perre, L.: A novel energy efficient wireless domain specific processor template. IEEE Trans. Circuits Syst. (TCAS) (2013, in preparation)
32. Fasthuber, R., Raghavan, P., Catthoor, F.: A scalable MIMO detector processor with near-ASIC energy-efficiency. IEEE Trans. Very Large Scale Integr. Syst. (TVLSI) (2013)
33. Fasthuber, R., Raghavan, P., Catthoor, F.: An enhancement for enabling variable multiplications on a general shift-add/sub datapath. (2013, in preparation)
34. Fasthuber, R., Raghavan, P., Van Thillo, W., Catthoor, F.: An energy-efficient software SIMD processor for very high throughput 60 GHz PHY FIR. J. Sig. Proc. Syst. (JSPS) (2013, in preparation)
35. Freescale Semiconductor Inc.: Star Core MSC8158 Product Brief, Rev. 0 (2010). http://www.freescale.com/
36. Glossner, J., Iancu, D., Moudgill, M., Schulte, M., Vassiliadis, S.: Trends in low power handset software defined radio. In: Workshop on Embedded Computer Systems: Architectures, Modeling, and Simulation (SAMOS) (2007)
37. Gonzalez, R.: Xtensa: a configurable and extensible processor. IEEE Micro **20**(2), 60–70 (2000). doi:10.1109/40.848473
38. Guan, X., Fei, Y., Lin, H.: Hierarchical design of an application-specific instruction set processor for high-throughput and scalable FFT processing. IEEE Trans. Very Large Scale Integr. Syst. (TVLSI) **20**(3), 551–563 (2012). doi:10.1109/TVLSI.2011.2105512
39. Gupta, R.: The variability expeditions: exploring the software stack for underdesigned computing machines. Qualcomm, UCSD (2011)
40. Hansen, C.: WiGiG: multi-gigabit wireless communications in the 60 GHz band. IEEE Wireless Commun. **18**(6), 6–7 (2011). doi:10.1109/MWC.2011.6108325

41. Instruments, T.: SPRU731A: TMS320C62x DSP CPU and instruction set reference guide (2010)
42. INVOMEC, Division in IMEC, B-3001 Leuven, Belgium
43. IRC: ITRS roadmap on design (2009)
44. IRC: ITRS roadmap on interconnect (2009)
45. IRC: ITRS roadmap on lithography (2009)
46. Itoh, M., Higaki, S., Sato, J., Shiomi, A., Takeuchi, Y., Kitajima, A., Imai, M.: PEAS-III: an ASIP design environment. In: International Conference on Computer Design (ICCD), pp. 430–436 (2000). doi:10.1109/ICCD.2000.878319
47. Janhunen, J., Silvén, O., Juntti, M.: Programmable processor implementations of K-best list sphere detector for MIMO receiver. Sig. Process. **90**(1), 313–323 (2010). doi:10.1016/j.sigpro.2009.06.026
48. Jones, H.: Feature dimension reduction slowdown. International Business Strategies Inc. EE Times (2012). http://www.eetimes.com/discussion/other/4238315/Feature-dimension-reduction-slowdown/
49. Kahng, A.: Design technology productivity in the DSM era. In: Asia and South Pacific Design Automation Conference (ASP-DAC), pp. 443–448 (2001). doi:10.1109/ASPDAC.2001.913348
50. Karuri, K., Leupers, R.: Application Analysis Tools for ASIP Design: Application Profiling and Instruction-Set Customization. Springer, New York (2011)
51. Keutzer, K., Malik, S., Newton, A.: From ASIC to ASIP: the next design discontinuity. In: IEEE International Conference on Computer Design (ICCD), pp. 84–90 (2002). doi:10.1109/ICCD.2002.1106752
52. Kneckt, J., Alanen, O., Chang, Z., Huovinen, T., Nihtila, T.: IEEE 802.11ac: Enhancements for very high throughput WLANs. In: IEEE International Symposium on Personal, Indoor and Mobile Radio Comm. (PIMRC), pp. 849–853 (2011). doi:10.1109/PIMRC.2011.6140087
53. Li, M.: Algorithm and architecture co-design for software defined radio baseband. Ph.D. thesis, KU Leuven (2010)
54. Li, M., Fasthuber, R., Novo, D., Van Der Perre, L., Catthoor, F.: Algorithm-architecture co-design of soft-output ML MIMO detector for parallel application specific instruction set processors. In: Design, Automatin and Test in Europe (DATE), pp. 1608–1613. IMEC (2009)
55. Li, M., Appeltans, R., Amin, A., Torrea-Duran, R., Cappelle, H., Hartmann, M., Yomo, H., Kobayashi, K., Dejonghe, A., Van Der Perre, L.: Overview of a software defined downlink inner receiver for category-E LTE-advanced UE. In: IEEE International Conference on Communication (ICC), pp. 1–5 (2011). doi:10.1109/icc.2011.5963387
56. Liu, D., Nilsson, A., Tell, E., Wu, D., Eilert, J.: Bridging dream and reality: programmable baseband processors for software-defined radio. IEEE Commun. Mag. **47**(9), 134–140 (2009). doi:10.1109/MCOM.2009.5277467
57. McKernan, E.: Intel's first 14nm Chip NOT an x86 processor. The semiconductor wiki project (2012). http://www.semiwiki.com/forum/content/1112-intel92s-first-14nm-chip-will-not-x86-processor.html
58. Meyr, H.: System-on-chip for communications: the dawn of ASIPs and the dusk of ASICs. In: IEEE Workshop on Signal Processing Systems (SIPS), pp. 4–5 (2003). doi:DOIurl10.1109/SIPS.2003.1235634
59. Miniskar, N.R., Hammari, E., Munaga, S., Mamagkakis, S., Kjeldsberg, P.G., Catthoor, F.: Scenario based mapping of dynamic applications on MPSoC: a 3D graphics case study. In: International Workshop on Embedded Computer Systems: Architectures, Modeling, and Simulation (SAMOS), pp. 48–57 (2009). doi:10.1007/978-3-642-03138-0_6
60. Mollick, E.: Establishing Moore's law. IEEE Ann. Hist. Comput. **28**(3), 62–75 (2006). doi:10.1109/MAHC.2006.45
61. Munaga, S., Catthoor, F.: Systematic design principles for cost-effective hard constraint management in dynamic nonlinear systems. Int. J. Adapt. Resilient Auton. Syst. (IJARAS) **2**(1), 18–45 (2011)

62. Nigam, T.: Scaling to the final frontier: reliability challenges in sub 20 nm technologies. In: IEEE International Integrated Reliability Workshop (IIRW), pp. xi–xi (2011). doi:10.1109/IIRW.2011.6142574
63. Niknejad, A.: Siliconization of 60 GHz. IEEE Microw. Mag. **11**(1), 78–85 (2010). doi:10.1109/MMM.2009.935209
64. Picochip: PC205 product brief (2009). http://www.picochip.com/
65. Portero, A., Talvera, G., Catthoor, F., Carrabina, J.: A study of a MPEG-4 codec in a multiprocessor platform. In: International Symposium on Industrial Electronics (ISIE), pp. 661–666 (2006). doi:10.1109/ISIE.2006.295539
66. Portero, A., Talavera, G., Moreno, M., Carrabina, J., Catthoor, F.: Methodology for energy-flexibility space exploration and mapping of multimedia applications to single-processor platform styles. IEEE Trans. Circuits Syst. Video Technol. (TCSVT) **21**(8), 1027–1039 (2011). doi:10.1109/TCSVT.2011.2129750
67. Rabaey, J.M., Rabaey, J.M.: Silicon platforms for the next generation wireless systems—What role does...In: International Workshop on Field-Programmable Logic and Applications (FPL), pp. 277–285 (2000)
68. Ramacher, U.: Software-defined radio prospects for multistandard mobile phones. Computer **40**(10), 62–69 (2007)
69. Sanyal, A., Rastogi, A., Chen, W., Kundu, S.: An efficient technique for leakage current estimation in nanoscaled CMOS circuits incorporating self-loading effects. IEEE Trans. Comput. (TC) **59**(7), 922–932 (2010). doi:10.1109/TC.2010.75
70. Sawahashi, M., Kishiyama, Y., Taoka, H., Tanno, M., Nakamura, T.: Broadband radio access: LTE and LTE-advanced. In: International Symposium on Intelligent Signal Processing and Communication Systems (ISPACS), pp. 224–227 (2009). doi:10.1109/ISPACS.2009.5383862
71. Silicon Hive: HiveFlex CSP family (2010). http://www.siliconhive.com/
72. Stotas, S., Nallanathan, A.: On the throughput and spectrum sensing enhancement of opportunistic spectrum access cognitive radio networks. IEEE Trans. Wireless Commun. (TWC) **11**(1), 97–107 (2012). doi:10.1109/TWC.2011.111611.101716
73. Strojwas, A.J.: Cost effective scaling to 22nm and below technology nodes. In: International Symposium on VLSI Technology, Systems and Applications (VTSA), pp. 1–2. IEEE (2011). doi:10.1109/VTSA.2011.5872265
74. Trends in Semiconductor Scaling. International business strategies. http://www.internationalbusinessstrategies.com/
75. van Berkel, C.: Multi-core for mobile phones. In: Design, Automation and Test in Europe (DATE), pp. 1260–1265 (2009)
76. Woh, M., Mahlke, S., Mudge, T., Chakrabarti, C.: Mobile supercomputers for the next-generation cell phone. Computer **43**(1), 81–85 (2010). doi:10.1109/MC.2010.16
77. Zhou, F., Lv, Y.: Analysis of the 4G technologic requirements and key technology. In: International Symposium on IT in Medicine and Education (ITIME), pp. 494–497 (2011). doi:10.1109/ITiME.2011.6132157

Chapter 2
Background and Related Work

2.1 Background on Wireless Communication Systems

This section describes the general digital wireless communication system, depicts the main reasons for the increasing system complexity and explains the physical layer signal processing of advanced receivers.

2.1.1 The General Digital Wireless Communication System

A communication system can generally be characterized in 7 abstraction layers, which are referred as Open Systems Interconnection (OSI) layers. Each of the layers, i.e. ranging from physical layer up to application layer, specifies parts of the communication system. The lowest layer is the PHYsical layer (PHY), which specifies all characteristics related to the raw transmission of bits over a physical link. In the rest of this book, if not otherwise mentioned, we refer to the digital signal processing of the physical layer.

As shown in Fig. 2.1, a general wireless communication system consists of a transmitter and a receiver. The data is communicated via a wireless link, i.e. the physical transmission medium is air. Because the wireless link can disturb the data, the transmitter encodes the data before transmission. The encoding adds redundant data to the data-stream, which ensures that the receiver can recover the disturbed data. After encoding, the data is modulated so that it can be transmitted in a certain frequency band. The modulation includes the separation of the data-stream and the mapping onto symbols, the filtering to limit the bandwidth of the signal, the digital-to-analog conversion and the frequency-shifting of the baseband signal to the transmit frequency. Then, the resulting passband/Radio Frequency (RF) signal is amplified and transmitted via the antenna. The antenna of the receiver captures the transmitted signal. The demodulation reverses the steps of the modulation. An analog filter selects the frequency band in which the signal was transmitted. Then, the passband signal is amplified and down-converted to baseband. After the

R. Fasthuber et al., *Energy-Efficient Communication Processors*,
DOI: 10.1007/978-1-4614-4992-8_2,

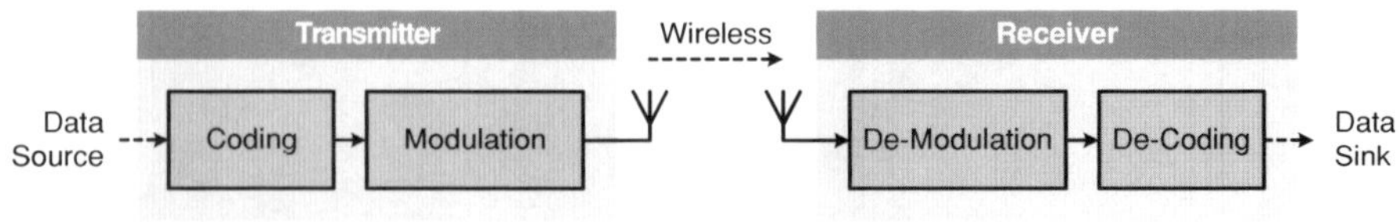

Fig. 2.1 The general wireless communication system

analog-to-digital conversion, an equalizer compensates the signal for certain undesired effects, such as phase distortion, time spreading and frequency selective fading, which have been introduced by the communication channel and the analog components. This compensation is essential to avoid inter-symbol interference and symbol distortion. The last step of the demodulation is the demapping, which converts the symbols back to a data-stream. The decoder leverages on the redundant bits to correct potential errors.

To enable a duplex communication, a mobile device includes a transmitter and a receiver, i.e. a transceiver. From the communication device of view, the wireless link for transmitting data is denoted as up-link, the wireless link for receiving data is denoted as down-link. For the up-link and the down-link in cellular networks, typically different frequency bands are used.

2.1.2 The Increasing Complexity of Wireless Communication Systems

For a wireless transmission, a certain frequency band is utilized for a certain time. With the wide variety of applications, e.g. ranging from radio broadcasting to radio communication for firefighters, the available frequency spectrum, which is most suited for wireless communication, has been fully allocated by now. For instance, for cellular communication networks in Europe, frequency bands between 700 MHz and 2.4 GHz have been assigned. Because of the increasing number of mobile users and the increasing channel bandwidths to enable higher data rates, the limited frequency spectrum is becoming more and more an issue [43]. For this reason, a broad research community has been focusing on new techniques to better utilize the available frequency spectrum and to approach the theoretical channel capacity defined by the Shannon bound. The most promising techniques are included in new communication standards. Thus, for each newly released communication standard, not only the data rates, but also the amount and the complexity of specified techniques increases. In general, the following main evolutions can be observed [162, 170]:

- From simple Quadrature Phase-Shift Keying (QPSK) to high order Quadrature Amplitude Modulation (QAM).

- From simple Time-Division Multiplexing (TDM) and Frequency-Division Multiplexing (FDM) to advanced Orthogonal Frequency-Division Multiplexing (OFDM) systems.
- From single channel to multi-channel use (channel bonding).
- From single antenna systems to systems with multiple transmit and receive antennas, i.e. Multiple-Input Multiple-Output (MIMO).
- From simple Viterbi error correction codes to complex Low-Density Parity-Check (LDPC) codes.
- From a static frequency spectrum use to a more dynamic one. Emerging systems, which sense and observe the availability of the frequency spectrum at run-time, and then adapt accordingly, are referred as Cognitive Radio (CR) systems.

Because of the trend towards higher data rates, more standards, more modes per standard and more complex techniques, the complexity of wireless communication devices is increasing dramatically [19, 162].

2.1.3 Physical Layer Signal Processing in an Advanced Receiver

In an advanced wireless communication system, the physical layer digital signal processing of the receiver is more complex than that of the transmitter. Therefore, the explanations and case studies of this book focus mainly on the receiver. Nevertheless, the proposed concepts of this work are also applicable to the transmitter.

Figure 2.2 shows the physical layer block diagram of a typical advanced wide-band OFDM receiver. The receiver consists of an Analog Front-End (AFE), an ADC, a Digital Frond-End (DFE), a digital inner modem and a digital outer modem. The AFE includes the antennas, the analog filters and the amplifiers, the mixers and the oscillators for the down-conversion. Once the analog/RF signal is down-converted to the baseband, the ADC quantizes the signal. The DFE performs certain compensations, such as DC offset and IQ imbalance compensation. In a wide-band receiver, not only the bandwidth of one channel, but a wide frequency spectrum is down-converted to the baseband. The wide frequency spectrum may be used for sensing and/or for the simultaneous reception of multiple channels. From the down-converted frequency spectrum, the actual channel selection and the adjustment of the sampling rate is also part of the DFE functionality. The main blocks of the inner modem are the FFT, which transforms the signal from the time-domain to the frequency-domain, the MIMO detector, which combines the signal information from multiple receive antennas and the demapper, which converts symbols to a bit-stream. The outer modem consists mainly of the Forward Error Correction (FEC) block, which performs the decoding. Beside the signal processing of the actual data, i.e. the payload data, the signal processing of the packet header information, i.e. pre-amble, is performed. The pre-amble functionality

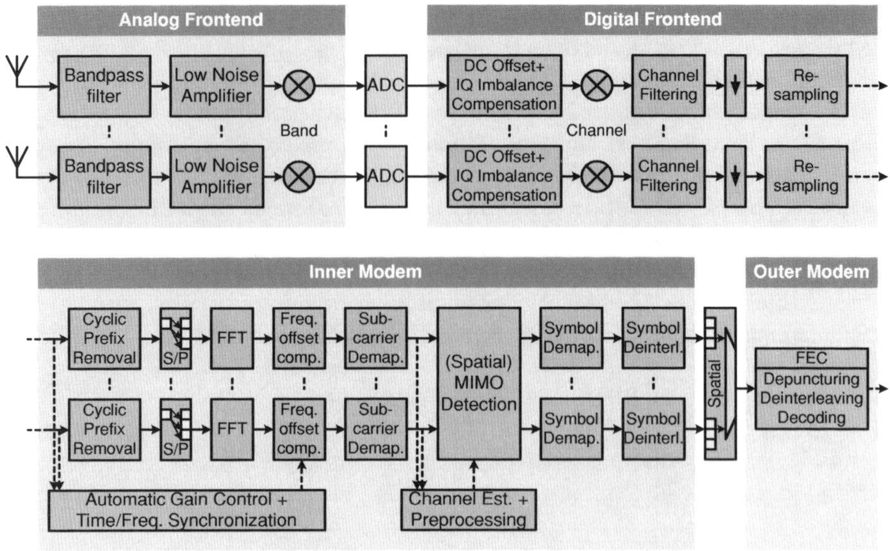

Fig. 2.2 Functional blocks of a typical wide-band OFDM MIMO receiver, which can perform the physical layer signal processing of modern wireless communication systems. The functional blocks with the highest computational complexity, from the digital domain, are typically the channel filter in the Digital Front-End (DFE), the Fast Fourier Transformation (FFT) and the MIMO detector in the inner modem and the Forward Error Correction (FEC) decoder in the outer modem

computes parameters, such as time and frequency offset, which are then used for the payload signal processing. In general, the payload processing has a higher computational complexity than the pre-amble processing. Nevertheless, the latter is typically more demanding in terms of latency. The functional blocks of the digital physical layer processing, often denoted as baseband blocks or kernels, are generally very heterogeneous in terms of computational complexity, functionality, employed type of algorithms and operations. This is a further reason for the high design complexity of an advanced wireless communication system.

2.2 Background on Architecture Styles for Wireless Communication Systems

For the implementation of the digital physical layer functionality, architecture styles, with different degrees of customization have been proposed. Ranging from very specific to very generic, we distinguish between ASICs and reconfigurable ASICs (rASIC)s, ASIPs (AP: Application Processor) and ASIPs (BP: Baseband Processor), DSIPs, DSPs, Field Programmable Gate Arrays (FPGA)s, Graphic Processing Units (GPU)s and general purpose Central Processing Units (CPU)s.

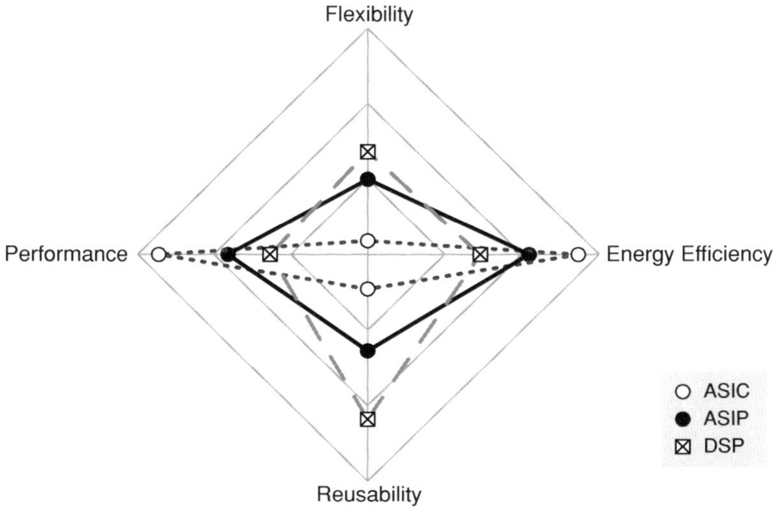

Fig. 2.3 Current architecture styles rASIC/ASIC, ASIP-AP/BP, DSIP/DSP offer different trade-offs in terms of performance, energy efficiency, flexibility and reusability. Nevertheless, considering the requirements, all of these metrics should be maximized

Because FPGAs, GPUs and general purpose CPUs consume clearly too much energy for the considered performance targets, they are basically only suited for prototyping in this domain [149]. Figure 2.3 characterizes the ASIC, ASIP and DSIP/DSP architecture styles in terms of performance, energy efficiency, flexibility and reusability, which are the most relevant metrics for wireless communication systems. Since metrics, such as performance and flexibility, oppose conflicting constraints, no architecture style is able to maximize all benchmark criteria. Thus, the architecture styles offer different trade-offs. As it can be seen in Fig. 2.4, the style of an architecture and the applied system design approach are closely linked. For instance, for each functional block, an individual ASIC will be designed. In contrast, several functional blocks will be implemented on a single ASIP (BP). ASICs are completely customized for the application. DSPs are typically not customized at all, i.e. they are directly purchased from vendors. In the following we provide a brief description of the considered architecture styles and the main differences regarding design.

2.2.1 ASIC/rASIC

ASICs are dedicated, non-flexible circuits that are designed for one specific functionality and for one specific algorithm. rASICs are like ASICs, with the difference that they can support a limited number of configurations.

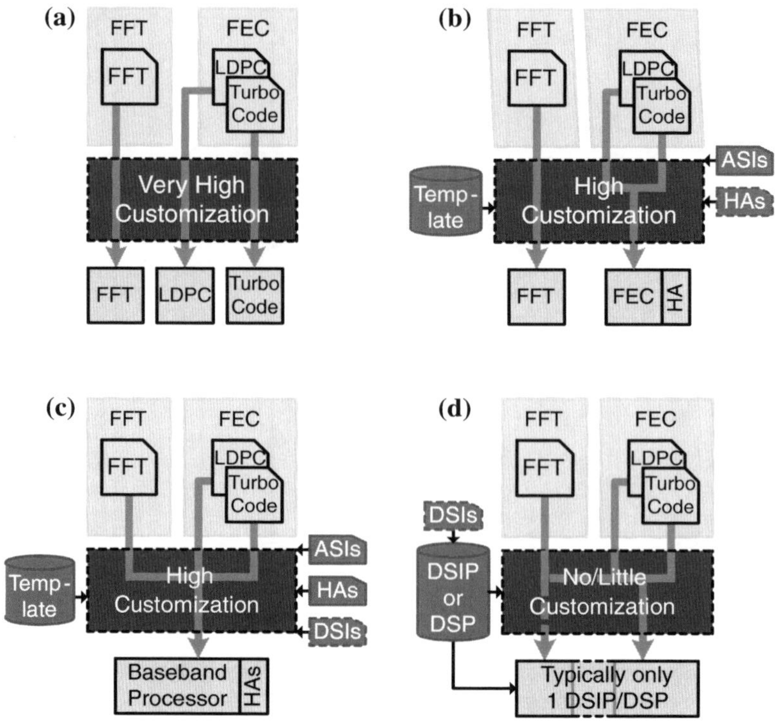

Fig. 2.4 Current architecture styles and design approaches for the the implementation of the digital physical layer functionality. The degree of customization varies across the different options. Input on the *top*: the functional blocks and the algorithms that need to be supported. Output: the implemented hardware blocks. **a** ASIC, rASIC. **b** ASIP (AP). **c** ASIP (BP). **d** DSIP, DSP

As shown in Fig. 2.4a, for each functional block of the system, and for each sub-functionality, an individual ASIC/rASIC is designed. The architectures of ASICs/rASICs are designed manually and/or with High-Level Synthesis (HLS) tools [57], which offer a trade-off between design time and implementation efficiency.

2.2.2 ASIP (Application Processor)

ASIPs (AP) are designed for one specific functionality and for one or multiple specific algorithms. Opposite to ASICs, ASIPs are programmable. Therefore they enable the execution of different algorithms with different control flows. The ASIP architecture style is generally attractive for functionality, which is based on rather stable algorithms.

As shown in Fig. 2.4b, for each functional block of the system, an individual ASIP (AP) is designed. Contrary to ASICs, sub-functionalities, which do not have to be executed at the same time, can be supported with one ASIP (AP) (time-sharing). ASIPs (AP) are customized with Application/Algorithm Specific Instructions (ASI)s which accelerate the execution of the program. For algorithms that cannot efficiently be mapped onto processors, also a customization with Hardware Accelerators (HA)s, which are ASIC/rASIC circuits, is possible. Because for ASIPs, a co-design and a compilation of software onto the hardware is needed, a manual design of ASIPs, especially when starting from scratch, demands a high investment, e.g. for building a compiler and instruction set simulation environment and for writing the Register Transfer Level (RTL) Hardware Description Language (HDL) code [119] of the architecture. Therefore, ASIPs are typically designed with processor design tools. The processor architecture is described in a certain tool-specific Architecture Description Language (ADL) [93]. To enable compiler and simulation support, the ADL and the processor design tool limit the design freedom to a certain extend. Although no clear boundary exists, a general distinction between the following two types of ADLs and processor design tools can be made [184]: (1) ADLs and processor design tools that target a specific customizable processor template. In this case, the architecture and the customization space is rather well defined. Thus, the design freedom in mainly limited to the selection of predefined template customization options. (2) ADLs and processor design tools that enable the design of a broader range of architectures. In this case, parts of the architecture, such as the instruction memory hierarchy, are also predefined, but especially the arithmetic datapath can be strongly customized (almost to the same extend as for ASICs). For instance, contrary to (1), it would be possible to employ multiple parallel Functional Units (FU)s which are customized with different instruction sets and with a different number of pipeline registers. In Sect. 2.8.2 the related ADLs and tools will be reviewed.

2.2.3 ASIP (Baseband Processor)

A single ASIP (BP) is designed for multiple specific functionalities with multiple specific algorithms. This type of processor is typically utilized to perform the computations of the complete baseband. Accordingly, ASIPs (BP) are often referred as (SDR) baseband processors.

As shown in Fig. 2.4c, multiple specific functional blocks of the system are implemented with one ASIP (BP). Functionalities and sub-functionalities, which do not have to be executed at the same time, can potentially be mapped onto the same hardware (time sharing). Like ASIPs (AP), ASIPs (BP) are customized with ASIs and with HAs. However, the instruction set of ASIPs (BP) is much richer and the customization with HAs is more common than for ASIPs (AP). Beside ASIs and HAs, ASIPs (BP) can also include Domain Specific Instructions (DSI)s. DSIs are instructions that are shared across functionality borders. For instance, complex multiplications, floating point and trigonometric operations. Like ASIPs (AP), ASIPs (BP) are usually designed with architecture templates and processor design tools. Considering the complexity, this is often the only way to keep design time and design cost acceptable.

2.2.4 DSIP

DSIPs are designed for a specific domain, but not for a specific functionality/algorithm. Contrary to ASIPs, DSIPs do not include ASIs/HAs which would limit the reusability. Nevertheless, DSIPs typically include DSIs.

As shown in Fig. 2.4d, typically multiple functional blocks of the system are implemented with one DSIP. Also DSIPs can leverage on time-sharing and time-multiplexing to save hardware area. Today's DSIP design approach does not foresee major customizations. Therefore today's DSIP architectures are slightly customized, e.g. number of cores, memory sizes, or not customized at all and directly purchased as fixed instances.

Compared to today's DSIP design approach, the proposed design approach of this book differs in the following main aspects: (1) Instead of multiple, typically only one functional block of the system is implemented with one DSIP. This leads to smaller design instances. Note that this is the same approach as for ASIPs (AP). (2) Each DSIP instance is stronger customized for the requirements of the functional block. Because the customizations are more functionality than algorithm specific, the same DSIP can be reused for different algorithms of the functionality-domain. For example, the FFT functionality requires block processing, while the functionality of a filter requires stream processing. The functionality for channel estimation requires a high fixed-point precision, while for FEC decoding a few bits are sufficient. Thus, without sacrificing the benefit of flexibility and reusability, the customization of instances is possible. However, contrary to the ASIP (AP) and ASIP (BP) architecture styles, the customization involves only the selection of pre-defined DSIP architecture template parameters and no ASIs/HAs are incorporated. Therefore the design time can remain significantly lower and a better link between architecture and technology can be established.

2.2.5 DSP

General purpose DSPs are designed and utilized for multiple application domains, e.g. for multimedia, medical, aerospace and wireless applications. Because it would limit the reusability across domains, DSPs do not even include DSIs. Like for ASIPs (BP) and DSIPs, typically multiple functional blocks of the system are implemented with one DSP. DSP architectures are basically not customized.

Instead, the common approach is to select the DSP, which has the most suited architecture, from a DSP vendor.

2.2.6 Other Styles

FPGAs, GPUs and general purpose CPUs are also sometimes used for wireless implementations (especially in academia and for prototyping). However, they are clearly not sufficiently energy effective and also not cost effective for the targeted physical layer consumer application domain. For this reason we do not further review and consider them in this book.

2.3 Background on the Physical Layer System Design

In this section the general physical layer system design flow is described. More details and more references will be provided in the Sects. 2.4–2.7.

The physical layer functionality of a digital wireless communication system is implemented in several design phases. As shown in Fig. 2.5, the following design phases can be defined: functionality design, algorithm design, partitioning, architecture design, software mapping and compilation and hardware implementation. The design flow starts with the system specification and ends with the implemented hardware and software. The system specification can generally be separated into functional system specification and hardware system specification. With the functional system specification, the targeted communication standards are specified. Since throughput and latency requirements are defined in standards, these requirements are implicitly included in the functional system specification. The hardware system specification includes constraints on chip area, energy consumption, technology, available hardware IP and interfaces.

2.3.1 Functionality Design

The first phase in the design flow is the functionality design, which starts with the functional system specification. The main task of this phase is the design of a simulatable standard-compliant model of the targeted wireless communication system. This task starts with the extraction of requirements from standards, e.g. modes, number of antennas, channel bandwidth, modulation scheme, techniques such as MIMO and OFDM, maximal bit-error rate, and ends with the implementation of the system functionality with several functions/functional blocks. Thereby the analog and the digital-domain of the physical layer are considered. Because accurate channel models and noise models of the AFE are normally not

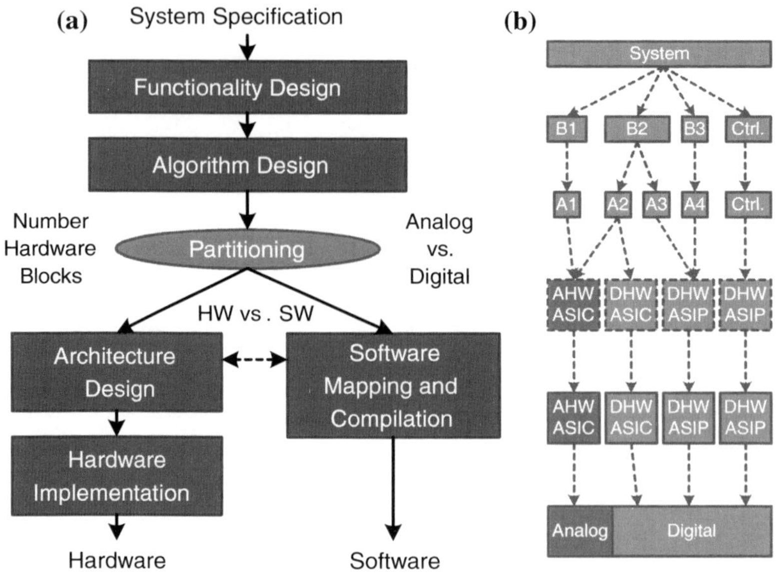

Fig. 2.5 **a** General system design flow and **b** illustrative example for the implementation of the physical layer functionality. Abbreviations: *B* Functional block, *A* Algorithm, AHW/DHW...Analog/Digital hardware block

available at this point, pessimistic assumptions are made. To limit the design complexity, the functional blocks are typically implemented with standard algorithms and floating-point arithmetic. At this phase, the link to hardware is rather weak, therefore throughput and latency requirements are largely neglected.

2.3.2 Algorithm Design

During this second phase, the functional blocks are implemented with algorithms that are better suited for implementation. Since the hardware platform has not been defined yet, several algorithm types, e.g. with different trade-offs regarding parallelization, memory and operation requirements, may be considered. Typically I/O-consistent and I/O-approximate algorithm transformations are performed [113]. I/O-consistent transformations, e.g. the conversion of a serial data-flow structure to a parallel one, change complexity trade-offs but do not influence the BER performance. In contrast, because of the approximations that are made, I/O-approximate transformations have an influence on the BER performance. An important I/O-approximate transformation, which is typically performed at this phase, is the floating-point to fixed-point conversion [143]. For functionality that is situated close to the analog/digital interface, the implementation in analog, the implementation in digital or the implementation as mix of both may be considered. After the algorithm design phase, for each digital functional block, one or several optimized and quantized algorithms are available.

2.3.3 Partitioning

In this third phase, the general hardware platform is designed. The partitioning of the algorithms/functionalities can be divided into three hierarchical sub-steps: (1) Separation into analog and digital functionality. (2) Separation of the analog/digital functionality into several hardware blocks. A hardware block could be implemented as an ASIC or a as programmable processor. (3) Separation of the digital functionality of a hardware block into software and/or hardware.

The separation into analog and digital functionality (sub-step 1) is mainly driven by hardware system specifications and by hardware cost models. Since the analog functionality has to be implemented with dedicated circuits, the separation into several hardware blocks and the software/hardware partitioning is rather obvious. In contrary, especially with the increasing complexity of wireless systems and the trend of implementing more functionality in digital, the partitioning of the digital functionally is very challenging [17]. The separation of the digital functionality into several hardware blocks (sub-step 2) is typically driven by (A) the system functionality, e.g. for which mode/at what time, which algorithm has to be executed, by (B) algorithm characteristics, e.g. type of operations, memory requirements and parallelization options, by (C) throughput and latency requirements, by (D) hardware system specifications and by (E) hardware cost models. For the software/hardware partitioning (sub-step 3), basically the same information as in sub-step 2 is used. The software/hardware partitioning, of the functionality of a digital hardware block, includes an implicit decision on the architecture style. For instance, for a hardware implementation, the ASIC architecture style has to be chosen. For the implementation of an algorithm in software, the DSIP architecture may be best suited. If only a sub-set of algorithm-specific operations should be implemented in hardware, the ASIP architecture style can be chosen. After the partitioning phase, for each digital hardware block, the architecture style, the type of assigned algorithms, the budget on latency, the flexibility requirements and the hardware system specifications are defined.

2.3.4 Architecture Design

In this phase, the architectures of the individual hardware blocks (which implement certain signal processing functionality) and the architectures for the platform communication and platform control are designed. Thereby the individually assigned algorithms, functionalities and specifications as well as the system specifications are considered. Depending on the chosen architecture styles, more or less customizations are performed. For an ASIP hardware block, typically the number of cores, the SIMD size, the subword size, the ASIs and/or HAs and the memory sizes are defined. The design approach, in which the architecture and the algorithm are co-optimized, is referred as algorithm-architecture co-design. This book focuses mainly on the architecture design phase.

2.3.5 Software Mapping and Compilation

During this phase, the algorithms, which should be executed in software, are mapped onto and compiled for the designed processor architectures. The software mapping refers to the coarse assignment of software parts to hardware parts and considers mainly Thread Level Parallelism (TLP). For instance, during the mapping, the computations for the DC-offset compensation could be assigned to core 1 and the computations for the IQ-imbalance compensation could be assigned to core 2. The mapping involves manual effort. The software compilation refers to the final assignment and scheduling of the software onto the hardware. For example, the computation $c = a + b$ is computed at time t on core 1, SIMD slot 1, the variables a and b are read from register R1 and R2 respectively and the sum c is stored in register R3. The compilation, which is performed by a compiler, is typically fully automated. In practice, mapping and compilation are performed together. Beside the software for the data computations, the platform level control software has to be designed.

2.3.6 Hardware Implementation

In this final phase, the hardware platform, which consists of the main hardware blocks and the glue logic, i.e. for data/control exchange and platform management, is implemented in hardware. Depending on the architecture style of a hardware block and the targeted degree of customization, different design flows are applicable. In general, analog blocks are manually designed and, for the integration into the hardware platform, provided as hard macros. For traditional digital standard-cell implementations, the circuits are first described in a Hardware Description Language (HDL), such as VHDL or Verilog. The HDL code can manually be written and/or automatically be generated. High-Level Synthesis (HLS) tools can generate the HDL code for digital ASICs, processor design tools can generate the HDL code for ASIPs. Once the HDL descriptions are available, logic synthesis can be performed. Thereby, under consideration of the hardware specifications, the circuits are mapped onto technology-dependent standard-cells and logic optimizations are applied. Afterwards, typically the individual physical layouts, i.e. hard macros, for the digital hardware blocks are made. The design of a physical layout includes floorplanning, e.g. placement of memory macros, placement of standard cells, clock tree synthesis and routing. As a final step, all hard macros of the platform, together with the remaining glue logic, are combined and the final chip is designed.

In general, the more phases of the design flow are considered together, the more cross-optimizations and thus, the more efficient designs can be made. However, in practice, because of the high complexity, a joint consideration of several phases is almost impossible. For this reason, it is important to leverage on appropriate abstraction layers and to leverage on appropriate cost models, which link the different phases of the design flow closely together.

2.4 Related Work on Functionality/Algorithm Design and Partitioning

In the Sects. 2.4–2.7, related work on design methodologies for the PHY system design is referred and reviewed. As previously mentioned, the intention of these sections is to provide a coarse overview and to point the reader to related literature and tools.

In the following we review related work on the functionality design, the algorithm design/optimization and the partitioning.

2.4.1 Functionality Design

Since the functionality design is rather far related to this book, we refer directly to other work for more information [144]. For the functionality design, mathematical tools that operate with floating-point numbers, such as Matlab, are used.

2.4.2 Algorithm Design and Optimization

The goal of the algorithm design and optimization phase is to obtain an algorithm or a set of algorithms which can efficiently be implemented. In general, the algorithm design and optimization phase is driven by the following two main tasks [111]: (1) Design/optimize an algorithm, for a certain BER performance requirement, which can most efficiently be implemented (with the considered hardware). (2) Combine different algorithms or enable the BER performance scalability of an algorithm to enable run-time trade-offs. The latter is only needed, if the concept of run-time scalability should be exploited. To enable an efficient implementation, the algorithm and the architecture has to be well-matched. A good match can be obtained by applying an algorithm-architecture co-design approach [112, 114]. This approach requires that the targeted architecture style is considered from the beginning on. The algorithm design and optimization flow starts generally with I/O-approximate transformations and ends with I/O-consistent transformations [113]. I/O-approximate transformations influence the BER performance while I/O-consistent transformations do not have an influence.

2.4.2.1 I/O-Approximate Transformations

In the following we consider the general design approach of [113], which includes the evaluation of multiple algorithm options. The design flow starts with the evaluation of possible algorithm candidates. For instance, to implement a MIMO detection functionality, the Minimum Mean-Squared Error (MMSE) algorithm, but also the Selective Spanning with Fast Enumeration (SSFE) algorithm could be used. The considered algorithms may offer different BER performance/complexity trade-offs. As second step, the algorithms are expanded and the full algorithm details are made visible. For instance, a matrix inversion function may be expanded to an Orthogonal Triangular Decomposition (QRD) based inversion algorithm. In general, the concept of hierarchical decomposition is applied [111]. The next step is to perform the floating-point to fixed-point conversion (quantization). Thereby also the numerical stability aspect has to be considered. In literature, many quantization methodologies have been proposed [143]. We refer to the PhD work of David Novo, which proposes a fine-grain quantization method [143]. Till here, instead of one, typically a set of algorithms is considered. By applying a High-Level Estimation (HLE) flow [22, 35] at this stage, the algorithms, which would anyway lead to inefficient implementations, can be detected and pruned, so that they are no longer considered. At this stage, also the combination of different algorithms, i.e. to enable different BER performance/complexity trade-offs, could be considered.

2.4.2.2 I/O-Consistent Transformations

The I/O-consistent transformations are stronger linked to the architecture. Here, not only the computational complexity, but also other criteria, such as data access behavior, address generation and compiler friendliness, are taken into account. Examples for optimization techniques are (1) strength reduction and coding to reduce the complexity of multiplications [38], (2) introducing more regularity and removing conditions to enable an efficient data-parallelization and (3) re-ordering of operations to reduce data storage requirements. Transformations are typically based on Control and Data Flow Graphs (CDFG)s. In literature, a wide range of I/O-consistent transformations have been proposed [13, 150, 165]. Different techniques for enabling run-time scalability are mentioned in [111].

2.4.3 Partitioning

As mentioned in Sect. 2.3.3, the general partitioning problem can be divided into three hierarchical sub-steps, which are (1) separation into analog and digital functionality, (2) separation of the analog/digital functionality into several hardware blocks and (3) separation of the digital functionality of a hardware block into software and/or hardware.

To reduce cost, analog and digital circuits are more and more integrated onto the same silicon. This high degree of integration enables new opportunities for analog-digital partitioning, i.e. for analog-digital co-optimizations. Because digital circuits benefit generally more from technology scaling and are easier to reproduce than analog ones, an increasing amount of analog functionality is implemented in the digital domain. Beside the concept of moving an analog functional block to the digital domain, the concept of digital compensation can be applied: By relaxing the quality factor of the analog circuits, they can be implemented with lower cost. However, the increased noise needs to be compensated. By compensating this noise with digital circuits, which may have a low implementation cost, the overall implementation cost (analog + digital) can be reduced. Related work on analog-digital partitioning can be found in [14, 73, 103, 203].

Especially with the increasing number of standards and modes, and the need for more energy and area efficient implementations, the partitioning of digital functionality across hardware blocks is becoming very challenging [17]. For this reason, a wide research community is focusing on systematic partitioning methodologies which reduce the design complexity. In literature, this problem is also often referred as heterogeneous MultiProcessor System-on-Chip (MPSoC) design. Related work can be found in [9, 17, 53, 54, 83]. In addition, we want to refer to the PhD work of Prashant Agrawal [3–5]. Here, also the work on HLE flows has to be mentioned [22, 35]. By applying a HLE flow at this early stage, the least promising options can be detected and pruned. Thus, the detailed exploration needs only to be performed on a smaller search space, which contains the most promising options.

The third partitioning problem is the partitioning between software and/or hardware, i.e. deciding, whether an algorithm or part of an algorithm should be computed in software or directly in hardware. At algorithm/functionality level, this is essentially the decision whether to employ a programmable processors or an ASIC. For parts of the algorithm, in case a programmable processor is employed, this relates basically to the decision whether ASIs/DSIs should be included or not. In literature, a wide variety of related exploration methodologies and algorithms have been proposed [110, 209, 213]. For more information on the design of ASIs/DSIs, we refer to Sect. 2.5.4.2.

2.5 Related Work on Architecture Design

In the following we review related work on the architecture design of the platform and of the ASICs and processors that are integrated in the platform.

2.5.1 Design of the Platform

The hardware blocks of the platform, i.e. ASIC and processor architectures, are first individually designed and then integrated in the hardware platform. The data communication between hardware blocks is typically enabled with point-to-point connections and/or with buses, e.g. with the AMBA bus from ARM. Network controllers, with Direct Memory Access (DMA) capability, control the data transfers. For the platform level control, typically Reduced Instruction Set Computer (RISC) architectures, e.g. from ARM, are employed. Beside the digital hardware, also the ADC and the analog circuits are integrated in the platform. The resulting mixed signal design leads to a high integration and thus, to low cost.

2.5.2 Design of an ASIC

The goal of this task is to design an efficient architecture for a given algorithm. Thereby hardware specifications and typically technology constraints are considered. The result is a verified RTL description in a HDL, like VHDL or Verilog, of the designed architecture.

After the algorithm design and optimization phase, the algorithm is typically represented with C code. To reduce design complexity and to avoid failures, especially for bigger designs, the architecture is first described in a HDL that enables a higher abstraction level. For instance, in SystemC, which is also based on C code. Then, in a second phase, the low-level RTL description is made. Thereby either a manual or an automated design approach is applied. For the latter, High-Level Synthesis (HLS) tools are used.

2.5.2.1 Manual Design

The manual ASIC design is the traditional approach. Thereby the RTL description of the architecture is written "by hand". In addition, to enable a verification of the the design, test-benches with representative test-vectors are made. For the RTL verification, HDL simulation tools, such as Mentor Graphics ModelSim [137], are utilized. This design approach enables maximal design freedom and can therefore lead to implementations with the highest efficiency. However, the design, maintenance and verification effort of this approach is very high. A small change in the algorithm requires the manual change of the design, e.g. in SystemC and in VHDL, which may even involve different design communities. In addition, especially for big designs, because a certain verification scenario has not been considered, the risk of failure is prevailing.

2.5.2.2 High-Level Synthesis

Because of the fore-mentioned issues of the manual design, which mainly effect the productivity, HLS tools are becoming more and more popular. HLS tools generate a synthesizable RTL code from an algorithm description in C, C++ or SystemC. The HLS design flow starts with the algorithm, with the hardware design specification and typically with a technology library. Based on this input, HDL tools perform generally the following steps fully automated [56]: (1) compilation and modeling, i.e. transform the input algorithm description into a formal representation and apply initial optimizations, (2) allocation, i.e. define the type and the number of hardware resources, (3) scheduling, i.e. assign cycles to all operations, (4) binding, i.e. bind variables, which are alive across cycles, to storage elements, and (5) RTL generation, i.e. generate a RTL model of the synthesized design. Often HLS tools are combined with logic synthesis tools so that technology-dependent decisions can be made [27, 134, 181]. In general, compared to the manual approach, the ASIC architecture design productivity by using HLS tools is much higher.

In recent years, a big research and design community has significantly improved the efficiency of HLS tools [56, 161]. State-of-the-art tools claim to generate RTL code that is as efficient as the manual written one [27]. However, current HLS tools do not incorporate enough technology information. For this reason, especially in future DDSM technologies, the energy efficiency of the resulting implementations will be low.

Popular HLS tools are Catapult C from Mentor Graphics [134], Symphony C Compiler from Synopsys [181], C-to-Silicon from Cadence Design Systems [27] PICO from Synfora (now Synopsys) [175] and the free and open-source tool ROCCC from Jacquard Computing [87].

2.5.3 Design of an Overall Processor Platform

As mentioned in Sect. 2.2, the architecture style has an influence on the processor architecture design. For instance, ASIPs (AP) are customized with ASIs, DSIPs are customized with DSIs and for DSPs the instruction set may not be changed at all. The processor design tools for ASIPs (AP) enable a higher degree of customization than the processor design tools for other processor architecture styles. Architecture-style specific processor design tools and templates will be reviewed in

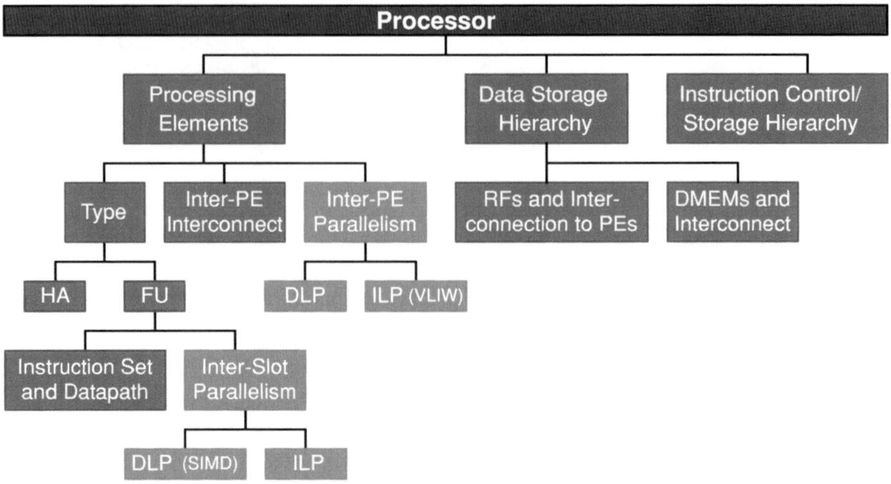

Fig. 2.6 Main architectural elements of a processor

Sect. 2.8. In the following we describe the processor architecture design in more general. The RTL description of the processor can be obtained with processor design tools, templates, soft macro IPs and/or by manual implementation.

As shown in Fig. 2.6, a processor generally consists of Processing Elements (PE)s, a Data Storage Hierarchy (DSH) and an Instruction Control/Storage Hierarchy (ICH/ISH). The PEs perform the actual arithmetic and logic computations. A PE can either be a Hardware Accelerator (HA) or a Functional Unit (FU). HAs are based on customized ASICs/rASICs that are designed for one fixed algorithm/ functionality. FUs are the flexible computational units of a processor. The instruction set of a FU defines the supported instructions, e.g. addition, multiplication. The datapath consists of datapath operators, e.g. adder and shifter, and of pipeline registers. A FU typically includes several parallel computational slots. These slots can execute the same instruction, i.e. Data Level Parallelism (DLP) can be exploited, or different individual instructions, i.e. Instruction Level Parallelism (ILP) can be exploited. The most common is DLP in form of a Single Instruction Multiple Data (SIMD) datapath. A processor consists typically of multiple PEs, which perform again DLP and/or ILP. Processors that perform ILP across FUs are often referred as Very Long Instruction Word (VLIW) processors. The PEs can be arranged and connected in form of an array, e.g. 3×3 or 4×4. Processors that include such arrays are referred as array processors. PEs could be interconnected with a bus [81, 146] or with a Network on Chip (NoC) [102, 108, 136]. The DSH generally consists of L0 Register Files (RF)s with the interconnection to PEs and the Data MEMory (DMEM) hierarchy with the interconnection to the RFs and to the hardware outside the processor. The ICH/ISH defines the overall control of the processor. It generally consists of a common instruction controller and/or distributed instruction controllers, optional L0 instruction buffers and the L1 and L2 Instruction MEMory (IMEM) hierarchy.

2.5.4 *Design of the Processing Elements of a Processor*

The design of the PEs can be separated into two abstraction layers.

2.5.4.1 Design at Processor Level

At processor level, essentially the number of PEs, the type of each PE, the interconnect and the employed parallelism is decided. This decisions are mainly driven by the characteristics of the considered algorithms, e.g. type of operations and exploitable parallelism, and by throughput requirements. HAs are employed to perform all operations or a big sub-set of operations of a certain algorithm. HAs are used for functionality that is rather stable, i.e. requires little flexibility, and for which an efficient implementation is essential. To decide on the type of interconnect between PEs, the amount of data transfers is considered. Since PEs are typically interconnected via RFs, the design of the inter-PE communication is jointly considered with the design of the data storage hierarchy. Multiple PEs enable typically ILP, often in form of a VLIW instruction organization.

2.5.4.2 Design at Processing-Element Level

For the design of HAs, we refer to the fore-mentioned literature on ASIC design. The instruction set of a FU can include general instructions that are used across application domains, Domain Specific Instructions (DSI)s and Application/Algorithm-Specific Instructions (ASI)s. As mentioned in Sect. 2.2, the employable instruction types are constraint by the architecture style. For instance, only ASIPs can be customized with ASIs. To decide on customized instructions, typically a code profiling of the targeted algorithms is performed. Thereby sub-graphs of CDFGs, which contribute significantly to the computational complexity, are identified [215]. By implementing the operations of the dominant sub-graphs directly in hardware, the throughput can substantially be accelerated. Nevertheless, the addition of an instruction is effecting the energy efficiency of the instructions that are already part of the instruction set. Reasons for that, amongst others, are the increased amount of multiplexing and the longer wire lengths. For the maximal acceleration of an algorithm, a very algorithm-specific ASI could be designed. Nevertheless, this high degree of customization will prevent the reusability, even if the algorithm changes only slightly. As it can be noticed, the design of instruction sets involves many trade-offs and can therefore be very challenging [215]. For this reason, a wide research community is investigating in this topic [83].

The insertion of pipeline registers, i.e. the computation of the instruction in several parallel stages, enables to increase the clock frequency.

The employed parallelism depends essentially on the characteristics of the considered algorithms. Most common is the support for DLP in form of SIMD

[194, 195, 207], i.e. the same operation is performed on multiple data words in parallel. This approach enables to share control cost. Slots, which can execute different instructions in parallel, enable ILP.

2.5.5 *Design of the Data Storage Hierarchy of a Processor*

The design of the DSH includes the design of L0 and the design of L1+.

2.5.5.1 Design of L0

This task involves (1) the design of the RFs, also denoted as foreground memories, (2) the design of the interconnect between RFs and PEs and generally also (3) the design of the interconnect between PEs, which are typically indirectly connected via the RFs. The PEs read the inputs from the RFs, perform the operations on the data, and then write the results back to the RFs. In general, the optimization goal is to minimize the number of RF words, the number of RF ports and the number of interconnects while avoiding data storage bottlenecks [106, 159]. RFs can be individually assigned to PEs or shared amongst several PEs. A technique to increase the bandwidth of a RF, without increasing the number of ports, is to split the RF into several parts [74]. Different types of RFs, such as hierarchical RFs [211], FIFO based RFs [191] and vector RFs [92, 104], may be considered. The latter is especially suited for FUs that employ SIMD. To reduce the number of RFs accesses, forwarding paths within a FU and amongst FUs can be made.

2.5.5.2 Design of L1+

This task involves (1) the design of L1 and L2 DMEMs, also referred as background memories, (2) the design of the internal interconnect, i.e. between L1 DMEMs and RFs, and between L1 and L2 DMEM, and (3) the design of the I/O data interconnect. The optimization goal is the same as for L0. Because the data flows in the PHY algorithms are rather deterministic and regular, typically software-controlled data memories, also referred as scratchpads, instead of hardware-controlled caches are employed. A scratchpad-based solution is typically more energy and performance efficient [16, 91], however, it requires a DMA controller to initiate the data transfers. To reduce the number of memory accesses and thus, the energy consumption, the concept of locality can be exploited. With this concept, the data is kept locally, e.g. in L1, and several operations are performed on it before it is written to the L2 memory (data reuse). To enable an exploitation of locality, the algorithm needs to be transformed and the data memory sizes must be adjusted accordingly. Data Transfer and

Storage Exploration (DTSE) [37, 39, 59, 145] methodologies, which analyze the possible reuse of data, can support design decisions.

As for the design of the PEs, the design of the storage hierarchy considers domain and algorithm specific characteristics. Code profiling tools, DTSE/HLE flows and hardware cost models can guide the decision making process.

2.5.6 Design of the Instruction C./S. Hierarchy of a Processor

The IMEMs contain the compiled program code, i.e. the instruction bits, which describe the tasks that have to be performed and the relative task execution order. To reduce storage space, the instruction bits are typically encoded and have to be decoded before execution. Beside the encoding, to further reduce the storage requirements, instruction and No OPeration (NOP) compression can be applied [23, 78, 193]. The instruction controller addresses the IMEMs and is responsible for the overall control of the processor. For the address computation, it leverages on control information from the IMEMs, from an optional branch unit and from external controllers. Like for the DSH, the ICH/ISH is composed of several hierarchy levels. The IMEMs in the bottom of the hierarchy are smaller and enable therefore a faster access. The IMEMs in the top of the hierarchy are bigger and enable therefore to store the whole program code. As for DMEMs, IMEMs can be hardware and/or software controlled. Hardware controlled IMEMs are referred as Instruction Caches (ICache)s. Software-controlled IMEMs, which are more commonly utilized for L0 and L1, are referred as configuration memories and loop buffers [90, 133]. The term configuration memory is used in the context of FPGAs and array processors. In the considered application domain, the program code is typically loop dominated, i.e. the most time is spend in the repetitive execution of a loop body. A loop buffer enables to store the instructions of a loop body. By reading the instructions from a small L0 loop buffer instead of reading them from a big L1 IMEM, a substantial amount of energy can be saved [154]. For this reason, ICH/ISHs with loop buffers have become more common. The organization of the hierarchy can be centralized and/or distributed [154]. Especially the separation of the L0 loop buffer into multiple loop buffers, and the distribution across PEs, can lead to significant energy savings [10]. Nevertheless, because of the increased compiler complexity, in todays designs, mainly centralized approaches are used.

The design of the ICH/ISH decides on the hierarchy levels and the organization, the IMEM types and sizes and the implementation of the controllers. Like for the DSH, the reduction of memory accesses, e.g. by exploiting locality, is important. The fact, that the number of read accesses is higher than the number of write accesses, can be considered for optimizations. Techniques to optimize the behavior of ICaches for loop code have been proposed in [98, 160]. Related work on centralized, clustered and distributed loop buffers can be found in [90, 94, 154].

2.6 Related Work on Software Mapping/Compilation

In the following we review related work on software mapping and compilation.

2.6.1 Software Mapping

Software mapping refers to the step of assigning and optimizing software code to/for certain hardware parts. The mapping of software is normally done in a hierarchical way. First, at hardware platform level, the system software code is separated and assigned into hardware blocks, i.e. to ASICs and processors. This is software mapping in the coarsest granularity, which is usually part of partitioning. Then, the software mapping is performed within a hardware block, i.e. within a processor. Thereby mainly tasks and/or algorithms, which have previously been assigned to the processor, are assigned to PEs. Beside the assignment of software code, software mapping includes also software optimizations which enable a more efficient mapping. For instance, loop unrolling can enable a better utilization of parallel datapaths. Software code optimizations are also performed within the compiler. However, to obtain efficient results, typically some manual software optimizations are needed. Often, the manual software optimizations leverage on compilation results. Data Transfer and Storage Exploration (DTSE) [36, 37, 39, 59, 145] and Instruction Transfer and Storage Exploration (ITSE) [65, 192] methodologies, which are also partly integrated as functionality in the compiler, can guide the software optimization process. Further information on transformation techniques can be found in [2, 131].

2.6.2 Compilation

Compilation is software mapping with the finest granularity. Thereby software code, e.g. operations and data variables, are assigned and scheduled to/for the hardware components. The software for the platform includes 1000s of operations, therefore the support of an automated compilation is essential. Commercially available processors and processor design tools include C compilers. A compiler performs typically the following steps (on the basis of the ADRES compiler) [133, 153]: (1) scratchpad locality optimization (e.g. DTSE), (2) SIMD/vectorization, (3) streaming data locality optimization, (4) platform independent instruction transfer and storage management (e.g. ITSE), (5) instruction selection for the address and the datapath, (6) scheduling and assignment for the address and the datapath, (7) platform dependent instruction transfer and storage management and finally (8) generation of assembly code.

To reduce design time, the software mapping/compilation phase and the hardware implementation phase are normally performed in parallel, typically with two different design communities.

2.7 Related Work on Hardware Implementation

In the following we review related work on the physical hardware implementation of the platform and on the physical hardware implementation of the blocks of the platform.

2.7.1 Implementation of the Platform

The implementation of the hardware platform typically starts with a rough chip planning, whereby the relative positions of the hardware blocks and the pin positions of the chip are considered. The obtained information, e.g. the optimal edges for I/O-connections of a hardware block, is then considered for the floorplanning within one hardware block. Only a rough chip planning is possible at this stage, because the exact sizes and dimensions of the hardware blocks are not yet known. Then, each hardware block is implemented individually. Thereby either a conventional standard-cell, a semi-custom and/or a full-custom design flow is applied. The different design flows will be explained below.

Once the netlists and the layouts of the individual analog and digital hardware blocks have been designed, they are combined and integrated to the platform level chip layout. The integration includes the design of the communication paths and the glue logic between the hardware blocks. As a final step, the hardware of the platform is verified. For the chip level design, commercial tools, such as Magma Hydra [125] are available.

2.7.2 Implementation of a Block

As previously mentioned, for the hardware implementation of a block, a conventional standard-cell, a semi-custom and/or a full-custom design flow is applied.

2.7.2.1 Conventional Standard-Cell Design

For the hardware implementation of a digital hardware block, i.e. an ASIC or a processor, traditionally the automated standard-cell design flow is applied. Thereby, the logic of the design is implemented with a set of primitive standard

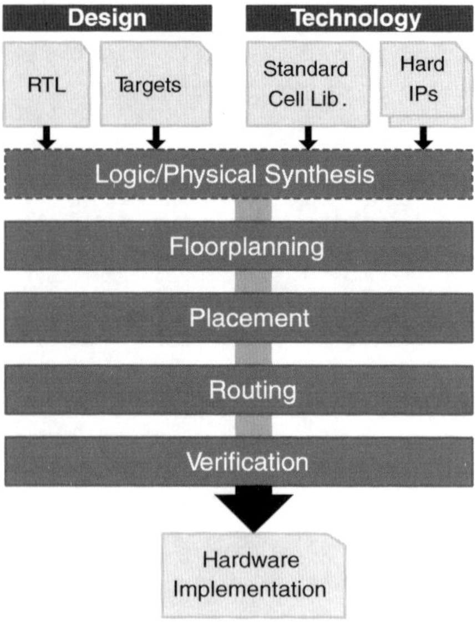

Fig. 2.7 Main steps of the traditional standard-cell design flow

cells, e.g. AND, OR, NAND, which are defined in a standard-cell library. The main advantage of this design flow is the high degree of automation, which results in shorter design times compared to custom-design flows. To enable the high degree of automation, clear restrictions on the standard cells and on the chip layout exist. For instance, all standard cells have a predefined physical height, the power lines are on the top/bottom edges of the cells and the transistors of the cells are only routed with poly and metal 1 wires. The physical width of the standard cells depends on the logic functionality and on the drive strength. The standard cells are placed within standard cell rows, which have a regular grid. In general, because of these restrictions, the efficiency of standard cell layouts is significantly lower compared to custom layouts [48].

Figure 2.7 shows the main steps of the traditional standard cell flow. The design flow starts with the functional RTL description of the design, e.g. in VHDL or Verilog, the design targets, e.g. clock frequency, area, power, pin positions and power islands, the standard cell libraries, e.g. for different design corners and threshold voltages, and the hard macros IPs, e.g. for memories.

During logic synthesis, the logic of the design is first translated to a technology-independent boolean description. Then, to reduce the design complexity, general boolean optimizations are performed. After that, the logic is mapped onto technology-dependent standards cells, test circuits are inserted and the design is further optimized. With the scaling towards DDSM technologies, the influence of wires on the delays is becoming very significant. Nevertheless, at this stage, the actual wire lengths and hence, the actual influence of wires is not yet known. Thus, to enable the essential logic optimizations, an accurate estimation of the wire influence is required. Until recently, wireload models have been employed to estimate the wire influence at this stage. Wireload models are look-up tables that define resistance and capacitance of wires based on design complexity and fan-outs of cells. Nevertheless, for DDSM technologies, the accuracy of these models is not sufficient. As a consequence, the design targets during the physical design phase cannot be met and frequent design iterations are needed. To overcome this limitation, wireload models are being replaced with actual placement and routing information. Synthesis tools that perform an initial placement and routing at this stage are referred as physical synthesis tools [61]. Commercial physical synthesis tools are Synopsys Design Compiler [177], Cadence Encounter RTL Compiler Physical [30] and Magma Talus Design [127]. Once technology-dependent optimizations have been performed, the gate-level netlist and the design reports, e.g. including standard-cell area and maximal delay, are generated.

After synthesis, the floorplan, i.e. the chip layout, is defined. Thereby the main wire connections, the area of the hard macro IPs, the area of the standard-cell logic and the pin positions, which have been obtained from the platform-level floorplanning, are considered. The result of floorplanning is a chip layout, where the hard macro IPs and the main pin positions are placed, and for the placement of standard cells, certain areas may be assigned/defined.

As third step, the placement of the standard cells takes place. This task is driven by coarse wire routings and by wire congestion analyses.

Then, Clock Tree Synthesis (CTS) is carried out. Thereby the placement of the sequential cells is refined, buffer cells are inserted and the clock-tree related cells are routed. After that, the routing of the hard macros and all other standard cells is performed. Usually in each routing layer only vertical or horizontal wires are made and a wire is typically only routed uni-directionally on each layer (e.g. the routing around a via is not possible). During the final optimization step, which is based on accurate RC-parasitic information, the drive strengths of the cells are adjusted and the placement and routing is refined. Commercial physical design tools are Synopsys IC Compiler [178], Cadence Encounter Digital Implementation System [28] and Magma Talus Design [127].

Finally, the design is verified for functional and electrical correctness. Popular commercial tools are Synopsys IC Validator [179], Synopsys PrimeTime [180], Cadence Encounter Timing System (ETS) [31], Cadence Encounter Power System (EPS) [29] and Cadence Physical Verification System [32].

2.7.2.2 Semi-custom Design

The semi-custom design flow is less automated, but enables to obtain more efficient layouts. The higher efficiency is mainly obtained because of the following two reasons: (1) Shorter wire lengths, especially of highly active wires. By constraining the cell/macro placement and the routing, the wire lengths can potentially be reduced. For instance, the cells of the datapath could be placed in a more regular way, e.g. in bit-slices. The routing could be influenced by defining the routing layers for certain nets and by manually placing pin positions. In addition, the design of custom cells can lead to shorter wire lengths. For instance, by implementing a complex logic within one cell, the wire lengths between transistors is reduced. Shorter wire lengths lead to lower RC delays and thus, to a lower amount of buffers. Because of the lower wire-switching capacities and the lower amount of buffers, the energy consumption is reduced. (2) A more fine-grain sizing of transistors, i.e. less over-sizing. To leverage on this, typically the design of a custom-cell library is required.

In the following we distinguish between two main semi-custom design flow types. For type A, the traditional standard-cell library is used. However, contrary to the traditional flow, the design steps are strongly influenced by manually defined constraints. Type B is like type A, but here also the cell library, which has no restrictions on cell dimensions, is designed manually.

Design flow type A: Here, the gain in efficiency origins from a reduction of wire lengths (reason 1). Most related work is focusing on a reduction of wire lengths by taking information of the datapath architecture into account. For instance, the information, that bits are processed in parallel and that the data flows from left to right. By making a regular bit-sliced datapath layout, wire lengths can be reduced. In this design flow, the steps are basically the same as for the traditional automated standard-cell flow. The main exception is the constraint for regularity, which is valid throughout the flow. Conventional synthesis tools optimize for criteria like minimum delay and minimum area, but not for regularity. Nevertheless, to enable a regular placement, the netlist needs to contain a form of regularity. Synthesis algorithm that optimize for regularity are proposed in [115, 205]. The work of [44, 64, 185, 206] proposes techniques to make regular cell placements. Related work on routing algorithms, which extract regularity and perform a regular routing, can be found in [60, 197].

Recent versions of commercial back-end tools, such as Cadence Encounter Digital Implementation System [28] and Magma Talus Design [127], support features for constraining the placement of standard cells. After defining the relative placement of cells, the DataPath Generator (DPG) of EECS Aachen [75, 198] can generate the corresponding layout. The datapath compilers of ChipEDA [11, 142] and MicroMagic [85] offer a similar functionality.

Design flow type B: Here, the gain in efficiency origins not only from a reduction of wire lengths (reason 1), but also from a more fine-grain sizing of transistors (reason 2). To exploit the latter, a cell library, which fits better for the targeted circuits, needs to be designed and characterized. Cells can be designed with analog design tools, such as Synopsys Custom Designer [176] or Cadence Virtuoso [33, 34], and libraries can be characterized with tools like Liberate from Altos [8] and SiliconSmart from Magma [126]. Related work on cell design can be found in [18, 157, 171, 190].

The design flow is basically similar as the one of type A. Nevertheless, the steps are more complicated to automate, because the cells can potentially have different heights/shapes.

As motivated in Sect. 1.3.3, to ensure that an architecture can efficiently be implemented in DDSM technologies, a strong link between architecture and technology/layout is required. To establish such a strong link, while keeping the design time acceptable, the application of a semi-custom design flow is essential. Semi-custom design flows, which target similar goals as we in this book (see Sect. 3.2.2, appendix of Chap. 3 and [132]) have been proposed in [64, 115, 157, 198, 206]. These design flows provide a good basis for our work. However, to achieve the targeted goals of this book in the best-suited way, we need to combine and extend different concepts of these semi-custom design flows. In the appendix of Chap. 3 we will propose a semi-custom design flow that is well-suited for the targeted DSIP template design approach. The main differences to the fore-mentioned related semi-custom design flows will be mentioned there as well.

2.7.2.3 Full-Custom Design

As the name suggests, with the full-custom design flow, the most customized layout can be made. Beside the technology-specific design rules, basically no restrictions exist. This design flow enables to obtain the most efficient layout, but the design effort is also the highest. A full-custom design of today's wireless systems, which contain multi-million devices, is completely infeasible. Nevertheless, this design flow is applied for analog circuits, which contain only up to a few hundreds of devices. The main steps of this design flow are (1) design of the netlist, (2) sizing of transistors, (3) placement and routing of transistors and (4) functional, electrical and design rule verification. Popular commercial design tools are Cadence Virtuoso [33, 34] Synopsys Custom Designer [176] and Magma Titan Mixed-Signal Design Platform [128].

Because of the considerable high effort to design, to characterize and to maintain a processor architecture, we do not adopt this design approach.

2.8 Related Work on Wireless Architectures and Templates

In this section we review related designs and templates for the wireless baseband domain, which are based on the ASIC, rASIC, ASIP, DSIP and DSP architecture style. Thereby the main drawbacks of existing solutions are highlighted.

2.8.1 ASIC/rASIC

Until recently, standard-cell based ASICs have been the traditional implementation choice. For this reason, ASICs and rASICs for many different functionalities, e.g. ranging from simple FIR filters [166, 187] to complex LDPC decoders [130], have been proposed. Especially for popular functionalities, such as MIMO detection, an immense number of ASIC [40, 58, 67, 97, 99, 121, 129, 147, 167, 172, 173, 183, 189] and rASIC [24, 25, 47, 66, 96, 116, 120, 124, 138, 139, 164, 199, 204, 208] solutions have appeared in literature.

Because of the high degree of customization, ASICs offer generally the highest energy and area efficiency. However, the ASIC design approach is becoming less and less attractive [135]: (1) The trend to multiple standards and the resulting flexibility requirements can only be fulfilled with multiple ASIC instances. Nevertheless, this translates to a decreasing area efficiency. (2) The high degree on customization causes typically the execution of the whole design flow even if only little changes are required. Especially when considering advanced process technologies, this long design cycles are becoming unacceptable. HLS tools can partly compensate for this issue [57]. (3) Because the functionality of ASICs is completely fixed, upgrades are not possible. This implies the risk of failure and limits the reusability. (4) To amortize the high NRE costs, high market volumes are needed. Considering the fast evolving requirements and the lack of reusability, high market volumes are becoming unfeasible. Therefore the costs for ASICs become unaffordable in DDSM technology nodes. Structured ASICs [210] can only marginally contribute in solving this problem. Reconfigurable ASICs (rASIC)s can partly compensate for the fore-mentioned issues but the flexibility and reusability requirements of future systems can still not be fulfilled. Because of these drawbacks that are clearly prevailing in the wireless domain, a shift from traditional ASIC/rASIC to ASIP/DSIP solutions can be observed [9, 95, 135].

2.8.2 ASIP (Application Processor)

ASIPs (AP: Application Processor) are customized with ASIs and/or with HAs for one specific functionality and for one or multiple specific algorithms. With the trend towards more programmability, many ASIP-AP designs for different functionalities appear in literature. Examples are the ASIPs for synchronization [1], for FFT [79], for interleaving [196] and for FEC [105].

As mentioned in Sect. 2.2, a general distinction between the following two types of ADLs and processor design tools can be made [184]: (1) ADLs and processor design tools that target a specific customizable processor template and (2) ADLs and processor design tools that enable the design of a broader range of architectures. Related work that falls under (1) are the Xtensa Customizable Processors from Tensilica [77] and PEAS-I/II [7, 84] (predecessor of ASIP Meister). Related ADLs and commercial processor design tools that fall under (2) are nML with the tools from Target Technologies [69, 186], ASIP Meister (formerly PEAS-III) with the tools from ASIP Solutions [12, 86, 184] LISA with the tools from CoWare (now Synopsys) [182, 214], EXPRESSION [80], MIMOLA [109], ArchC [158] and NoGap [93].

ASIPs (AP) are often based on a Reduced Instruction Set Computing (RISC) architecture with customized instruction set [184]. However, also ASIPs with data-flow architectures have been proposed. Examples are the Transport Triggered Architecture (TTA) [55] and the Synchronous Transfer Architecture (STA) [50]. For these architectures, the functionality of the units and their interconnects are typically customized [89].

The ASIP (AP) design approach offers a trade-off between flexibility and efficiency. Nevertheless, this design approach has the following drawbacks: (1) ASIs, HAs and data-flow architectures are often difficult to design and the effort can hardly be reused. (2) If ASIPs (AP) are designed from scratch, the design time can be very high. If they are designed with templates, then certain optimizations may not be possible. For instance, utilizing several data memories to reduce data storage access cost. (3) Traditional ASIPs (AP) are significantly less energy efficient than ASICs. (4) Since ASIPs (AP) are designed for a particular functionality, also here the market volume may not be very high.

2.8.3 ASIP (Baseband Processor)

ASIPs (BP: Baseband Processor) are customized with ASIs and/or with HAs for multiple specific functionalities and for multiple specific algorithms. Besides, they can employ DSIs. Like for ASIPs (AP), ASIPs (BP) are typically designed based on architecture templates and processor design tools. However, mainly because of the high design complexity, the design freedom is typically more restricted. Examples for templates are the baseband processor CSP20x1, which can be customized with ASIs [168], and the configurable logic platform [169], both from Silicon Hive. Many related processors, which fit in the definition of ASIP (BP), have been proposed in literature [9]. In general, a distinction between (1) processors, which are mainly based on a SIMD, a Very Long Instruction Word (VLIW) and/or a vector architecture and (2) processors, which are mainly based on an array of interconnected FUs, can be made.

The following related processors fall under (1): The Linkoping University has proposed SIMT, which includes HAs for FEC processing and ASIs for FFT computations [141]. The Tomahawk SDR processor platform has been proposed by the Technology University of Dresden [117]. The vector and scalar units are supported with several HAs for filtering and decoding. ARM and the University of Michigan have proposed Ardbeg, which is a commercial prototype of SODA [202]. It includes a HA for Turbo Decoding and it supports Thread Level Parallelism (TLP). Infineon has introduced the MuSIC [155] and the X-Gold SDR20 baseband processor platform [152, 156]. Both include configurable HAs for filtering and decoding. Sandbridge Technologies has proposed Sandblaster 2 for 4G baseband processing [140]. This architecture supports TLP and includes ASIs for FFT and FEC computations. Also ASIs for advanced CORDIC operations can be supported [163]. The HiperSonic processor, which includes OnDSP for inner modem processing, has been introduced by Systemonic [100]. The CEVA processor XC321, which targets 4G, employs ASIs for matrix processing, filtering,

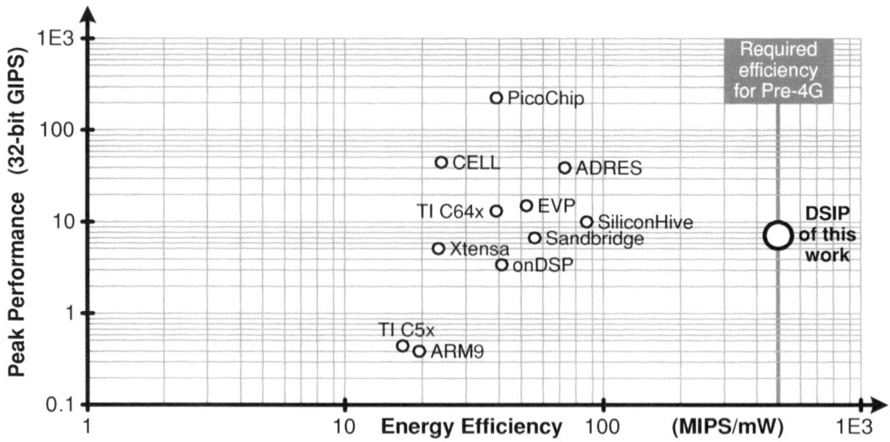

Fig. 2.8 Computational performance and energy efficiency of state-of-the-art processors. Normalized to 40 nm technology. Figure based on [38, 200]. The required energy efficiency of 100 GOPS/0.2 W for Pre-4G is specified in [19]

MIMO detection and so on [42]. Freescale offers several processors that fall in this category. One of them is the StarCore MSC8158, which includes the MAPLE-B2 HA to improve the efficiency for baseband processing [70]. An SDR processor with ASIs for FFT and demapping has been proposed in [46].

The following array processors fall under (2): The Intel RCA Coarse Grain Array (CGA) processor has been introduced with HAs in [49]. PicoChip's Pico-Array is a Fine Grain Array (FGA) processor. The processor PC205 includes HAs for FEC and FFT computations and targets physical layer and MAC layer processing [148].

Todays ASIPs (BP) have basically similar drawbacks as ASIPs (AP). The customization limits the reusability and hence, the market volumes. The high design complexity in combination with the limited design time restricts the space for optimizations. Besides, as shown in Fig. 2.8, the energy efficiency of current ASIPs (BP) constraints the employment in future Pre-4G and 4G systems. Reasons for the low energy efficiency are indicated in Table 2.1.

2.8.4 DSIP

DSIPs are designed for a specific domain, but not for a specific functionality or algorithm. Contrary to ASIPs, DSIPs do not include ASIs or HAs, but typically they include DSIs. Like for ASIPs (BP), a general distinction between (1) SIMD/VLIW/vector processors and (2) array processors can be made.

Table 2.1 Reasons for the low energy efficiency of existing ASIPs (BP) and DSIPs. We refer to Sect. 3.3 to better understand the mentioned reasons

Ref.	Processor	Reasons for low energy efficiency
[168]	SiHive CSP20x1	only 16 bits supported, conventional narrow-ported register file (e.g. 32 bits), centralized instruction hierarchy
[141]	SIMT	only 16 bits supported, suboptimal instruction hierarchy (long physical communication path because of sharing, loop characteristic not exploitable), rather big shuffling network, data exchange between heterogeneous units only via shuffling network
[117]	Tomahawk	only 16 bits and/or floating point hardware supported, conventional narrow-ported register file
[202]	SODA/ Ardbeg	only 16 bits supported (SODA) rather big shuffling networks, centralized instruction hierarchy
[156]	X-Gold SDR20	only 16 and 32 bits supported, big instruction set (210 instr.), shared memory with many ports
[107]	SODA-II	only 16 bits supported, big I/O-multiplexer of computation unit (activated for every operation), 3 ALUs present (but not very utilized)
[15]	ELM	conventional ALU, rather generic
[20]	EVP	only 16 (and 8) bits supported, data communication between units only via shared multi-ported register file, centralized program memory
[21]	MEPHISTO	only 16 bits supported, MAC datapath with hardware multiplier, conventional narrow-ported regfile/RAM, instruction cache, no customization of instances
[26]	ADRES	only 16 bits supported, conventional FUs, rather complex inter-FU interconnect
[88]	GENEPY/ SMEP	only 16 bits supported, conventional DSP FUs

The following proposals, which typically also support DLP and ILP, fit in the first category: The highly flexible processors SODA [118], SODA-II [107] and AnySP [201], which have been introduced by the University of Michigan. Sandbridge Technologies and others have proposed Sandblaster, which supports TLP [76]. The University of Standford has proposed ELM [15], which is a rather general processor that includes distributed data and instruction register files. EVP is a vector processor that has been proposed by Philips Research (NXP) and the University of Eindhoven [20]. Icera has introduced DXP [101]. MEPHISTO, which has been utilized in the MAGALI platform/chip [51, 52], has been proposed by LETI, CEA [21]. Other architectures are [123, 212]. Also, our previous work, the FEENECS architecture template, fits in this category [38].

DSIP CGA processors are the Intel's RCA [45], IMEC's ADRES [26, 174], which has been utilized in the SDR MPSoC platform/chip proposed in [62, 63], MONTIUM [82] from the University of Twente, Ninesilica [6] from the University of Tampere and GENEPY/SMEP [88] from LETI, CEA. Beside in CGA mode, the ADRES can also operate in VLIW mode. PicoChip's PicoArray [151] and XiRisc [122], from the University of Bologna, are FGA processors with DSIP characteristic.

ASIPs are customized with ASIs and/or HAs. Because of the limited reusability for new algorithms/requirements, this design approach causes a big issue for DDSM technologies. Nevertheless, since DSIPs include only DSIs, they can potentially support new disruptive communication standards without requiring architecture changes. Therefore, this architecture style is generally a good choice for the fast evolving market in which high volumes are becoming a necessity. However, the main drawback of state-of-the-art DSIPs is the energy efficiency, which is generally too low for emerging Pre-4G and 4G systems. Some reasons for that are indicated in Table 2.1. Besides, as it is also the case for other architecture styles, the proposed architectures are not well suited for the scaling in advanced DDSM technologies. An exception is the FEENECS architecture template [38] from our previous work, which generally considers the combination of energy efficiency and advanced technologies. However, for the employment in the wireless baseband domain, the template needs to be customized and extended to cope efficiently with the requirements of this domain. Besides, certain architectural concepts that are key elements for high energy efficiency, such as software SIMD or distributed loop buffers, have been proposed in [38], but not implemented and evaluated. We use the FEENECS architecture template as a basis for the proposed wireless domain-specific DSIP architecture template of this book.

2.8.5 DSP

Since general purpose DSPs are not customized with ASIs, HAs or DSIs, they can be employed in multiple application domains. The following commercial DSPs are often employed in the wireless domain: TMS320c6X from Texas Instruments [188], MSC8256 from Freescale [71], TigerSHARC from Analog Devices [72] and X1641 from CEVA [41].

DSPs offer the highest reusability, but generally also the lowest energy efficiency [68]. Because of the latter, DSPs are hardly considered for the wide employment in advanced battery-powered wireless communication devices.

2.9 Summary and Conclusions

This chapter can be summarized and concluded as follows:

As mentioned in Sect. 2.1, the need for higher data rates, the need for higher connectivity, the need for more efficient frequency spectrum use and the need for legacy support are the main reasons for the increasing flexibility requirements and the increasing complexity of wireless systems. Thus, only programmable and highly energy-efficient solutions are suited for this application domain.

As explained in Sect. 2.1.3, the digital physical layer functionality of wireless systems has the following two characteristics, which can be exploited to obtain a more efficient implementation: (1) the complete functionality can be separated into sub-functionalities, i.e. functional blocks (FFT, MIMO detector, etc.). (2) the functional blocks are rather heterogeneous. Thus, instead of utilizing one big hardware block, which needs to support all heterogeneous sub-functionalities, it is better to utilize multiple hardware blocks which are individually customized for the requirements of the sub-functionality.

As mentioned in Sects. 2.2 and 2.8, several architecture styles for the implementation of the digital physical layer functionality have been proposed. All architecture styles offer different trade-offs in terms of performance, energy efficiency, flexibility and reusability. The main issue of ASICs/rASICs is the lack of flexibility and reusability. The latter translates to high NRE costs in DDSM technology nodes. ASIPs are customized for certain algorithms, which decreases the reusability. Beside the reusability, the rather low energy efficiency of today's ASIPs is an issue. Existing DSIPs and DSPs offer high flexibility and reusability, but currently, the low energy efficiency prevents them from a wide employment by industry for high data rate standards. Instead of employing only one, the system could be implemented with a mix of different architecture styles. Nevertheless, because of the different design approaches and tools, the design and integration complexity may increase. Thus, the current combinations of architecture styles and associated design approaches are not well suited for the considered domain.

As it can be noticed from Sect. 2.3 and from the Sects. 2.4–2.7, the phases of the system design flow are strongly linked together and it is almost impossible to clearly split them. Besides, the design space and the cross-optimization space is huge. Thus, design approaches, that introduce abstraction layers, but where the abstraction layers do not hamper to achieve optimal implementations, are needed.

As mentioned in Sect. 2.7, because of the increasing wire influence in DDSM technologies, the traditional standard-cell hardware implementation flow is starting to break-down. Thus, a need for semi-custom design flows, which establish a stronger link between the application/architecture and the technology, is prevailing.

Considering the fore-mentioned issues and requirements, a clear need for new disruptive solutions exists. The proposed DSIP approach of this book, which combines (1) innovative architecture concepts with a strong link to functionality requirements and technology and (2) the DSIP template concept with a good mix of reusability and customization, shows a promising direction towards a pareto-optimal solution.

References

1. Abdelall, M., Shalash, A.F., Fahmy, H.A.H.: A reconfigurable baseband processor for wireless OFDM synchronization sub-system. In: IEEE International Symposium on Circuits and System (ISCAS), pp. 2385–2388. IEEE (2011). doi:10.1109/ISCAS.2011.5938083
2. Absar, J., Lambrechts, A., Min, L., Jayapala, M., Raghavan, P., Vandecappelle, A.: Locality optimization in wireless applications. In: International Conference on Hardware/Software Codesign and System, Synthesis (CODES+ISSS), pp. 125–130 (2007)

3. Agrawal, P., Fasthuber, R., Raghavan, P., Van der Aa, T., Ahmad, U., Van der Perre, L., Catthoor, F.: High level analysis of trade-offs across different partitioning schemes for wireless applications. In: IEEE Workshop on Signal Processing System (SIPS) (2011)
4. Agrawal, P., Raghavan, P., Hartmann, M., Sharma, N., Van der Perre, L., Catthoor, F.: Early exploration for platform architecture instantiation with multi-mode application partitioning. In: Design Automation Conference (DAC) (2013)
5. Agrawal, P., Sugand, K., Palkovic, M., Raghavan, P., Van der Perre, L., Catthoor, F.: Partitioning and assignment exploration for multiple modes of IEEE 802.11n modem on heterogeneous MPSoC platform. In: Euromicro Conference on Digital System Design (DSD), pp. 608–615 (2012). doi:10.1109/DSD.2012.102
6. Airoldi, R., Garzia, F., Anjum, O., Nurmi, J.: Homogeneous MPSoC as baseband signal processing engine for OFDM systems. In: International Symposium on System on Chip (ISSOC), pp. 26–30 (2010). doi:10.1109/ISSOC.2010.5625562
7. Alomary, A., Nakata, T., Honma, Y., Sato, J., Hikichi, N., Imai, M.: PEAS-I: A hardware/software co-design system for ASIPs. In: European Design Automation Conference (EURO-DAC), pp. 2–7 (1993). doi:10.1109/EURDAC.1993.410608
8. Liberate, Altos Design Automation. http://www.altos-da.com/
9. Anjum, O., Ahonen, T., Garzia, F., Nurmi, J., Brunelli, C., Berg, H.: State of the art baseband DSP platforms for Software Defined Radio: a survey. EURASIP J. Wirel. Commun. Networking **2011**(1), 5 (2011). doi:10.1186/1687-1499-2011-5
10. Artes, A., Huisken, J., Ayala Rodrigo, J.L., Fasthuber, R., Catthoor, F.: Energy impact of loop buffer schemes used in embedded systems. IEEE Embedded System Letters (ESL) (Submitted, 2012)
11. Ashkar, M.: Integrate custom layout with ASIC back-end design flow for high performance datapath design. In: International Conference on Solid-State and Integrated-Circuit Technology (ICSICT), pp. 1901–1904 (2008). doi:10.1109/ICSICT.2008.4734931
12. ASIP Solutions, Inc.: ASIP Meister. URL http://www.asip-solutions.com/
13. Auger, F., Lou, Z., Feuvrie, B., Li, F.: Multiplier-free divide, square root, and log algorithms [DSP Tips and Tricks]. IEEE Signal Process. Mag. **28**(4), 122–126 (2011). doi:10.1109/MSP.2011.941101
14. Badaroglu, M., Desset, C., Ryckaert, J., De Heyn, V., Der Plas, G., Wambacq, P., Van Poucke, B.: Analog-digital partitioning for low-power UWB impulse radios under CMOS scaling. EURASIP J. Wirel. Commun. Networking pp. 1–8 (2006). doi:10.1155/WCN/2006/72430
15. Balfour, J., Dally, W.J., Black-Schaffer, D., Parikh, V., Park, J.: An energy-efficient processor architecture for embedded systems. Comput. Archit. Lett. **7**(1), 29–32 (2007)
16. Banakar, R., Steinke, S., Balakrishnan, M., Marwedel, P.: Scratchpad memory: a design alternative for cache on-chip memory in embedded systems. In: International Symposium on Hardware/Software Codesign (CODES), pp. 73–78. ACM Press (2002). doi:10.1109/CODES.2002.1003604
17. Baruah, S.: Task partitioning upon heterogeneous multiprocessor platforms. In: IEEE Real-Time and Embedded Technology and Applications Symposium (RTAS), pp. 536–543 (2004). doi:10.1109/RTTAS.2004.1317301
18. Bekiaris, D., Papanikolaou, A., Stamelos, G., Soudris, D., Economakos, G., Pekmestzi, K.: A standard-cell library suite for deep-deep sub-micron CMOS technologies. In: International Conference on Design and Technology of Integrated Systems in Nanoscale Era (DTIS), pp. 1–6 (2011). doi:10.1109/DTIS.2011.5941445
19. van Berkel, C.: Multi-core for mobile phones. In: Design, Automation and Test in Europe (DATE), pp. 1260–1265 (2009)
20. van Berkel, K., Heinle, F., Meuwissen, P.P.E., Moerman, K., Weiss, M.: Vector processing as an enabler for software-defined radio in handheld devices. J. Adv. Signal Process. (EURASIP) **2005**(16), 2613–2625 (2005). doi:10.1155/ASP.2005.2613
21. Bernard, C., Clermidy, F.: A low-power VLIW processor for 3GPP-LTE complex numbers processing. In: Design, Automation and Test in Europe (DATE), pp. 1–6 (2011)

22. Berset, T., Catthoor, F.: Technical note TN-11-WATS-TIP2: high level estimation for independent component analysis. Technical report, Holst Centre, IMEC-NL (2012)
23. Beszédes, A., Ferenc, R., Gyimóthy, T., Dolenc, A., Karsisto, K.: Survey of code-size reduction methods. ACM Comput. Surv. **35**(3), 223–267 (2003). doi:10.1145/937503.937504
24. Bhagawat, P., Dash, R., Choi, G.: Dynamically reconfigurable soft output MIMO detector. In: International Conference on Computer Design (ICCD), pp. 68–73 (2008). doi:10.1109/ICCD.2008.4751842
25. Bhagawat, P., Dash, R., Choi, G.: Array like runtime reconfigurable MIMO detectors for 802.11n WLAN: a design case study. In: Asia and South Pacific Design Automation Conference (DAC-ASP), pp. 751–756 (2009). doi:10.1109/ASPDAC.2009.4796570
26. Bougard, B., De Sutter, B., Rabou, S., Novo, D., Allam, O., Dupont, S., Van der Perre, L.: A coarse-grained array based baseband processor for 100Mbps+ software defined radio. In: Design, Automation and Test in Europe (DATE), pp. 716–721. IMEC (2008). doi:10.1109/DATE.2008.4484763
27. Cadence, C-to-Silicon. http://www.cadence.com/
28. Cadence, Encounter Digital Implementation System. http://www.cadence.com/
29. Cadence, Encounter Power System. http://www.cadence.com/
30. Cadence, Encounter RTL Compiler. http://www.cadence.com/
31. Cadence, Encounter Timing System. http://www.cadence.com/
32. Cadence, Physical Verification System. http://www.cadence.com/
33. Cadence, Virtuoso Analog Design Environment. http://www.cadence.com/
34. Cadence, Virtuoso Layout Suite. http://www.cadence.com/
35. Catthoor, F.: Slide set that explains the high-level estimation flow, IMEC. Available on request
36. Catthoor, F., Danckaert, K., Kulkarni, K., Brockmeyer, E., Kjeldsberg, P.G., van Achteren, T., Omnes, T.: Data Access and Storage Management for Embedded Programmable Processors. Springer (2010)
37. Catthoor, F., Danckaert, K., Wuytack, S., Dutt, N.: Code transformations for data transfer and storage exploration preprocessing in multimedia processors. IEEE Des. Test Comput. **18**(3), 70–82 (2001)
38. Catthoor, F., Raghavan, P., Lambrechts, A., Jayapala, M., Kritikakou, A., Absar, J.: Ultra-Low Energy Domain-Specific Instruction-Set Processors, 1st edn. Springer (2010)
39. Catthoor, F., Wuytack, S., de Greef, G., Banica, F., Nachtergaele, L., Vandecappelle, A.: Custom Memory Management Methodology: Exploration of Memory Organisation for Embedded Multimedia System Design. Springer (1998)
40. Cerato, B., Viterbo, E.: Hardware implementation of a low-complexity detector for large MIMO. In: IEEE International Symposium on Circuits and System (ISCAS), pp. 593–596 (2009). doi:10.1109/ISCAS.2009.5117818
41. CEVA Inc.: CEVA DSP core X1641 Product Note (2008). http://www.ceva-dsp.com/
42. CEVA Inc.: CEVA-XC321 Product Note (2009). http://www.ceva-dsp.com/
43. Chandra, A.: Spectrum management for future generation wireless based technology. In: European Wireless Technology Conference (EuWIT), pp. 201–205 (2009)
44. Changqi, Y., Xianlong, H., Yici, C., Wenting, H., Tong, J., Weimin, W.: Standard-cell based data-path placement utilizing regularity. In: International Conference on ASIC (ICASIC), pp. 97–100 vol. 1 (2003). doi:10.1109/ICASIC.2003.1277499
45. Chen, I., Chun, A., Tsui, E., Honary, H., Tsai, V.: Overview of intels reconfigurable communication architecture. In: Workshop on Application Specific Processor, vol. 3, pp. 95–102. Intel (2004)
46. Chen, K.C.: A green software-defined communication processor for dynamic spectrum access. In: IEEE International Symposium on Personal, Indoor and Mobile Radio Communication (PIMRC), pp. 774–779 (2010). doi:10.1109/PIMRC.2010.5671956

47. Chien-Jen, H., Chung-Wen, Y., Hsi-Pin, M.: A power-efficient configurable low-complexity MIMO detector. IEEE Trans. Circuits and Syst. **56**(2), 485–496 (2009). doi:10.1109/TCSI.2008.2001368
48. Chinnery, D., Keutzer, K.: Closing the Power Gap between ASIC & Custom: Tools and Techniques for Low Power Design. Springer (2007)
49. Chun, A., Tsui, E., Chen, I., Honary, H., Lin, J.: Application of the intel reconfigurable communications architecture to 802.11a, 3G and 4G standards. In: Proceedings of the 6th Circuits and Systems Symposium on Emerging Technologies (CASSET), vol. 2, pp. 659–662. Intel (2004)
50. Cichon, G., Robelly, P., Seidel, H., Matus, E., Bronzel, M., Fettweis, G.: Synchronous Transfer Architecture (STA). In: Lecture Notes in Computer Science. Mobile Communcations Chair, TU-Dresden, pp. 193–207. Springer (2004). doi:10.1007/978-3-540-27776-7_36
51. Clermidy, F., Bernard, C., Lemaire, R., Martin, J., Miro-Panades, I., Thonnart, Y., Vivet, P., Wehn, N.: A 477mW NoC-based digital baseband for MIMO 4G SDR. In: IEEE International Solid-State Circuits Conference (ISSCC), pp. 278–279 (2010). doi:10.1109/ISSCC.2010.5433920
52. Clermidy, F., Bernard, C., Lemaire, R., Martin, J., Miro-Panades, I., Thonnart, Y., Vivet, P., Wehn, N.: MAGALI: a network-on-chip based multi-core system-on-chip for MIMO 4G SDR. In: International Conference on Integrated Circuit Design and Technology (ICICDT), pp. 74–77 (2010). doi:10.1109/ICICDT.2010.5510291
53. Cong, J., Gururaj, K., Han, G., Kaplan, A., Naik, M., Reinman, G.: MC-Sim: an efficient simulation tool for MPSoC designs. In: IEEE/ACM International Conference on Computer-Aided Design (ICCAD), pp. 364–371 (2008). doi:10.1109/ICCAD.2008.4681599
54. Cong, J., Sarkar, V., Reinman, G., Bui, A.: Customizable domain-specific computing. IEEE Des. Test Comput. **28**(2), 6–15 (2011). doi:10.1109/MDT.2010.141
55. Corporaal, H.: Microprocessor Architectures: From VLIW to TTA. Wiley (1998)
56. Coussy, P., Gajski, D.D., Meredith, M., Takach, A.: An introduction to high-level synthesis. IEEE Des. Test Comput. **26**(4), 8–17 (2009). doi:10.1109/MDT.2009.69
57. Coussy, P., Morawiec, A.: High-Level Synthesis from Algorithm to Digital Circuit, 1 edn. Springer, Dordrecht (2008)
58. Cupaiuolo, T., Siti, M., Tomasoni, A.: Low-complexity high throughput VLSI architecture of soft-output ML MIMO detector. In: Design, Automation and Test in Europe (DATE), pp. 1396–1401 (2010)
59. Danckaert, K., Catthoor, F., De Man, H.: Platform independent data transfer and storage exploration illustrated on a parallel cavity detection algorithm. In: Conference on Parallel and Distributed Processing Techniques and Applications (PDPTA), pp. 1669–1675 (1999)
60. Das, S., Khatri, S.: An efficient and regular routing methodology for datapath designs using net regularity extraction. IEEE Trans. Comput. Aided Des. **21**(1), 93–101 (2002). doi:10.1109/43.974141
61. De Micheli, G.: Logic synthesis and physical design: Quo vadis? In: Design, Automation and Test in Europe (DATE), pp. 1–1. EPFL (2011)
62. Dejonghe, A.: R &D fact sheet: digital solutions for cognitive Radio. Technical report, IMEC (2010)
63. Derudder, V., Bougard, B., Couvreur, A., Dewilde, A., Dupont, S., Folens, L., Hollevoet, L., Naessens, F., Novo, D., Raghavan, P., Schuster, T., Stinkens, K., Weijers, J.W., Van der Perre, L.: A 200Mbps + 2.14nJ/b digital baseband multi processor system-on-chip for SDRs. In: Symposium on VLSI Circuits, pp. 292–293 (2009)
64. Eleyan, N.N., Ken, L., Kamal, M., Mohammed, B., Bassett, P.: Semi-custom design flow: leveraging place and route tools in custom circuit design. In: IEEE IC Design and Technology (ICICDT), pp. 143–147. Qualcomm, DSP core design (2009)
65. Falk, H., Marwedel, P.: Source Code Optimization Techniques for Data Flow Dominated Embedded Software. Springer (2004)

66. Fasthuber, R., Li, M., Novo, D., Raghavan, P., Van der Perre, L., Catthoor, F.: Energy-efficient run-time scalable soft-output SSFE MIMO detector architectures. In: Transaction on High-Performance Embedded Architectures and Compilers (HiPEAC), Special Issue SAMOS 2009, vol. 5, no. 3, pp. 1–20 (2011)
67. Fasthuber, R., Li, M., Novo, D., Van Der Perre, L., Catthoor, F.: Novel energy-efficient scalable soft-output SSFE MIMO detector architectures. In: International Conference on Embedded Computer Systems (IC-SAMOS). IMEC (2009). doi:10.1109/ICSAMOS.2009.5289228
68. Fasthuber, R., Li, M., Novo, D., Van Der Perre, L., Catthoor, F.: Exploration of soft-output MIMO detector implementations on massive parallel processors. J. Signal Process. Syst. **64**(1), 75–92 (2010). doi:10.1007/s11265-010-0499-0
69. Fauth, A., Van Praet, J., Freericks, M.: Describing instruction set processors using nML. In: Design, Automation and Test in Europe (DATE), pp. 503–507 (1995). doi:10.1109/EDTC.1995.470354
70. Freescale Semiconductor Inc.: Star Core MSC8158 Product Brief, Review (2010). http://www.freescale.com/
71. Freescale Semiconductor Inc.: Star Core MSC8256 Data Sheet, Rev. 1(2010). http://www.freescale.com/
72. Fridman, J., Greenfield, Z.: The TigerSHARC DSP architecture. IEEE Micro **20**(1), 66–76 (2000). doi:10.1109/40.820055
73. Ganesan, S., Vemuri, R.: Analog-digital partitioning for field-programmable mixed signal systems. In: Conference on Advanced Research in VLSI (ARVLSI), pp. 172–185 (2001). doi:10.1109/ARVLSI.2001.915559
74. Gangwar, A., Balakrishnan, M., Kumar, A.: Impact of inter-cluster communication mechanisms on ILP in clustered VLIW architectures. ACM Trans. Des. Autom. Electron. Syst. **12**(1) (2007)
75. Gansen, M., Richter, F., Weiss, O., Noll, T.: A datapath generator for full-custom macros of iterative logic arrays. In: IEEE International Conference on Application-Specific Systems, Architectures and Processors (ASAP), pp. 438–447 (1997). doi:10.1109/ASAP.1997.606849
76. Glossner, J., Chirca, K., Schulte, M., Wang, H., Nasimzada, N., Har, D., Wang, S., Hoane, J.A., Nacer, G., Moudgill, M., Vassiliadis, S.: Sandblaster Low Power DSP. In: IEEE Custom Integration Circuits Conference (CICC), pp. 575–581. Sandbridge Technologies, Delft University of Technologies, University of Wisconsin (2004)
77. Gonzalez, R.: Xtensa: a configurable and extensible processor. IEEE Micro **20**(2), 60–70 (2000). doi:10.1109/40.848473
78. Gordon-Ross, A., Vahid, F.: Dynamic loop caching meets preloaded loop caching-a hybrid approach. In: IEEE International Conference on Computer Design (ICCD), pp. 446–449 (2002). doi:10.1109/ICCD.2002.1106810
79. Guan, X., Fei, Y., Lin, H.: A hierarchical design of an application-specific instruction set processor for high-throughput FFT. IEEE International Symposium on Circuits and System (ISCAS) (2009)
80. Halambi, A., Grun, P., Ganesh, V., Khare, A., Dutt, N., Nicolau, A.: EXPRESSION: a language for architecture exploration through compiler/simulator retargetability. In: Design, Automation and Test in Europe (DATE), pp. 485–490 (1999). doi:10.1109/DATE.1999.761170
81. Heyrman, K.: Control of sectioned on-chip communication. Ph.D. thesis, University of Gent, Belgium (2009)
82. Heysters, P.M., Smit, G.J.M.: Mapping of DSP algorithms on the MONTIUM architecture. In: International Parallel and Distributed Processing Symposium (IPDPS). University of Twente (Netherlands) (2003)
83. Ienne, P., Leupers, R.: Customizable Embedded Processors: Design Technologies and Applications. Morgan Kaufmann (2006)

84. Imai, M., Takeuchi, Y., Ohtsuki, N., Hikichi, N.: Compiler generation techniques for embedded processors and their application to HW/SW codesign. System-level, synthesis pp. 293–320 (1999)
85. Inc., M.M.: DataPath compiler datasheet (2006). http://www.micromagic.com/
86. Itoh, M., Higaki, S., Sato, J., Shiomi, A., Takeuchi, Y., Kitajima, A., Imai, M.: PEAS-III: an ASIP design environment. In: International Conference on Computer Design (ICCD), pp. 430–436 (2000). doi:10.1109/ICCD.2000.878319
87. ROCCC, Jacquard Computing. http://www.jacquardcomputing.com/roccc/
88. Jalier, C., Lattard, D., Sassatelli, G., Benoit, P., Torres, L.: A homogeneous MPSoC with dynamic task mapping for software defined radio. In: IEEE Computer Society Symposium on VLSI (ISVLSI), pp. 345–350 (2010). doi:10.1109/ISVLSI.2010.110
89. Janhunen, J., Pitkanen, T., Silven, O., Juntti, M.: Fixed- and floating-point processor comparison for MIMO-OFDM detector. IEEE J. Sel. Top. Signal Process. **5**(8), 1588–1598 (2011). doi:10.1109/JSTSP.2011.2165830
90. Jayapala, M., Barat, F., Catthoor, F., Corporaal, H., Deconinck, G.: Clustered loop buffer organization for low energy VLIW embedded processors. IEEE Trans. Comput. **54**(6), 672–683 (2005)
91. Kandemir, M., Kadayif, I., Choudhary, A., Ramanujam, J., Kolcu, I.: Compiler-directed scratch pad memory optimization for embedded multiprocessors. IEEE Trans. Very Large Scale Integr. VLSI Syst. **12**(3), 281–287 (2004). doi:10.1109/TVLSI.2004.824299
92. Kapasi, U., Rixner, S., Dally, W., Khailany, B., Mattson, P., Owens, J.: Programmable stream processors. Computer **36**(8), 54–62 (2003). doi:10.1109/MC.2003.1220582
93. Karlstrom, P., Zhou, W., Wang, C.h., Liu, D.: Design of PIONEER: A case study using NoGap. In: Asia Pacific Conference on Postgraduate Research in Microelectronics and Electronics (PrimeAsia), pp. 53–56 (2010). doi:10.1109/PRIMEASIA.2010.5604962
94. Kavvadias, N., Nikolaidis, S.: Zero-overhead loop controller that implements multimedia algorithms. IEEE Proc. Comput. Digital Tech. **152**(4), 517 (2005). doi:10.1049/ip-cdt:20041187
95. Keutzer, K., Malik, S., Newton, A.: From ASIC to ASIP: the next design discontinuity. In: IEEE International Conference on Computer Design (ICCD), pp. 84–90 (2002). doi:10.1109/ICCD.2002.1106752
96. Kim, E.P., Shanbhag, N.R.: An energy-efficient multiple-input multiple-output (MIMO) detector architecture. In: IEEE Workshop on Signal Processing System (SIPS), pp. 239–244 (2011). doi:10.1109/SiPS.2011.6088981
97. Kim, T.H., Park, I.C.: Small-area and low-energy K-best MIMO detector using relaxed tree expansion and early forwarding. IEEE Trans. Circuits Syst. **57**(10), 2753–2761 (2010). doi:10.1109/TCSI.2010.2046249
98. Kin, J., Gupta, M., Mangione-Smith, W.: Filtering memory references to increase energy efficiency. IEEE Trans. Comput. **49**(1), 1–15 (2000). doi:10.1109/12.822560
99. Knagge, G., Bickerstaff, M., Ninness, B., Weller, S.R., Woodward, G.: A VLSI 8x8 MIMO Near-ML decoder engine. In: IEEE Workshop on Signal Processing System (SIPS), pp. 387–392 (2006). doi:10.1109/SIPS.2006.352614
100. Kneip, J., Weiss, M., Drescher, W., Aue, V., Strobel, J., Oberthuer, T., Bolle, M., Fettweis, G.: Single chip programmable baseband ASSP for 5 GHz wireless LAN applications. IECICE Trans. Electron. **E85-C**(2), 359–367 (2002)
101. Knowles, S.: The SOC future is soft. In: IEEE Cambridge Branch Seminar (2005)
102. Kogel, T., Leupers, R., Meyr, H.: Integrated System-Level Modeling of Network-on-Chip enabled Multi-Processor Platforms. Springer (2006)
103. Kokkeler, A.B.J.: Analog-digital codesign using coarse quantization. Ph.D. thesis, University of Twente, Enschede, The Netherlands (2005)
104. Kozyrakis, C., Patterson, D.: Scalable vector processors for embedded systems. IEEE Micro **23**(6), 36–45 (2003). doi:10.1109/MM.2003.1261385

105. Kunze, S., Matus, E., Fettweis, G.: ASIP decoder architecture for convolutional and LDPC codes. In: IEEE International Symposium on Circuits and System (ISCAS), i, pp. 2457–2460 (2009). doi:10.1109/ISCAS.2009.5118298
106. Lapinskii, V., Jacome, M., de Veciana, G.: Application-specific clustered VLIW datapaths: early exploration on a parameterized design space. IEEE Trans. Comput. Aided Des. **21**(8), 889–903 (2002). doi:10.1109/TCAD.2002.800451
107. Lee, H., Chakrabarti, C., Mudge, T.: A low-power DSP for wireless communications. IEEE Trans. Very Large Scale Integr. Syst. **18**(9), 1310–1322 (2010)
108. Leroy, A., Milojevic, D., Verkest, D., Robert, F., Catthoor, F.: Concepts and implementation of spatial division multiplexing for guaranteed throughput in networks-on-chip. IEEE Trans. Comput. **57**(9), 1182–1195 (2008). doi:10.1109/TC.2008.82
109. Leupers, R., Marwedel, P.: Retargetable code generation based on structural processor descriptions. In: Design Automation for Embedded Systems, pp. 1–36. Kluwer Academic Publishers (1998)
110. Li, L., Shi, M.: Software-hardware partitioning strategy using hybrid genetic and Tabu search. In: International Conference on Computer Science and Software Engineering (CSSE), pp. 83–86 (2008). doi:10.1109/CSSE.2008.488
111. Li, M.: Algorithm and architecture co-design for software defined radio baseband. Ph.D. thesis, KU Leuven (2010)
112. Li, M., Amin, A., Appeltans, R., Torrea, R., Cappelle, H., Fasthuber, R., Dejonghe, A., Van der Perre, L.: Instruction set support and algorithm-architecture for fully parallel multi-standard soft-output demapping on baseband processors. In: IEEE Workshop on Signal Processing System (SIPS), pp. 140–145. IMEC (2010). doi:10.1109/SIPS.2010.5624777
113. Li, M., Appeltans, R., Amin, A., Torrea-Duran, R., Cappelle, H., Hartmann, M., Yomo, H., Kobayashi, K., Dejonghe, A., Van Der Perre, L.: Overview of a software defined downlink inner receiver for category-E LTE-advanced UE. In: IEEE International Conference on Communication (ICC), pp. 1–5 (2011). doi:10.1109/icc.2011.5963387
114. Li, M., Fasthuber, R., Novo, D., Van Der Perre, L., Catthoor, F.: Algorithm-architecture co-design of soft-output ML MIMO detector for parallel application specific instruction set processors. In: Design, Automation and Test in Europe (DATE), pp. 1608–1613. IMEC (2009)
115. Li-ya, L., Peng, L.: Low power implementation of datapath using regularity. J. Zhejiang Univ. Sci. A **6**(6), 596–600 (2005). doi:10.1007/BF02841771
116. Liao, C.H., Wang, T.P., Chiueh, T.D.: A 74.8 mW soft-output detector IC for 8 x 8 spatial-multiplexing MIMO communications. IEEE J. Solid-State Circuits **45**(2), 411–421 (2010). doi:10.1109/JSSC.2009.2037292
117. Limberg, T., Winter, M., Bimberg, M., Klemm, R., Matus, E., Tavares, M.B., Fettweis, G., Ahlendorf, H., Robelly, P.: A fully programmable 40 GOPS SDR single chip baseband for LTE/WiMAX terminals. In: European Solid-State Circuits Conference (ESSCIRC), pp. 466–469. Technische Universtaet Dresden and Vodafone Chair Mobile Communications Systems (Fettweis) (2008)
118. Lin, Y., Lee, H., Woh, M., Harel, Y., Mahlke, S., Mudge, T.: SODA: A low-power architecture for software radio. In: International Symposium on Computer Architecture (ISCA), vol. 33, pp. 89–101. University of Michigan; Arizona State University, ARM Ltd., IEEE (2006). doi:10.1109/ISCA.2006.37
119. Liu, D.: Embedded DSP Processor Design: Application Specific Instruction Set Processors. Morgan Kaufmann (2008)
120. Liu, L., Ye, F., Ma, X., Zhang, T., Ren, J.: A 1.1-Gb/s 115-pJ/bit configurable MIMO detector using 0.13um CMOS technology. IEEE Trans. Circuits Syst. **57**(9), 701–705 (2010). doi:10.1109/TCSII.2010.2058494
121. Liu, T.H., Jiang, J.Y., Chu, Y.S.: A Low-Cost MMSE-SIC detector for the MIMO system: algorithm and hardware implementation. IEEE Trans. Circuits Syst. **58**(1), 56–61 (2011). doi:10.1109/TCSII.2010.2092819

122. Lodi, A., Cappelli, A., Bocchi, M., Mucci, C., Innocenti, M., Bartolomeis, C.D., Ciccarelli, L., Giansante, R., Deledda, A., Campi, F., Toma, M., Guerrieri, R.: XiSystem: a XiRisc-based SoC with reconfigurable IO module. IEEE J. Solid-State Circuits **41**(1), 85–96 (2006)
123. Lu, W., Zhao, S., Zhou, X., Ren, J., Sobelman, G.: Reconfigurable baseband processing architecture for communication. IET Comput. Digit. Tech. **5**(1), 63 (2011). doi:10.1049/iet-cdt.2009.0121
124. Luo, D., Tsui, C.Y.: A hybrid algorithm and its re-configurable architecture for MIMO detector. In: IEEE International Midwest Symposium on Circuits and System (MWSCAS), pp. 297–300 (2009). doi:10.1109/MWSCAS.2009.5236095
125. Magma, Hydra. http://www.magma-da.com/
126. Magma, SiliconSmart. http://www.magma-da.com/
127. Magma, Talus Design. http://www.magma-da.com/
128. Magma, Titan Mixed-Signal Design Platform. http://www.magma-da.com/
129. Mahdavi, M., Shabany, M.: Ultra high-throughput architectures for hard-output MIMO detectors in the complex domain. In: IEEE International Midwest Symposium on Circuits and System (MWSCAS), pp. 1–4 (2011). doi:10.1109/MWSCAS.2011.6026425
130. Mansour, M., Shanbhag, N.: High-throughput LDPC decoders. IEEE Trans. Very Large Scale Integr. Syst. **11**(6), 976–996 (2003). doi:10.1109/TVLSI.2003.817545
131. Marwedel, P.: Embedded System Design. Springer (2005)
132. Medra, A.: Semi-custom design to Reduce the interconnect energy in advanced technologies. Master thesis, Nile University, Egypt (2012)
133. Mei, B., Vernalde, S., Verkest, D., De Man, H., Lauwereins, R.: ADRES: an architecture with tightly coupled VLIW processor and coarse-grained reconfigurable matrix. Lect. Notes Comput. Sci. **2778**, 61–70 (2003)
134. Catapult C, Mentor graphics. http://www.mentor.com/esl/catapult/overview
135. Meyr, H.: System-on-chip for communications: the dawn of ASIPs and the dusk of ASICs. In: IEEE Workshop on Signal Processing Systems (SIPS), pp. 4–5 (2003). doi:10.1109/SIPS.2003.1235634
136. Micheli, G.D., Benini, L.: Networks on Chips: Technology and Tools (Systems on Silicon). Morgan Kaufmann (2006)
137. ModelSim, Mentor Graphics. http://model.com/
138. Moezzi-Madani, N., Thorolfsson, T., Chiang, P., Davis, W.R.: Area-efficient antenna-scalable MIMO detector for K-best sphere decoding. J. Signal Process. Syst. pp. 1–12 (2011). doi:10.1007/s11265-011-0595-9
139. Mohammed, K., Mohamed, M.I.A., Daneshrad, B.: A Parameterized Programmable MIMO decoding architecture with a scalable instruction set and compiler. IEEE Trans. Very Large Scale Integr. Syst. **19**(8), 1485–1489 (2011). doi:10.1109/TVLSI.2010.2049592
140. Moudgill, M., Glossner, J., Agrawal, S., Nacer, G.: The Sandblaster 2.0 architecture and SB3500 implementation. In: SDR Forum. Sandbridge Technologies (2008)
141. Nilsson, A., Tell, E., Liu, D.: An 11 mm2, 70 mW fully programmable baseband processor for mobile WiMAX and DVB-T/H in 0.12um CMOS. IEEE J. Solid-State Circuits **44**(1), 90–97 (2009). doi:10.1109/JSSC.2008.2007167
142. Nofal, F.A., Nofal, M.R.: Top Down SoC Floor planning with ReUse. ChipEDA. EE Times, White paper (2008)
143. Novo, D.: Exploiting adaptive precision in software defined radios. Ph.D. thesis, KU Leuven (2010)
144. Nsenga, J.: Design of an air interface for 60 GHz multi-antenna systems and study of non-idealities impact on system performance. Ph.D. thesis, KU Leuven (2009)
145. Panda, P.R., Catthoor, F., Dutt, N.D., Danckaert, K., Brockmeyer, E., Kulkarni, C., Vandercappelle, A., Kjeldsberg, P.G.: Data and memory optimization techniques for embedded systems. ACM Trans. Des. Autom. Electron. Syst. **6**(2), 149–206 (2001). doi:10.1145/375977.375978
146. Papanikolaou, A.: Application-driven software configuration of communication networks and memory organizations. Ph.D. thesis, University of Gent, Belgium (2006)

147. Patel, D., Smolyakov, V., Shabany, M., Gulak, P.G.: VLSI implementation of a WiMAX/LTE compliant low-complexity high-throughput soft-output K-Best MIMO detector. In: IEEE International Symposium on Circuits and System (ISCAS), pp. 593–596 (2010). doi:10.1109/ISCAS.2010.5537524
148. Picochip: PC205 Product Brief (2009). http://www.picochip.com/
149. Portero, A., Carrabina, J., Catthoor, F.: HW/SW Implementation Trade-Offs of MPEG-4 Data-Flow Algorithm. LAP LAMBERT Academic Publishing (2010)
150. Potkonjak, M., Rabaey, J.: Optimizing resource utilization using transformations. IEEE Trans. Comput. Aided Des. Integr. Circuits Syst. **13**(3), 277–292 (1994). doi:10.1109/43.265670
151. Pulley, D., Baines, R.: Software defined baseband processing for 3G base stations. In: International Conference on 3G Mobile Communication Technologies, vol. 4, pp. 123–127. Picochip (2003). doi:10.1049/cp:20030350
152. Raab, W., Berthold, J., Hachmann, U., Langen, D., Schreiner, M.: Low power design of the X-GOLD SDR 20 baseband processor. In: Design, Automation and Test in Europe (DATE), pp. 792–793. Infineon Technologies AG, Germany and TU Dresden, Germany (2010)
153. Raghavan, P.: Low energy VLIW architecture extensions and compiler plug-ins for embedded systems. Ph.D. thesis, KU Leuven, IMEC (2009)
154. Raghavan, P., Lambrechts, A., Jayapala, M., Catthoor, F., Verkest, D.: Distributed loop controller for multi-threading in Uni-threaded ILP architectures. IEEE Trans. Comput. **58**(3), 311–321 (2009)
155. Ramacher, U.: Software-defined radio prospects for multistandard mobile phones. Computer **40**(10), 62–69 (2007)
156. Ramacher, U., Raab, W., Hachmann, U., Langen, D., Berthold, J., Kramer, R., Schackow, A., Grassmann, C., Sauermann, M., Szreder, P., Capar, F., Obradovic, G., Xu, W., Bruls, N., Lee, K., Weber, E., Kuhn, R., Harrington, J.: Architecture and implementation of a software-defined radio baseband processor. In: IEEE International Symposium on Circuits and System (ISCAS), pp. 2193–2196. Infineon (2011). doi:10.1109/ISCAS.2011.5938035
157. Ren, B., Wang, A., Bakshi, J., Liu, K., Dai, W.: A domain-specific cell based ASIC design methodology for digital signal processing applications. In: Design, Automation and Test in Europe (DATE), pp. 280–285. IEEE Computer Society (2004). doi:10.1109/DATE.2004.1269251
158. Rigo, S., Araujo, G., Bartholomeu, M., Azevedo, R.: ArchC: a systemC-based architecture description language. In: Symposium on Computer Architecture and High Performance Computing (SBAC-PAD), pp. 66–73 (2004). doi:10.1109/SBAC-PAD.2004.8
159. Rixner, S., Dally, W., Khailany, B., Mattson, P., Kapasi, U., Owens, J.: Register organization for media processing. In: International Symposium on High-Performance Computer Architecture (HPCA), pp. 375–386 (2000). doi:10.1109/HPCA.2000.824366
160. Rotenberg, E., Bennett, S., Smith, J.: Trace cache: a low latency approach to high bandwidth instruction fetching. In: IEEE/ACM International Symposium on Microarchitecture (MICRO), pp. 24–34 (1996). doi:10.1109/MICRO.1996.566447
161. Sarkar, S., Shashank, D., Mitra, R.S., Tiwari, P.K.: Lessons and experiences with high-level synthesis. IEEE Des. Test Comput. **26**(4), 34–45 (2009)
162. Sawahashi, M., Kishiyama, Y., Taoka, H., Tanno, M., Nakamura, T.: Broadband radio access: LTE and LTE-advanced. In: International Symposium on Intelligent Signal Processing and Communication Systems (ISPACS), pp. 224–227 (2009). doi:10.1109/ISPACS.2009.5383862
163. Senthilvelan, M., Sima, M., Iancu, D., Hormigo, J., Schulte, M.: CORDIC instruction set extensions for matrix decompositions on Software Defined Radio processors. In: Asilomar Conference on Signals, Systems and Computers (ACSSC), pp. 1792–1797. IEEE (2009). doi:10.1109/ACSSC.2009.5470207
164. Shabany, M., Gulak, P.G.: A 675 Mbps, 4x4 64-QAM K-Best MIMO detector in 13m CMOS. IEEE Trans. Very Large Scale Integr. Syst. **20**(1), 135–147 (2010). doi:10.1109/TVLSI.2010.2090367

165. Shanbhag, N.R.: Algorithms transformation techniques for low-power wireless VLSI systems design. Int. J. Wirel. Inf. Netw. **5**, 147–171 (1998)
166. Sheikh, F., Mill, M., Richards, B., Markovic, D., Nikolic, B.: A 1–190 MSample/s 8–64 tap energy-efficient reconfigurable FIR filter for multi-mode wireless communication. In: Symposium on VLSI Circuits, pp. 207–208 (2010). doi:10.1109/VLSIC.2010.5560297
167. Shiun Lin, J., Hao Fang, S., Hsuan Jen, Y., Der Shieh, M.: Design of high-throughput MIMO detectors using sort-free and early-pruned techniques. In: IEEE TENCON Conference, pp. 1513–1516 (2010). doi:10.1109/TENCON.2010.5686145
168. Silicon Hive: HiveFlex CSP Family (2010). http://www.siliconhive.com/
169. Silicon Hive: HiveLogic Configurable Parallel Processing Platform (2010). http://www.siliconhive.com/
170. Stotas, S., Nallanathan, A.: On the throughput and spectrum sensing enhancement of opportunistic spectrum access cognitive radio networks. IEEE Trans. Wirel. Commun. **11**(1), 97–107 (2012). doi:10.1109/TWC.2011.111611.101716
171. Strojwas, A.J.: Cost effective scaling to 22nm and below technology nodes. In: International Symposium on VLSI Technology, Systems and Applications (VTSA), pp. 1–2. IEEE (2011). doi:10.1109/VTSA.2011.5872265
172. Studer, C., Fateh, S., Seethaler, D.: A 757Mb/s 1.5 mm2 90 nm CMOS soft-input soft-output MIMO detector for IEEE 802.11n. In: ESSCIRC, pp. 530–533 (2010). doi:10.1109/ESSCIRC.2010.5619760
173. Sun, Y., Cavallaro, J.R.: High-throughput soft-output MIMO detector based on path-preserving trellis-search algorithm. IEEE Trans. Very Large Scale Integr. Syst. **20**(7), 1235–1247 (2012). doi:10.1109/TVLSI.2011.2147811
174. Suzuki, T., Yamada, H., Yamagishi, T., Takeda, D., Horisaki, K., Vander Aa, T., Fujisawa, T., Van der Perre, L., Unekawa, Y.: High throughput and low power software defined radio using dynamically reconfigurable baseband processor. IEEE Micro **31**(6), 19–28 (2011). doi:10.1109/MM.2011.95
175. Synfora/Synopsys, PICO. http://www.synopsys.com/
176. Synopsys, Custom Designer. http://www.synopsys.com/
177. Synopsys, Design Compiler. http://www.synopsys.com/
178. Synopsys, IC Compiler. http://www.synopsys.com/
179. Synopsys, IC Validator. http://www.synopsys.com/
180. Synopsys, PrimeTime. http://www.synopsys.com/
181. Synopsys, Symphony C Compiler. http://www.synopsys.com/Systems/
182. Synopsys Inc.: Processor Designer Datasheet (2010). http://www.synopsys.com/
183. Tae-Hwan, K., In-Cheol, P.: A 2.6Gb/s 1.56mm2 near-optimal MIMO detector in 0.18μm CMOS. In: IEEE Custom Integration Circuits Conference (CICC), pp. 1–4 (2010). doi:10.1109/CICC.2010.5617463
184. Takeuchi, Y., Sakanushi, K., Imai, M.: Generation of application-domain specific instruction-set processors. In: International SoC Design Conference (ISOCC), pp. 75–78 (2010). doi:10.1109/SOCDC.2010.5682970
185. Tao Ye, T., De Micheli, G.: Data path placement with regularity. In: IEEE/ACM International Conference on Computer Aided Design (ICCAD), pp. 264–270 (2000). doi:10.1109/ICCAD.2000.896484
186. Target ASIP development toolsuite (inc. IP Designer, IP Programmer), Target Compiler Technologies. http://www.retarget.com/
187. Tecpanecatl-Xihuitl, J., Aguilar-Ponce, R.M., Ismail, Y., Bayoumi, M.A.: Efficient mutliplierless polyphase FIR Filter based on new distributed arithmetic architecture. In: Asilomar Conference on Signals, Systems and Computers (ACSSC), pp. 958–962. IEEE (2007). doi:10.1109/ACSSC.2007.4487361
188. Texas Instruments: TMS320C6416 Datasheet, Rev. M (2009). http://www.ti.com/
189. Tsai, P.Y., Chen, W.T., Lin, X.C., Huang, M.Y.: A 4x4 64-QAM reduced-complexity K-best MIMO detector up to 1.5Gbps. In: IEEE International Symposium on Circuits and System (ISCAS), pp. 3953–3956 (2010). doi:10.1109/ISCAS.2010.5537675

190. Tsujihashi, Y., Matsumoto, H., Kato, S., Nakao, H., Kitada, O., Okazaki, K., Shinohara, H.: A high density data path generator with stretchable cells. In: IEEE Custom Integrated Circuits Conferefnce (CICC), pp. 11.3.1–11.3.4 (1992). doi:10.1109/CICC.1992.591283
191. Tyson, G., Smelyanskiy, M., Davidson, E.: Evaluating the use of register queues in software pipelined loops. IEEE Trans. Comput. **50**(8), 769–783 (2001). doi:10.1109/TC.2001.947006
192. Vander Aa, T.: Instruction transfer and storage exploration for low energy embeded VLIWs. Ph.D. thesis, KU Leuven (2005)
193. Veidenbaum, A., Nicolau, A.: Reducing power with an L0 instruction cache using history-based prediction. In: International Workshop on Innovative Architecture for Future Generation High-Performance Processors and Systems (IWIA), pp. 11–18 (2002). doi:10.1109/IWIA.2002.1035013
194. van de Waerdt, J., Vassiliadis, S., Das, S., Mirolo, S., Yen, C., Zhong, B., Basto, C., van Itegem, J., Amirtharaj, D., Kalra, K., Rodriguez, P., van Antwerpen, H.: The TM3270 media-processor. In: IEEE/ACM International Symposium on Microarchitecture (MICRO), pp. 331–342 (2005). doi:10.1109/MICRO.2005.35
195. Wang, J., Sohl, J., Kraigher, O., Liu, D.: ePUMA: A novel embedded parallel DSP platform for predictable computing. In: International Conference on Education Technology and Computer (ICETC), pp. V5-32–V5-35 (2010). doi:10.1109/ICETC.2010.5529952
196. Wani, M., Miljanic, Z., Spasojevic, P., Redington, J.: ASIP data plane processor for multi-standard Interleaving and De-Interleaving. In: Asilomar Conference on Signals, Systems and Computers (ACSSC), pp. 1259–1263 (2010). doi:10.1109/ACSSC.2010.5757733
197. Wei, Z., Qiang, Z., Yici, C., Xianlong, H.: A datapath routing algorithm using bit regularity extraction. In: International Conference on ASIC (ASICON), vol. 2, pp. 820–823 (2005). doi:10.1109/ICASIC.2005.1611453
198. Weiss, O., Gansen, M., Noll, T.: A flexible datapath generator for physical oriented design. In: European Solid-State Circuits Conference (ESSCIRC), pp. 393–396 (2001)
199. Witte, E.M., Borlenghi, F., Ascheid, G., Leupers, R., Meyr, H.: A scalable VLSI architecture for soft-input soft-output single tree-search sphere decoding. IEEE Trans. Circuits Syst. **57**(9), 706–710 (2010). doi:10.1109/TCSII.2010.2056014
200. Woh, M., Mahlke, S., Mudge, T., Chakrabarti, C.: Mobile supercomputers for the next-generation cell phone. Computer **43**(1), 81–85 (2010). doi:10.1109/MC.2010.16
201. Woh, M., Sangwon, S., Mahlke, S., Mudge, T., Chakrabarti, C., Flautner, K.: AnySP: anytime anywhere anyway signal processing. IEEE Micro **30**(1), 81–91 (2010)
202. Woh, M., Yuan, L., Sangwon, S., Mahlke, S., Mudge, T., Chakrabarti, C., Bruce, R., Kershaw, D., Reid, A., Wilder, M., Flautner, K.: From SODA to Scotch: The evolution of a wireless baseband processor. In: IEEE/ACM International Symposium on Microarchitecture (MICRO), vol. 41, pp. 152–163. Department of Electrical Engineering, Arizona State University, Tempe, AZ and ARM, Ltd., Cambridge, United Kingdom (2008)
203. Wolff, F., Knieser, M., Weyer, D., Papachristou, C.: Codesign paradigm in digital/analog tradeoffs. In: IEEE International ASIC/SOC Conference, pp. 76–80 (1999). doi:10.1109/ASIC.1999.806478
204. Wu, D., Eilert, J., Asghar, R., Liu, D.: VLSI implementation of a fixed-complexity soft-output MIMO detector for high-speed wireless. J. Wirel. Commun. Netw. **2010**, 1–14 (2010). doi:10.1155/2010/893184
205. Ye, T., Chaudhuri, S., Huang, F., Savoj, H., De Micheli, G.: Physical synthesis for ASIC datapath circuits. In: IEEE International Symposium on Circuits and Systems (ISCAS), vol. 3, pp. 365–368 (III) (2002). doi:10.1109/ISCAS.2002.1010236
206. Yian Mei, L., Rosdi, B.A.B., Cheen Kok, L.: A methodology for automation structured datapath placement In VLSI design. In: IEEE Symposium on Industrial Electronics and Applications (ISIEA), pp. 273–278 (2011). doi:10.1109/ISIEA.2011.6108714
207. Yifan, H., Yu, P., Zhenyu, Y., Londono, S., Kleihorst, R., Abbo, A., Corporaal, H.: Xetal-Pro: An ultra-low energy and high throughput SIMD processor. In: ACM/IEEE Design Automation Conference (DAC), pp. 543–548. Eindhoven University (2010)

208. Yoshizawa, S., Ikeuchi, H., Miyanaga, Y.: Scalable pipeline architecture of MMSE MIMO detector for 4x4 MIMO-OFDM receiver. In: IEEE International Symposium on Circuits and System. (ISCAS), pp. 2534–2537 (2010). doi:10.1109/ISCAS.2010.5537108
209. Youness, H., Hassan, M., Sakanushi, K., Takeuchi, Y., Imai, M., Salem, A., Wahdan, A.M., Moness, M.: A high performance algorithm for scheduling and hardware-software partitioning on MPSoCs. In: International Conference on Design and Technology of Integrated Systems in Nanoscale Era (DTIS), pp. 71–76 (2009). doi:10.1109/DTIS.2009.4938027
210. Yu-Wen, T., Kun-Chen, W., Hui-Hsiang, T., Rung-Bin, L.: Using structured ASIC to improve design productivity. In: International Symposium on Integrated Circuits (ISIC), pp. 25–28 (2009)
211. Zalamea, J., Llosa, J., Ayguadé, E., Valero, M.: Two-level hierarchical register file organization for VLIW processors. In: ACM/IEEE International Symposium on Microarchitecture (MICRO), pp. 137–146 (2000). doi:10.1145/360128.360143
212. Zhang, B., Liu, H., Zhao, H., Mo, F., Chen, T.: Domain specific architecture for next generation wireless communication. In: Design, Automation and Test in Europe (DATE), pp. 1414–1419. Group from China (2010)
213. Zheng, S., Zhang, Y., He, T.: The application of genetic algorithm in embedded system hardware-software partitioning. In: International Conference on Electronic Computer Technology (ICECT), pp. 219–222 (2009). doi:10.1109/ICECT.2009.132
214. Zivojnovic, V., Pees, S., Meyr, H.: LISA-machine description language and generic machine model for HW/SW co-design. In: VLSI Signal Processing (VLSISP), pp. 127–136 (1996). doi:10.1109/VLSISP.1996.558311
215. Zuluaga, M., Topham, N.: Design-space exploration of resource-sharing solutions for custom instruction set extensions. Trans. Comput. Aided Des. **28**(12), 1788–1801 (2009). doi:10.1109/TCAD.2009.2026355

Chapter 3
The Proposed DSIP Architecture Template for the Wireless Communication Domain

3.1 An Effective Architecture Template for the Wireless Communication Domain

In the following the considered domain and the proposed DSIP platform design approach, which combines all 5 design requirements discussed in Sect. 1.2, will be explained.

3.1.1 Considered Domain

This book proposes a DSIP architecture template for the digital physical-layer baseband processing of high-performance multi-standard/mode wireless communication systems. We consider the data-processing of the whole baseband domain, i.e. ranging from the digital front-end to the outer modem. However, the control-related processing of the domain (middleware layer) is outside the scope of this book. As mentioned in Sect. 2.2.4 and shown in Fig. 2.4, the complete system functionality can be separated into multiple sub-functionalities, i.e. functional blocks. Each of the functional blocks, e.g. channel filter, FFT, FEC decoder, performs the data-processing of a certain sub-functionality.

Although the focus of this book is on the fore-mentioned domain, it should be mentioned that the proposed DSIP architecture template can be largely reused for other domains. For instance, for the domain of digitally-assisted RF/analog [44] and for the baseband domain of low-performance wireless communication (e.g. Bluetooth, ZigBee). Nevertheless, to cope with the specific requirements and characteristics of these domains, some modifications will be necessary.

R. Fasthuber et al., *Energy-Efficient Communication Processors*,
DOI: 10.1007/978-1-4614-4992-8_3,

3.1.2 Proposed Design Approach

3.1.2.1 Conceptual Difference to Traditional Approaches

As mention in Sect. 2.2, in the traditional DSIP platform design approach, typically many functional blocks are implemented with a single DSIP hardware block. Therefore a traditional DSIP platform consists of very few hardware blocks. Because the functional blocks are rather heterogeneous and the DSIP has to support many of them on the same hardware (time-multiplexing and time-sharing), the FUs of the traditional DSIPs need to be kept rather generic and sufficient memory storage needs to be present. Besides, in order to enable real-time processing of several functional blocks, the DSIP has to employ massive parallelism. Thus, a traditional DSIP is not only rather generic but also rather big in area. This combination is the main reason for the low energy efficiency.

Contrary to a traditional DSIP platform, a traditional ASIP (AP)platform consists of many hardware blocks. This enables a better customization of each hardware block. Because of the inefficiency of today's ASIPs (AP), a traditional ASIP (AP) platform consists not only of processors, but also of ASIC HAs. The employed processors may be designed with the same tool framework and therefore the general structures of the resulting ASIP (AP) architectures can be similar. However, because of customizing each processor with individual ASIs, especially the datapaths of the FUs are different. Thus, due to the use of ASIC HAs and the customization with ASIs (which are designed at processor design time and not before), the reusability in the traditional ASIP (AP) platform design approach is limited.

As mentioned in Sect. 2.2, also due to the design of ASIs/HAs, the reusability of ASIPs (BP) is an issue. Nevertheless, the main issue of ASIPs (BP), which is the same as for traditional DSIPs and DSPs, is the low energy efficiency. Again, the low energy efficiency is mainly a consequence of implementing several heterogeneous functional blocks, i.e. algorithms, on the same, i.e. rather generic, hardware.

In this book, we combine the idea of domain-specificness from the DSIP architecture style with the idea of utilizing multiple customized hardware blocks from the ASIP (AP) architecture/platform style. Figure 3.1 shows the proposed DSIP design approach indexDesign approach of this book. As it can be seen, we implement only one or very few functional blocks with a single processor. Thus, to implement the whole functionality of the domain, several processors are employed. Each of these processors is customized for the corresponding sub-domain/sub-functionality. Because all processors are based on the same architecture template and the customizations are only done within a common pre-defined space (already defined before designing the actual instance), high reusability is ensured. Because of the domain-specificness of the template and the additional customization of each processor instance, at a medium-grain granularity, high energy efficiency is obtained. As shown in Fig. 3.1, we make a clear distinction between (a) the template design and (b) the design of template instances.

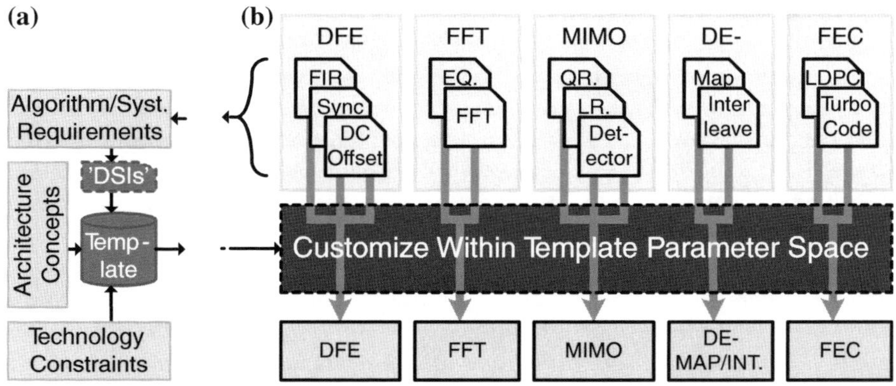

Fig. 3.1 Proposed DSIP template design approach. For the design of the template, the algorithm/system requirements, architecture concepts and technology constraints are considered. Contrary to conventional DSIP solutions, the proposed DSIP platform consists of several architecture instances which are all customized (medium grain) for the sub-domain/sub-functionality. Contrary to conventional ASIP solutions which customize with ASIs/HAs, we customize the architecture instances only within the pre-defined template parameter space. This ensures high reusability. The traditional design approaches can be seen in Fig. 2.4. **a** Template. **b** Instantiation of template

3.1.2.2 Template Design

During the design of the template, to ensure that the joint requirements of high performance, high energy efficiency, high flexibility, high reusability and technology scaling-friendliness can be fulfilled (i.e. are well balanced), the following main aspects are taken into account:

1. *Algorithmand system requirements from the considered domain.* This involves the analysis of algorithms from different/all functional blocks. It is important that the algorithms can efficiently be mapped onto the DSIP architecture instances. Besides, also system requirements, such as data rates and latencies need to be considered.
2. *Innovative architecture concepts.* It is essential to evaluate existing and to propose new innovative concepts to cope with the set of requirements.
3. *Scaled technology constraints.* This is important to ensure an efficient hardware implementation in DDSM technologies. Here, we focus mostly on the wire influence on the template scalability aspect.

To enable high reusability and to reduce the time for the design of template instances, the customization of template instances has to be done within a pre-defined (=pre-evaluated) architecture design space. Thus, during the template design, this template parameter/scalability space needs to be defined. This can basically be done by evaluating different implementation options and by pruning design points/choices which are not pareto-optimal. For instance, to support variable multiplications, different hardware multipliers could be considered. Since certain algorithms could require multiplications with up to 32 bits, the parameter space could include hardware multipliers with up to 32 bits. However, because (1) either algorithm optimizations can be applied to avoid 32-bit multiplications completely and/or because (2) this operation occurs very seldom, so that an iterative computation with a smaller multiplier would not harm the overall performance, the choice for a 32-bit hardware multiplier can be pruned. Note, this is clearly beneficial, because the layouts of parallel hardware multipliers are difficult to combine with the layouts of linear operators (see Sect. 3.2.2). To define the parameter/scalability space, also the wire delay has to be taken into account. For instance, it can be beneficial from a mapping/scheduling point-of-view to share a single register file amongst 20 FUs. However, the enormous capacitive load on the read port, i.e. caused by many/long wires and the input logic of the FUs, results in a very high delay. Thus, because of the significantly decreased clock frequency, this option is clearly not pareto-optimal. Therefore, by evaluation such options at template design time, the design space can be pruned. As a consequence, during the design of template instances, only pareto-optimal options have to be considered and therefore the design time can be much shorter. During the design of the template, also cost models should be made. This cost models, which capture important trade-offs, can then be used during the instance design. In general, it would be beneficial to have a defined template parameter space and corresponding cost models which are technology independent. However, because most trade-offs are influenced by technology properties, it seems that this is not possible.

3.1.2.3 Design of Template Instances

As illustrated in Fig. 3.1b, for the implementation of the platform, the functional blocks of the platform are first assigned to multiple sub-domains. This step, which is part of partitioning, considers several aspects, such as algorithm characteristics and concurrency of execution. Then, for each of the sub-domains, an individual DSIP architecture template instance is designed. Thereby, each architecture instance is only customized within the pre-defined parameter space. During the design of the architecture instance, not only the architecture is customized for the algorithms, but also the algorithms for the architecture. This algorithm-architecture co-design approach is essential to obtain high efficiency. The cost models of the template will guide design decisions.

Figure 3.2 shows an example of the envisioned DSIP platform. As it can be seen, the platform is implemented with 5 DSIP architecture template instances. Because they are all based on the same template, the architecture structures of all instances are the same. However, each instance is customized for the requirements of the sub-domain/sub-functionality. The example illustrates this customization with different memory sizes, different number of clusters/processing engines and different interconnections. However, as shown in Fig. 3.16, the template customization space is clearly not limited to that. This is a major difference compared to conventional DSIP architecture templates which only enable a coarse

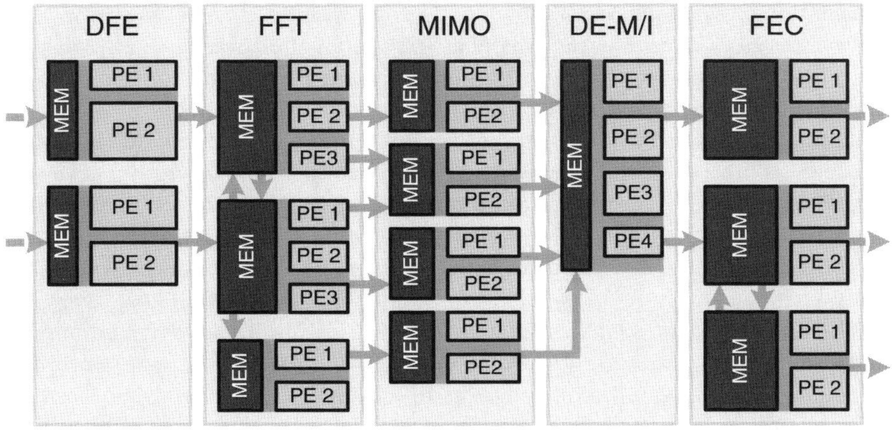

Fig. 3.2 Example of the envisioned DSIP platform, which consists of several DSIP architecture instances. Each DSIP architecture instance is customized for the requirements of the sub-domain/sub-functionality. This is illustratively shown with different memory sizes, different number of clusters/processing engines and different interconnections. However, as shown in Fig. 3.16, the template customization space is clearly not limited to that

customization. From Fig. 3.2 it can be seen that the data flow in the platform is unidirectional from left to right. This indicates that we clearly separate between transmitter and receiver hardware and that we exploit the data-streaming nature of the domain. This is important to achieve high energy efficiency at platform level.

3.2 Applied Design Approach to Define Architecture Template

In the following we explain the general design approach that was applied to obtain the proposed DSIP architecture template. It can be separated into 4 consecutive steps, which are (1) analysis and definition of system and algorithm requirements, (2) analysis and definition of technology constraints, (3) evaluation and selection of architectural concepts and (4) definition of the architecture template.

3.2.1 Analysis and Definition of System and Algorithm Requirements

3.2.1.1 System Requirements

As previously mentioned, this book targets advanced high-performance multi-stand-ard/mode wireless communication systems. We consider 4G standards, such as LTE, WLAN and IEEE 802.15.3c (60 GHz), as the main drivers. The DSIP architecture template must be well-suited to cope with the system requirements of the considered domain. As motivated in the introduction, we generally need to target high performance, high energy efficiency, high flexibility, high reusability and technology scaling-friendliness.

High performance refers to (1) high system (BER) performance, (2) high data rates and (3) low latencies. To obtain high BER performance, a focus on advanced algorithms, e.g. the K-Best instead of the zero-forcing MIMO detection algorithm, is needed. Depending on the considered standard, the maximal required data rates range from 300 Mbps for LTE, via 600 Mbps for IEEE 802.11n to several Gbps for 60 GHz standards such as IEEE 802.15.3c. The maximal allowed Short InterFrame Space (SIFS) latencies range from 16 μs for IEEE 802.11n (5 GHz band), via 6 μs for IEEE 802.15.3c (default) to 3 μs for IEEE 802.11ad. To enable high data rates, i.e. high throughputs, and low latencies, i.e. short delays between the availability of input data and the availability of computed output data, a high degree of parallel processing and high clock frequencies are needed.

As motivated in the introduction and shown in Fig. 2.8, to meet the targets on energy efficiency, computations have to be performed with less than 1–2 pJ while maintaining high performance [14]. For optimizing energy, techniques, such as near/sub-threshold voltage operation, have been proposed. Because of the low voltage, the achievable clock frequency is very low, e.g. several MHz or even lower. To still enable high data rates and low latencies, an enormous degree of parallelization needs to be exploited. Nevertheless, the efficiently exploitable amount of parallelism in the considered domain is limited. In addition, because of the massive amount of parallel hardware, the lengths of wires would potentially be very long, which is clearly a drawback for DDSM technologies. The increased chip area would lead to a higher unit cost, which is clearly also a disadvantage. For these reasons, near/sub-threshold voltage operation is currently not an option for the considered domain (in future technologies, e.g. when FinFET devices are available, this option should be re-evaluated). We focus on architectures which can operate at high clock frequencies to meet the data rate and latency requirements. Nevertheless, for low-performance modes, which can be enabled at run-time, we still consider optimization techniques, such as Dynamic-Voltage and Frequency Scaling (DVFS) and power gating. In this book, we focus mainly on architectural concepts, on algorithm-architecture co-design and on short lengths of active wires to optimize for energy. All these measures are exploitable for the high-performance modes, which are causing the biggest challenges.

To cope with the flexibility and reusability requirements, we employ the proposed DSIP architecture template design approach. The processor style enables high flexibility, the domain-specific nature enables high reusability. The actual required flexibility depends on the requirements of the targeted standards/modes and the requirements of each functional block. During the design of each architecture template instance, with the degree of customization, the flexibility/efficiency trade-off can be selected. At the platform level, the support for multi data-streams (because of multiple antennas) need to be enabled. In this book, we focus mostly on the implementation of specific architecture instances and consider the platform design as future work.

Regarding technology scaling-friendliness, we refer to Sect. 3.2.2.

3.2.1.2 Algorithm Requirements

To obtain an efficient implementation, it is very important that the algorithms and the architectures are well matched. In case of a mismatch, the algorithms cannot be well mapped/scheduled and therefore the achievable hardware utilization is low. As a consequence, because of the low area/energy efficiency of the implementation, the implementation targets cannot be met. To avoid such as mismatch, we analyze and consider the characteristics of domain-specific algorithms during the design of the architecture template. The results of this analysis will be shown in Sect. 3.3.

3.2.2 Analysis and Definition of Technology Constraints

3.2.2.1 Focus on Wire Influence

As mentioned in Sect. 1.2.2, with the scaling towards DDSM technologies, new technology-related effects, such as decreasing reliability [42], increasing variability [12, 21], and increasing influence of wires [22], are appearing. To cope with these effects efficiently, considerations and countermeasures not only at technology, but also at architectural level are required. We consider countermeasures for effects, such as reliability and variability, which can be seen as part of another problem/research field, out of scope. However, because (1) wires will dominate the delay and the energy consumption of designs in DDSM technologies, and because (2) wires and architectures cannot be considered independently from each other, we take the influence of wires strongly into account. Moreover, in this book, we explicitly target an architecture template that can efficiently be implemented in future technologies. This is a major difference compared to traditional architectures which often neglect technology constraints completely.

The main target of the architecture design, in regard to wires, is to ensure that the most influencing wires, i.e. wires in the critical path and highly active wires, are kept short and that wire congestions are generally avoided. In addition, it is also important to ensure that the critical path is in the most important part of the architecture. For instance, if the most important/utilized FU could operate at 1 GHz but the delay of the control unit limits the clock frequency to 400 MHz, the efficiency of the design is significantly reduced. Furthermore, for the architecture template, the architecture scalability and the influence on wires need to be considered. For example, when increasing the number of FUs per register file read port, there will be a moment when the critical path moves from the datapath within a FU to the datapath between the register file and a FU. Clearly, as mentioned above, this is not desired.

3.2.2.2 Design Flow for Hardware Implementation

As explained in Sect. 2.7.2, by using the traditional standard-cell flow, the mapping of the architecture description onto the technology and the design of the layout can be done fully automated. This high degree of automation is clearly beneficial, however, the main issue of this flow is that the designer has little influence on the resulting layout, i.e. wire lengths. Thus, even if the architecture designer targets a certain layout, which ensures that the wires are well considered, it is most-likely that the automation tool will implement it in a completely different way (because it does not see the same context). Because of this weak link between architecture and layout and the resulting unpredictability, the design of an efficient scalable architecture for DDSM technologies is very difficult. For this reason, we propose a semi-custom design flow in the appendix of this chapter that overcomes these issues. We assume that this flow, which translates the layout considerations of the architecture designer to the actual layout, will be applied to implement the proposed DSIP architecture template.

3.2.2.3 Defined Layout Constraints/Guidelines

We define several layout/architecture constraints (or at least guidelines) that are considered during the architecture design. As stated above, the main goal of this constraints is to control the wire lengths, i.e. to ensure scalability and high energy/area efficiency. In the following we summarize these constraints, which are illustrated in Fig. 3.3:

1. *Basis for layout*: As basis for the layout, we leverage on the standard-cell grid (which enables automation with existing tools) and on the (1a) bit-slice scheme. For every bit, a standard-cell row is utilized. As an example, a 16-bit adder would use 16 standard-cell rows. The bit-slice scheme has an inherent regularity and provides therefore a good basis for scalability. In addition, we define the main (1b) data flow/wires to be horizontally, e.g. from left to right, and the control flow/wires vertically, e.g. from top to bottom. This rule considers the fact, that the width of most logic standard cells is smaller than the height (the power lines are at the top/bottom of a cell) and therefore the data-wires can be short. The considered domain is clearly data dominated, i.e. much more data wires than control wires, therefore it is important to mainly optimize for short data wires. To compensate for the delays of long wires in the control path, pipeline registers could be inserted. Because the control flow is rather deterministic, additional pipeline registers would typically not harm the computational/scheduling performance. Nevertheless, because of the additional energy consumption in the pipeline registers, this approach should only be applied for control wires in the critical path. In addition, the duplication of control registers, i.e. the load is distributed amongst several buffers which can then physically be placed at suited locations, could be considered.

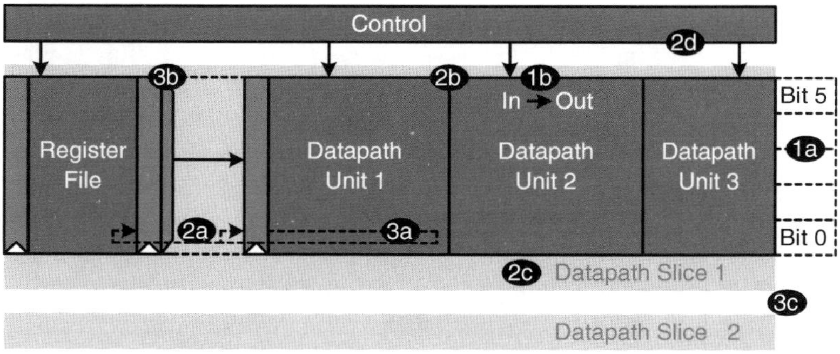

Fig. 3.3 Considered layout constraints/guidelines to control the wire lengths, i.e. to ensure scalability and high energy/area efficiency: **1a** Leverage on bit-slice standard-cell layout. **1b** Target mainly horizontal data and vertical control wires. **2a** Ensure short wires in critical path (relative order). **2b** Place frequently communicating components/operators next and horizontally to each other. **2c** Group components/operators to datapath slices and place the slices vertically to each other. **2d** Place the local control on *top/bottom* of a datapath slice. **3a** Employ linear/iterative arithmetic/logic operators to enable horizontal abutment. **3b** Enable the abutment of arithmetic/logic datapaths with register files/data memories. **3c** Prefer homogeneous datapath slices to ensure high area utilization

2. *Relative order of architectural components/operators*: To ensure that the wires of the critical path and the wires that are most active (high toggling frequency) are as short as possible, the relative order of architectural components (e.g. different units, register file, control) and operators (e.g. multiplexers, adders) is important. Considering the horizontal data-flow direction, components/operators of the critical path and components/operators that communicate frequently have to be placed horizontally and close to each other. For instance, (2a) if the critical path of the most important unit includes the connection to the register file, than the most important unit has to be placed next to this register file. For example, (2b) if unit 1 and unit 2 in Fig. 3.3 communicate more frequently with each other than with unit 3, than unit 1 and unit 2 should be placed next to each other. Because of the high amount of required data parallelization and data sharing, a layout, in which all datapath components are placed in a single row fashion (slice), would not be very efficient. For instance, the aspect ratio could become 100:1 (x:y) and therefore the horizontal data wires, which lead from the left most to the right most side of the layout, would be very long. For this reason, (2c) datapath components need to be grouped to datapath slices and these slices need to be placed vertically to each other. Note, a datapath slice contains components that communicate frequently with each other but not so frequently with other architectural parts. Beside the relative order of the datapath slices/components/operators, also the order of the control relative to the datapath is important. Because of the vertical control wires, (2d) we place especially the local L0 control (assume a control hierarchy is present), which has a higher toggling frequency than e.g. L1/L2, above or below and close to the datapath slices.

3. *Abutment of architectural components/operators*: To ensure short wire lengths between architectural components/operators, i.e. to avoid empty space, the components/operators must be abutment-able. Considering the bit-slice scheme and the horizontal alignment, we must employ linear operators to enable abutment. We refer to linear operators as operators, for which the output word length is the same as the input/internal word length. This means, the number of utilized standard-cell rows within the operator is constant. Nevertheless the number of inputs and outputs can be different. Basic elements, such as multiplexers, Flip-Flops (FF)s and adders, fulfill this constraint. However, this constraint is not fulfilled by parallel/array operators, such as parallel/array multipliers. Thus, we need to focus on (3a) linear/iterative arithmetic/logic operators during the architecture design. An exception can be made for operators with small word length that still fit in the overall scheme, e.g. a single 8-bit multiplier within a 64-bit datapath scheme. Not only the components/operators within the arithmetic/logic datapath, but also the (3b) components/operators between the arithmetic/logic datapath and the register file/data memory should be abutment-able. Note, components, which are further away from the arithmetic/logic datapath, e.g. L2 data/control memory, are typically not very active, therefore longer wire lengths are generally more acceptable. To ensure a high area utilization, (3c) the width of different datapath slices with the corresponding local control should not be very different. Thus, homogeneous rather than heterogeneous datapath slices are preferred.

Despite of applying the fore-mentioned constraints, certain long wires, especially of connections with low activity, can still be present. To avoid that these long "unimportant" wires are responsible for the critical path, their corresponding paths can potentially be declared as "multi-cycle delay paths". Furthermore, to avoid an unindented toggling of long wires, i.e. to reduce energy, the insertion of AND gates between the wire drivers/buffers and the beginnings of the long wires can be considered.

3.2.3 Evaluation and Selection of Architectural Concepts

After defining the system and algorithm requirements as well as the technology/layout requirements, the architecture template can be designed. The design starts by evaluating and by selecting architectural concepts that are best suited for filling the joint requirements.

First, during a literature study, the most promising concepts are identified. Then, further evaluations, which consider all the requirements that have been defined in the previous steps, are performed to select the best concepts for the proposed architecture template. Thereby, also new concepts, which complement existing ones, are introduced.

Section 3.4 will highlight the main architectural concepts that have been selected for the proposed architecture template. Supporting evaluation results, which motivate the design decisions, are provided. For the evaluation, we have implemented the architectural components in TSMC 40 nm standard-cell technology. The results have been obtained by performing automatic synthesis with Cadence RTL Compiler and automatic placement/routing with Cadence Encounter. Although the resulting layouts do not match the defined layout style, the results provide a good indication. Note, for a small design, e.g. only one architecture component implemented, the wire influence is also rather small.

3.2.4 Definition of the Architecture Template

As a last step, the architectural concepts are combined to define the complete DSIP architecture template. The architecture template will be explained in Sect. 3.5. By considering the required scalability, the template parameters can be defined. The template parameters are summarized in Sect. 3.6. For compliance with the proposed design approach, the parameter ranges have to be evaluated, defined, and characterized with cost models. Thereby, to ensure scalability together with predictability/consistency, a semi-custom design flow, as e.g. proposed in the appendix of this chapter, has to be applied. However, because of the effort and the lack of automation, we consider this step as future work.

In general, because it is infeasible to consider all future requirements, the architecture template will have to be refined over time. In this book, we have considered the experiences from the case studies of Chaps. 4, 5 and 6 to refine the template.

3.3 Requirements from Algorithm Perspective

The objective of the proposed architecture template is to support all domain-specific algorithms (sub-functionalities) very efficiently. To achieve this objective, the algorithm-related requirements (characteristics) need to be known and considered during the architecture template design. In this section, we analyze and determine this algorithm-related requirements and their consequences on the architecture template. Thereby we organize the main common domain-specific requirements in terms of (1) data representation, (2) arithmetic and logic operations, (3) parallelization, (4) data transfer operations, (5) data storage and (6) instruction control. A summary is provided in Sect. 3.3.7.

As previously motivated in 1.3.5, to achieve high energy efficiency, a good match between algorithms and architectures is absolutely crucial.

3.3.1 Data Representation

Algorithms are usually designed with floating point precision. However, the support of floating-point operations in hardware is more costly than the support of fixed-point operations [25]. A conversion from floating point to fixed point representation introduces quantization noise, i.e. reduces the BER performance. Nevertheless, because of the existing margin, a minor BER performance degradation is typically acceptable. Based on these reasons, we consider only fixed-point hardware in the architecture template.

The actual required fixed-point word length, i.e. the minimal fixed-point word length to still meet the BER performance constraint, varies across domain-specific algorithms. For instance, the algorithm for Low-Density Parity-Check (LDPC) decoding in [30] requires only 5 bits while the algorithm for matrix inversion in [52] is more sensitive to quantization noise and requires therefore up to 24 bits. The word-lengths requirements are not only varying from algorithm to algorithm, but also from signal to signal within an algorithm [10, 43]. Considering these requirements, the architecture template has to support a variety of different word lengths very energy-efficiently [43, 62]. In addition to different word lengths, the support of real and complex data is required.

3.3.2 Arithmetic and Logic Operations

Algorithms in the considered domain employ operations on real and on complex numbers. Nevertheless, the latter can be decomposed into a sequence of real number operations. For instance, a complex multiplication requires 4 real multiplications, 1 real addition and 1 real subtraction. Throughout this book, we distinguish between the following domain-specific operation types:

- *Basic arithmetic operations* (*BAO*s): E.g. addition, subtraction, shift and abs. The latter can be decomposed to a negation and an addition (2's complement).
- *Multiplications with a constant operator (MCO)s*: This type of operation can be implemented by using only basic arithmetic operations [29]. Compilers and logic synthesis tools leverage on various optimization techniques to reduce the cost of constant multiplications.
- *Multiplications with variable operators and a multiplier which can adopt a Wide value range (MVOW)s*: An example for this type of operation is a 16×12-bit multiplication in which the multiplier can adopt all 2^{12} different values. Note, the word length does not necessarily determine the value range since the application can imply restrictions.
- *Multiplications with variable operators and a multiplier which can adopt only a narrow value range (MVON)s*: A 16×3-bit multiplication, where the multiplier can only adopt maximal $2^3=8$ different values, falls in this category. Another example is a 16×10-bit multiplication, where the application restricts the

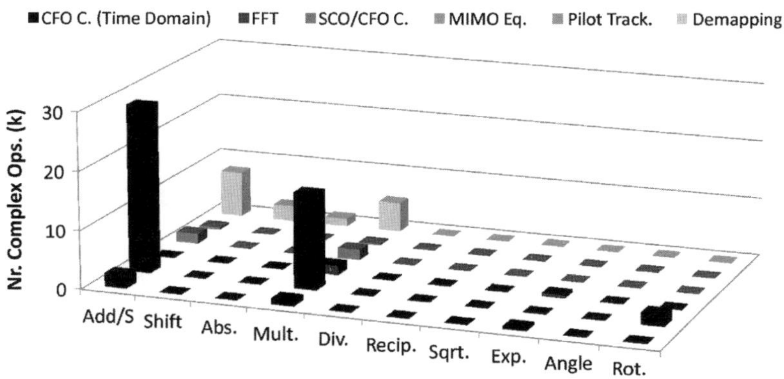

Fig. 3.4 Operation breakdown, based on a high-level profiling of the Matlab code, for the IEEE 802.11n MIMO 4×4 inner modem baseband payload processing. The functional blocks are Carrier-Frequency Offset (CFO) compensation in the time domain, FFT, Sampling Clock Offset (SCO)/CFO compensation in the frequency domain, MIMO equalization, pilot tracking and demapping. Computations per symbol and antenna. BAOs and multiplications are clearly dominant. TSOs, such as reciprocal and square root, are only required by few functional blocks

multiplier value range to only 10 different values (e.g. when one-hot encoding is applied).

- *Trigonometric and special operations (TSO)s*: E.g. cos, sin, log, $\frac{1}{x}$ and $\sqrt{x}$. Operations of this type can be approximated in several ways. For instance, by leveraging on the iterative CORDIC algorithm, by using the Dichotomous Coordinate Descent (DCD) algorithm [3] and/or by using a Look-Up Table (LUT) approach with pre-computed values [24, 52]. The CORDIC algorithm does not required specialized hardware, since it leverages only on BAOs.
- *Boolean logic and comparison operations (BLCO)s*: E.g. AND, OR, $\geq$, maximum value of a vector. Comparison operations are essential to enable non-deterministic data-flow structures (branches, jumps).

Because each of the fore-mentioned operation types has certain characteristics, which can exploited to enable an efficient implementation, a distinction between these operation types is important. As indicated above, complex operations can always be decomposed to sequences of simple/basic operations. This is an important observation, because the need for high reusability and high energy-efficiency implies the need for reusable hardware components with low complexity.

Figure 3.4 shows the relative operation breakdowns for different functional blocks, which are based on the profiling of IEEE 802.11n MIMO 4×4 high-level Matlab code. A profiling of the 60 GHz baseband processing functionality is reported in [53]. Also Table 3.1 shows, for typical domain-specific algorithms, an estimate of the required operation types and the relative duty cycle between these operation types. From these 3 profiling results it can be observed that the operation-type requirements and the relative duty cycles are strongly algorithm/

Table 3.1 Estimated operation breakdown of domain-specific algorithms. Numbers are based on real operations. Control operations are neglected

Ref.	Algorithm	Operation breakdown (%)			
		BAO	MCO	MVO	BLCO
[41]	Quadrature mixer (CORDIC)	100	–	–	–
[18]	Auto correlation (synchronization)	>60	–	<40	–
[32]	FIR filter	>50	<50	–	–
[15]	FFT	>75	<25	–	–
[36]	MIMO detection (low-level optimized)	>80	–	<10	<10
[34]	De-mapping (low-level optimized)	<40	>20	<40	–
[30]	LDPC (Offset min-sum)	>60	–	–	<40

functionality dependent. This information is essential for designing area-efficient architecture template instances. For example, as it can be seen in Table 3.1, an architecture instance, which targets FIR and FFT algorithms, does not need to include hardware for the support of variable multiplications (MVO)s and logic/comparison operations (BLCO)s. Also TSOs are only present in few functional blocks. Figure 3.4 and Table 3.1 show that BAOs (together with MCOs) are typically computationally dominant. Because the relative duty cycle of variable multiplications is rather low, they could be supported with area-efficient multi-cycle instead of single-cycle multipliers. Thus, the knowledge of the observed properties, i.e. support of certain operation types not always needed and computational dominance of BAOs/MCOs, should be exploited by the architecture template to increase the overall efficiency.

3.3.3 Parallelization

As mentioned before, upcoming 4G wireless standards demand data rates beyond 1 Gbps with strict latency requirements in the order of several μs [19]. Since the clock frequency can basically not be increased anymore, the enormous computational complexity has to be solved with massive parallel computing. In general, Data Level Parallelism (DLP), Instruction Level Parallelism (ILP) and Thread Level Parallelism (TLP) can be utilized.

To better classify the exploitation of parallelization, we define two main classes that are relevant for wireless applications:

1. *Fine-grain*: A relatively small Control Data Flow Graph (CDFG) of the application for which DLP, ILP and TLP can be exploited. The amount of communication/sharing between the different threads is medium/large. Examples are the parallelization across different sub-carriers in a OFDM receiver and the parallelization across different check-nodes in a LDPC decoder.
2. *Coarse-grain*: A relatively large CDFG of the application for which DLP, ILP and TLP can be exploited. The amount of communication/sharing between the

different threads is low. Examples are the parallelization across different functional blocks and the parallelization across different antennas in a MIMO receiver.

This classification provides a general understanding. The distinction between fine-grain and coarse-grain is important for the design of the architecture and for the mapping.

Algorithms are traditionally designed in a sequential and in a platform-independent manner. Thus, to obtain efficient implementations, algorithm transformations, which enable massive parallelization and which consider the underlying platform, are a necessity [34, 35, 61]. Optimized algorithms have typically a more deterministic data-flow structure and offer a parallelization with reduced dependency. In the considered domain, DLP in the form of SIMD is most commonly applied. Nevertheless, with the increasing number of computational units and the limitations of single-threaded data parallelization, ILP and TLP are becoming more important [1, 54]. Thus, to enable the required high degree of parallelization, all 3 types may need to be combined. Therefore, the architecture template has to provide support for DLP, ILP and TLP.

In general, the degree of exploitable parallelization, i.e. how many parallel units can efficiently be utilized, varies from algorithm to algorithm. Also the level on which a parallelization can occur, i.e. coarse-grain or fine-grain, is algorithm dependent. When designing an architecture, the exploitable parallelization of the targeted algorithms should be considered. An example: Two algorithms, one that allows a DLP of 4 and one that allows a DLP of 128, have to run on the same hardware. In this case, it is not optimal to design a datapath for a DLP of 128 since the first algorithm cannot leverage on it. Therefore, a better option can be to design multiple parallel datapaths (up to 32), each supporting a DLP of 4, and across these datapaths, ILP is supported. With the additional support for ILP, a parallelization across loop iterations and/or within a loop body, of the first algorithm would be enabled. Thus, to achieve an efficient implementation, typically a combination of fine-grain and coarse-grain parallelization need to be exploited.

Parallelization has implications on control, on storage and on data transfer operations. The implications are generally type (DLP, ILP or TLP) and class (coarse or grain) dependent. An architecture can offer several options for the exploitation of parallelization. Each option can be characterized with a cost. During the mapping phase, the costs of the different options can be compared and the cheapest solution can be chosen.

3.3.4 Data Transfer Operations

Data transfer operations are basically required because of the following 4 reasons: (1) operation dependencies that are inherently present in the algorithm, (2) data storage hierarchies and data storage related transfers, (3) data parallelization and

(4) data representation related. The first cause depends only on the algorithm (pattern explicitly defined) while the others depend additionally on the architecture and on the mapping.

Examples for inherent data transfer operations are the word-level shuffling in the LDPC algorithm, the bit-level shuffling in the Viterbi algorithm and the sorting in the K-Best MIMO algorithm.

A data transfer between the L1 data memory and the L0 register file is a data transfer that is caused by the storage hierarchy (between levels). The same is valid for a data transfer from a register file to a pipeline register. A data transfer from a data memory to a data memory falls also in this category (same level).

Data parallelization has several causes for data transfer operations: (1) packing of scalar to vector operands and vice versa, (2) exchange of data between parallel units (e.g. previously computed results, synchronization) and (3) distribution of shared data (e.g. constants). The frequency of data transfer operations depends essentially on the employed type (DLP/ILP/TLP) and class (fine/coarse grain) of parallelization. For instance, a fine-grain parallelization requires typically more frequent data transfers than a coarse-grained parallelization.

As mentioned in Sect. 3.3.1, the actual required word length varies typically from signal to signal within an algorithm. In case this word-length variations are exploited, i.e. multiple word lengths are supported on the architecture, data-transfer operations for fixed-point alignments are required.

In general, the architecture template has to be optimized for different types and characteristics of data transfer operations. For instance, between data memories and register files, wide instead of small data transfers could be exploited [50]. Paths, with frequent data communications, e.g. a path between register file and adder, should be kept simple to reduce energy cost (e.g. avoid many multiplexers). Algorithm transformations, efficient data layouts and the right choices of parallelization, e.g. employ sub-word parallelization if possible, can reduce the number of data transfers, the requirements on the hardware [9, 35] and therefore the implementation cost.

3.3.5 Data Storage

The data storage of variables in an algorithm can generally be classified as follows: (1) size, i.e. scalar vs. array types, (2) life-time, e.g. temporary variables with a few cycles of life-time vs. arrays with 1000 s of cycles of life-time, (3) spatial locality of arrays, i.e. access order of variables in an array, and (4) temporary locality or reuse of arrays, i.e. access frequency of the same array/variable. The actual requirements depend essentially on the characteristics of the algorithms and on the employed parallelization schemes, e.g. the mix of ILP, DLP and TLP.

For instance, in the case of a FIR filter, the coefficients, the input and the output array exhibit spatial locality and reuse. However, this may not be the case for algorithms which include conditions. From an application perspective, a general distinction between streaming-based and block-based algorithms can be made. Common algorithms with streaming nature are FIR filter, channel mixer and interpolator. FFT and LDPC are well-known block-based algorithms. Block-based algorithms require more data storage and the data has a longer life-time compared to streaming-based algorithms. At the platform level, the communication between functional blocks follows always a streaming nature. In general, the architecture has to provide data storage for different types of requirements (size, life-time, etc.) and has to enable the efficient mapping of streaming-based and block-based algorithms.

The employed parallelization scheme can have a significant influence on the data storage requirements. For instance, multiple units may need copies of the same data to perform computations independently. Often data has to be buffered before it can be processed in a vector fashion. Unrolled loops can increase the amount of required temporal data storage and also the code-size. Fortunately, to cope with the fore-mentioned issues, i.e. to reduce data storage size and access frequency, code transformations can be applied [9, 45, 59].

3.3.6 Instruction Control

Algorithms with a deterministic data-flow structure can efficiently be parallelized and require little control. Most algorithms of the considered domain have either inherently a fully deterministic data-flow structure (e.g. FIR and FFT) or can be transformed to make the data-flow structure more deterministic [35]. Because of a if-then-else condition, a path inside a loop of an algorithm can be non-deterministic. However, it can be made deterministic by using techniques like predication. In some cases, the exact number of loop iterations is unknown at design time. This is particularly the case for iterative algorithms, such as Turbo Code. Nevertheless, the maximum number of iterations is clearly defined, e.g. with a stop-criterion, since this is an essential requirement for meeting real-time performance constraints. The control complexity for supporting a variable number of iterations is typically rather low, since only a decision between the continuation of a loop execution or the processing of a new input packet needs to be made.

Algorithms of the wireless domain are typically loop-dominated, i.e. most of the execution time the same sequence of instructions is executed [57]. This characteristic should be exploited to further reduce the energy consumption of the control.

A cost-effective implementation has to support multiple standards and modes with the same hardware (time multiplexing). To support all different modes, parameterized and/or different algorithms are employed. At run-time, the control must be able to change the execution flow and/or the complete algorithm in an efficient manner. However, because of the availability of the source code, this can be fully planned at design time.

As mentioned in Sect. 3.3.3, to enable an efficient parallelization and to meet the performance constraint, the support of ILP is essential. Thus, the control needs to be able to provide different instructions to multiple homogeneous and heterogeneous units/architecture elements efficiently. Thereby incompatible loop body's need to be handled. Especially when considering architecture instances that should run multiple functionalities concurrently, the support for TLP is crucial. Hence, instructions for different threads/algorithms have to be provided concurrently and synchronization mechanisms must be enabled [10].

3.3.7 *Summary*

In the following the main conclusions/requirements are summarized:

- *Data representation*: Employ fixed-point hardware and support multiple word lengths.
- *Arithmetic and logic operations*: Decompose complex operations to simple ones and employ basic and reusable operators. Exploit the relative duty cycle of operation types. The efficient support of BAOs/MCOs is most important.
- *Parallelization*: Support TLP/DLP/ILP and fine/coarse-grain parallelization. Evaluate best option during architecture design/mapping.
- *Data transfer operations*: Perform wide data transfers and ensure little multiplexing/short wires for most frequently used connections. Apply algorithm transformations to reduce amount of data transfer operations.
- *Data storage*: Support streaming and block-based algorithms and their corresponding requirements on data storage. Apply code transformations to reduce data storage size and amount of data transfer operations.
- *Instruction control*: Apply algorithm transformations to make data-flow structures deterministic. Enable efficient loop support. Ensure that algorithm parameters and algorithms can efficiently be changed at run-time. Employ distributed control, with non-compatible loop-nest organization per instruction field, to support ILP efficiently.

Related work can be found in [2, 20, 37, 57].

3.4 Employed Architectural Concepts

In this section we highlight the main architectural concepts that have been selected for the proposed architecture template. Supporting evaluation results, which motivate the design decisions, are provided. The concepts are structured as in Sect. 3.3.

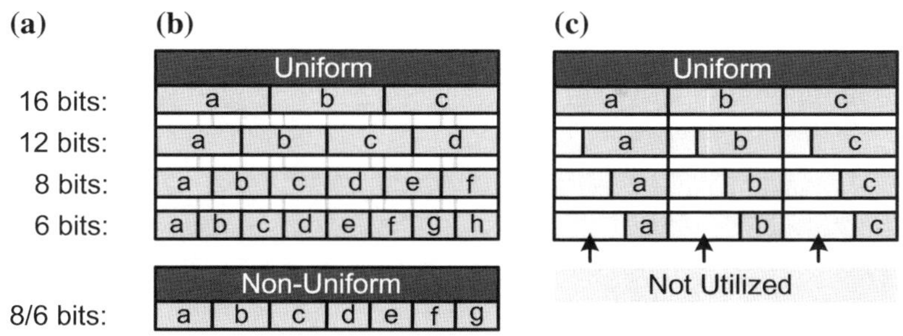

Fig. 3.5 Subword organizations for SoftSIMD and HardSIMD. SoftSIMD enables to fully utilize the datapath word (here 48 bits) and non-uniform subword sizes are generally possible. **a** Subwords. **b** SoftSIMD. **c** HardSIMD

3.4.1 Data Representation

As explained in Sect. 3.3.1, in general, the architecture template needs to enable the support for a wide range of different word lengths. To full this requirement, in the conventional approach, fixed-point HardSIMD units together with optional floating-point units are employed. In contrast, we propose the alternative approach of exploiting fixed-point SoftSIMD units (for DLP) [10]. The choice, whether HardSIMD or SoftSIMD units are used in an architecture instance, depends on the actual word-lengths requirements of the targeted algorithms.

A HardSIMD datapath consists of multiple parallel processing paths which are separated into hardware, i.e. the carry-chains between subwords are not connected. Figure 3.5c shows the subword organization for a 3-way HardSIMD datapath with 16-bit subwords. As it can be seen, if 16-bit subwords are used, the full datapath is utilized. However, if the word length of the subwords is smaller, the subwords still need to be "encapsulated" in 16-bit containers. During the operation, only a fraction of the hardware is and can be used. If a wide variety of different word lengths should be support on the same hardware, HardSIMD is not a good option. In this case, SoftSIMD, which enables a better utilization of the datapath, is a better choice [4, 31, 46]. As shown in Fig. 3.5b, instead of multiple narrow, SoftSIMD employs only 1 wide processing path. The subwords that need to be processed are placed next to each other to form a datapath word. Since subwords are no longer separated in hardware, a wide range of subword organizations are possible. In general, also subwords with different word lengths can be combined and computed in a single cycle. Hence, SoftSIMD enables much more flexibility than HardSIMD, which is clearly beneficial for achieving a high hardware utilization. Nevertheless, possible overflows, e.g. because of an addition, need to be analyzed at compile/design time. The potential "pollution" of adjacent subwords can be prevented by adding guard bits.

Figure 3.6 shows a comparison between HardSIMD and SoftSIMD for various distributions of signal word-lengths requirements. Thereby a datapath word length

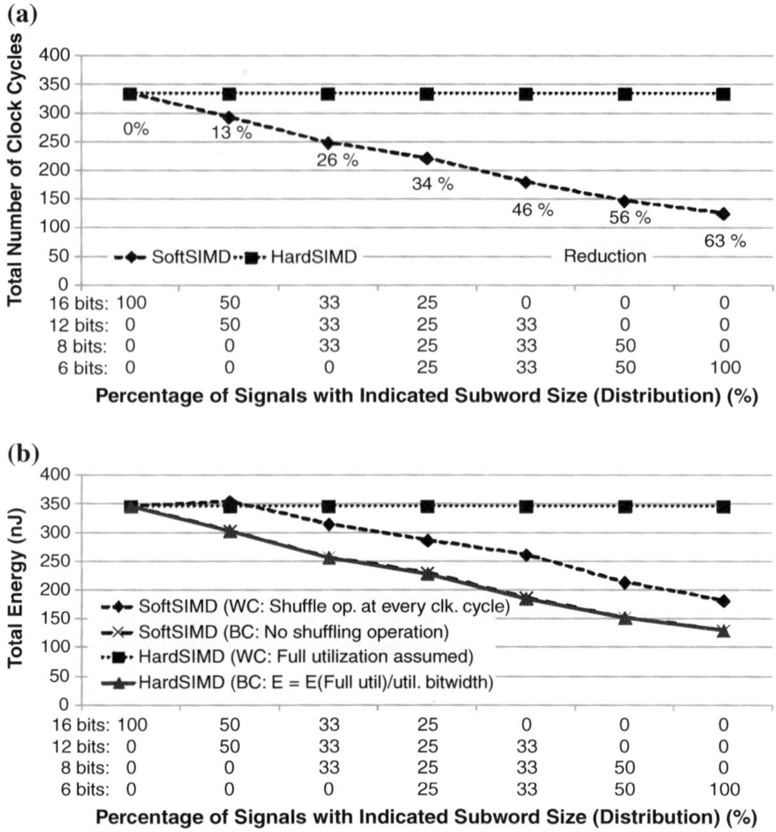

Fig. 3.6 Comparison of SoftSIMD and HardSIMD for an example with 1000 operations and different distributions of subword sizes. **a** Number of required clock cycles. **b** Energy consumption

of 48 bits is assumed (3×16 bits for HardSIMD). In Fig. 3.6a the comparison in terms of required clock cycles is shown. Since packing and unpacking operations can typically be hidden, i.e. this operations can be performed in advance on a different unit, they have been neglected. As it can be seen, SoftSIMD has a large gain potential, especially when small subword sizes are used (in this example up to 63 %). Figure 3.6b shows the comparison in terms of energy with different bounds for packing/unpacking (shuffling) requirements. In the Best Case (BC), there is not need for any shuffling operation and in the Worst Case (WC), a shuffling operation has to be performed at every clock cycle. For HardSIMD, rather the WC boundary can be assumed (small subword sizes are typically not exploited). The results clearly shows that the SoftSIMD approach can lead to substantial energy gains (in this example up to 63 %). Besides, considering the reduction in clock cycles, SoftSIMD implies also fewer instructions, fewer load/store operations and fewer register file accesses. Hence, the total energy gain, when considering the whole processor, would be even higher. Nevertheless, whether HardSIMD, SoftSIMD or even a hybrid solution is most efficient, depends on many factors, such as number of exploitable subword sizes and number of operations per subword size. For most cases, probably a hybrid solution is the best choice. To leverage on SoftSIMD in an effective manner, a fine-grain quantization is required.

3.4.2 Arithmetic and Logic Operations

As explained in Sect. 3.3.2, common domain-specific operations can be separated into the following types: BAOs, MCOs, MVOWs, MVONs, TSOs and BLCOs. Traditionally, an Arithmetic Logic Unit (ALU) is employed to perform all/most of these operations. Instead of employing one big unit, we propose to distribute the functionality across multiple smaller/specific units. Each of the small units supports only a small set of operations. Importantly, all supported operations of a unit have similar characteristics and can therefore share most of the hardware. For instance, an addition and a subtraction are both based on an adder circuit. Contrary to the ALU approach where one type of operation is executed at a time, these small units can work in parallel and therefore compute different types of operations concurrently. Depending on the required types of operations, an architecture instance can include several or only one of these small units. This enables a trade-off between flexibility and area-efficiency.

3.4.2.1 Basic Arithmetic Unit

As mentioned in Sect. 3.3.2, the most important/dominant operations in the considered domain are BAOs and MCOs. Therefore it is important to support these types of operations very efficiently. BAOs are shift, addition, subtraction and abs (=negation+addition).

MCOs can iteratively be computed with shifts and additions. For a conventional constant multiplication, the maximum number of required iterations corresponds to the word length of the coefficient minus 1 (e.g. for a 16-bit constant, 15 iterations are needed). However, a multiplication with this high number of iterations is not very efficient. Hence, to make an iterative computation feasible, the number of iterations need to be further reduced. Instead of binary representation, the constant can be coded in Canonical Signed Digit (CSD) format [10]. Since the number in CSD format is represented with additions and subtractions, the maximal number of non-zero bits is reduced by a factor of 2. Hence, maximal $WL/2-1$ iterations are required to perform a multiplication. In the considered domain, full computation accuracy is typically not required and minor noise, e.g. introduced by fixed-point quantization, can be accepted. Therefore, also the values of the constants can potentially be slightly changed in order to optimize the number of non-zero bits. As a result, constant multiplications can typically be realized with only 2 iterations

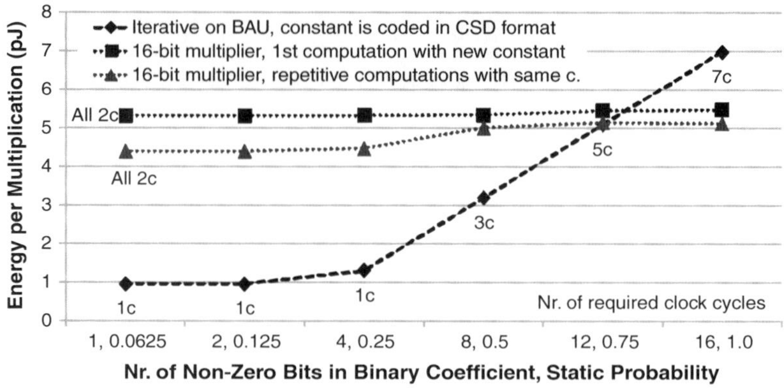

Fig. 3.7 Comparison of energy costs for performing MCOs on different units: BAU versus single-precision multiplier

in average (this was experienced with several driver examples). Figure 3.7 shows a comparison between an iterative CSD-based multiplication and a multiplier-based multiplication in terms of energy cost. As it can be seen, for typical cases (average), the proposed iterative approach is much cheaper.

Since MCOs and BAOs share the same set of primitive operations, they can both be computed on the same hardware unit. We denote this unit, which is capable of computing these shared set of primitive operations, as Basic Arithmetic Unit (BAU). Depending on the word-length requirements, this unit is either implemented with HardSIMD or SoftSIMD. Because of the growth in word length, performing multiplications in SoftSIMD is generally an issue. However, the right-shift method, as proposed in [10], can account for that.

The BAU computes the dominant operations. To achieve low latency and to obtain high area/energy efficiency, we ensure that this unit can operate at a very high frequency. This is done by taking the following two main measures: (1) The datapath is kept simple. In fact, the datapath consists basically only of a shifter and an adder/subtracter. (2) The critical path of all other units are equal or smaller than the one of the BAU. Hence, the BAU, which is the most important unit, determines the (global) clock frequency. Note, in traditional designs, the clock frequency is typically influenced by hardware which is not very high utilized. For instance, often the critical path involves circuits that are part of under-utilized ASIs. A traditional ALU has multiple computation paths. Depending on the instruction, the required output is selected. This selection involves a multiplexer which increases the delay of the critical path. Also by adding ASI/DSIs, the delay of the critical path is increased. In our approach, this degradation by other components, which are less important, is avoided. Because the BAU is very simple, the number of control bits is very low.

In order to support multiple subword sizes with the BAU, e.g. when exploiting SoftSIMD, the BAU will be tightly coupled with a Subword Shuffler Unit (SSU), which performs the resizing/alignment of subwords.

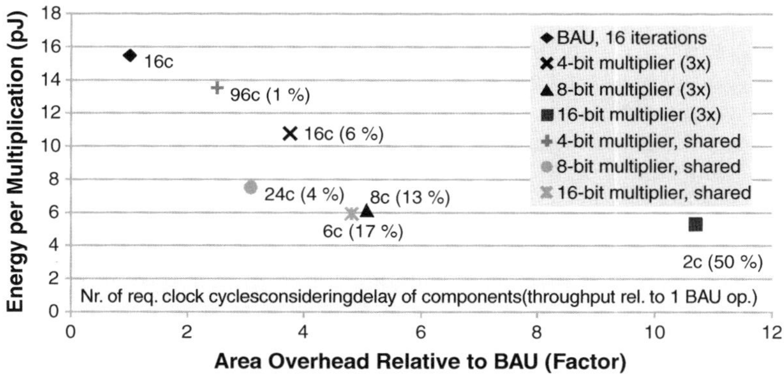

Fig. 3.8 Evaluation of different implementation options for enabling variable 16-bit multiplications on a 48-bit wide datapath: Word-serial multiplication on the BAU, single-precision and multi-precision multiplier. The multiplier is either shared or duplicated for all 3 HardSIMD slots

3.4.2.2 Multi-Precision Multiplier Unit

Constant multiplications (MCO)s are supported by the BAU. However, depending on the algorithms, also the support for variable multiplications (MVO)s can be required. As shown in Sect. 3.3.2, MVOs contribute typically to less than 40 % in the operation breakdown. By a joint consideration of all domain-specific algorithms, the contribution is typically even lower. We leverage on this information to maximize the area efficiency of the proposed architecture template.

In a conventional HardSIMD processor with N slots, the datapath typically enables to perform a variable multiplication of the whole vector with one instruction that takes only 1 or few clock cycles. To achieve this, N parallel full precision multipliers are employed. However, considering the typical operation breakdown, these multipliers are mostly idle. Hence, this solution is not area efficient for the considered domain. In addition, multiplications that could be computed with less precision still need to use the full precision multiplier.

We evaluate the following implementation options that could be used for supporting MVOs in the architecture template: (1) full-precision multiplier, (2) multi-precision multiplier and (3) performing binary multiplications with shift and additions on the BAU. For a SIMD datapath with N slots, we consider either N parallel multipliers or one multiplier that is shared across all slots. Figure 3.8 quantifies the different options in terms of energy cost and area overhead relatively to the BAU. Since the different options require different number of clock cycles, we also indicate the relative throughput of each option with reference to BAU operations. Thereby we take the critical path of the hardware multipliers into account. As it can been seen, the options imply different trade-offs. Hence, the best option depends on the area constraints, on the operation breakdown, on whether the multiplication falls in the category of wide (MVOW) or narrow (MVON) and on possible other constraints that would prevent an optimal utilization of the hardware (scheduling). Nevertheless, in general we find the option of employing N parallel multi-precision multipliers (e.g. 8 bits) most attractive for the considered domain. This approach enables to perform sub-precision multiplications as well as full-precision multiplications. Hence, different word lengths requirements can be better exploited. To increase the throughput, pipelining could be considered. We denote the unit that supports MVOs as Multiplier Unit (MU).

3.4.2.3 Look-Up Table Unit

As mentioned in Sect. 3.3.2, trigonometric and other special operations (TSO)s can be implemented by using a LUT approach [24, 52]. The storage for the constant data values is assigned to the Look-Up Table Unit (LUTU).

3.4.2.4 Logic Unit

To enable logic and comparison operations (BLCO)s, a further unit is needed. We denote the unit, which can perform BLCOs, as Logic Unit (LU). This unit is coupled to the control for condition handling. Since the complexity of this unit is rather low and optimization opportunities are limited, we do not further elaborate on it.

3.4.3 Parallelization

As mentioned in Sect. 3.3.3, to efficiently account for the different requirements, the architecture template has to support DLP, ILP and TLP at different granularities (fine and coarse grain). For this purpose, we compose the architecture template in a hierarchical and scalable way. Each hierarchy level is characterized by the employed elements and by the best exploitable type of parallelization. For instance, fine-grain parallelization implies high band-width for communication whereas coarse-grain parallelization requires no or little bandwidth.

Figure 3.9 shows the employed hierarchy. At the top level, DLP, ILP and TLP can be applied amongst multiple clusters. Inside a cluster, multiple engines can again perform DLP, ILP and TLP. At the engine level, DLP can be exploited amongst slices. A slice can contain multiple units, which can exploit DLP and ILP. Inside a unit, multiple slots strongly exploit DLP. The granularity increases from top to bottom. We extend the work of [10] by defining a scalable hierarchical architecture with clearly defined structure.

Fig. 3.9 Proposed architecture hierarchy for enabling scalability and the support for different parallelization requirements

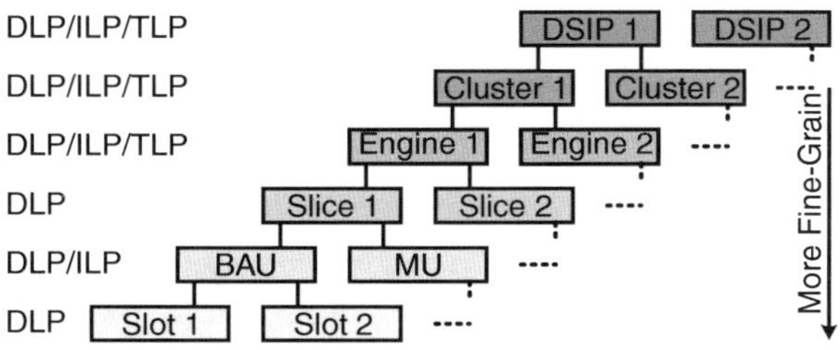

3.4.4 Data Transfer Operations

In conventional processors, the contribution of interconnect to the total energy cost is significant [6]. In fact, this is one of the main reasons for the energy-efficiency gap between processors and ASICs. With the trend towards more parallelization, the demand for more flexibility rises. Since the cost of a multiplexer increases non-linearly with the number of inputs (same delay) and because the cost of wires increases due to more routing congestions, the total cost of interconnect is becoming ever more expensive. Traditional processors typically enable a substantial amount of flexibility to account for all possible data flows [6, 51]. This provides high flexibility, but it is very expensive in terms of energy and area.

In the proposed architecture template, we design the interconnect based on the targeted algorithm characteristics. We analyze and optimized the interconnect for typical communication patterns. Data flows that are seldomly occurring are still enabled. However, in a way in which they do not have a significant impact on the most important ones (e.g. via detours). In that way, many direct connections are pruned while the general flexibility is still enabled.

Figure 3.10 shows the utilized interconnection and the typical communication frequency. Because of coarse-grain parallelization, the requirements on inter-cluster and inter-engine/slice communication are low to medium. We connect multiple clusters via the Very Wide Registers (VWR)s and multiple engines/slices via the VWRs and/or a Word Shuffler Unit (WSU). Inside an engine/slice, multiple heterogeneous units perform fine-grain parallelization and therefore the data exchange frequency is rather high. For instance, each Multiply-and-ACcumulate (MAC) operation requires a data transfer from the MU (multiplication) to the BAU (addition).

In conventional processors, typically all FUs are very tightly connected and direct data exchanges between all FUs are possible. However, because of the amount of wires and multiplexing, this approach is very costly. In addition, when adding one more unit, because of the new additional connections, all other units are affected. Nevertheless, by considering the relative importance of units and the fore-mentioned weighting of communication paths (optimizing for typical case), the cost of the interconnect can significantly be reduced. In general, based on the algorithm characteristics of the considered domain, the employed data flow is rather comparable to the ones that are applied in streaming processors [58]. In the

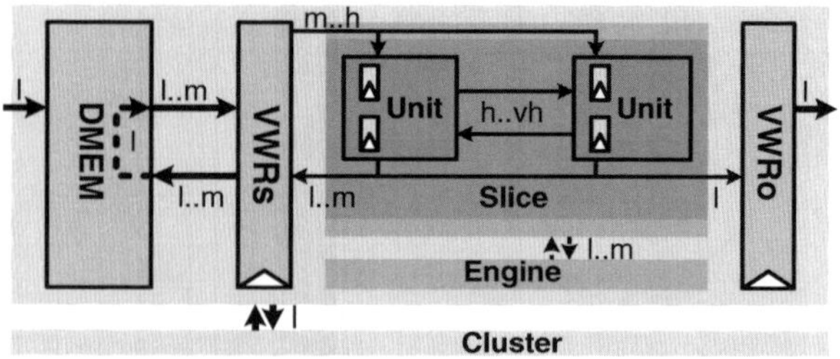

Fig. 3.10 Data storage hierarchy and typical communication frequency amongst components: (v)ery (l)ow, (m)edium, (h)igh

architecture template, examples for the scalability of the interconnect are the choice of different Very Wide Register (VWR) types and the choice of different stages for the Subword Shuffler Unit (SSU).

3.4.5 Data Storage

As mentioned in Sect. 3.3.5, the architecture has to provide data storage for variables with different characteristics. Conventional architectures employ data memories for arrays with long life-time and register files for data with temporary locality and medium to short life-time. Inside cores, pipeline registers are employed. However, these registers are typically not directly addressable and transparent for the compiler. The word length of the interface between register file, data memory and core is typically the same. The word length of pipeline registers is typically higher, since multiple operands have to be stored.

In the proposed architecture template, we construct the data storage hierarchy with Data MEMories (DMEM)s, with Very Wide Registers (VWR)s [50] and with local registers. This hierarchy is shown in Fig. 3.10. Contrary to conventional architectures, the interface between data memory and VWR is much wider than the interface between VWR and engines. This very wide interface leads to lower energy consumption, since the DMEM energy access cost (per bit) decreases with the increasing word length of the interface. Figure 3.11 shows this relationship for different storage types. In addition, because less instructions are required, also the control cost is lower. As indicated in [50], a physical abutment/alignment of VWR and data memory is possible. Thus, the wires between these two architecture elements can be made short. Unlike pipeline registers, the employed local registers are directly accessible and have the same word length as the datapath. Local registers are associated to units. Units, which may not concurrently be active, can be grouped so that they can share the same local registers. This reduces the area cost. Since the main data flow occurs typically between local registers, the energy cost is lower compared to conventional register file approaches. As it can be noted, contrary to traditional processors, we do not employ a dedicated load/store unit. Because of the rather deterministic data flow and the distributed control (see below), the data transfers between DMEMs and VWRs can typically be hidden.

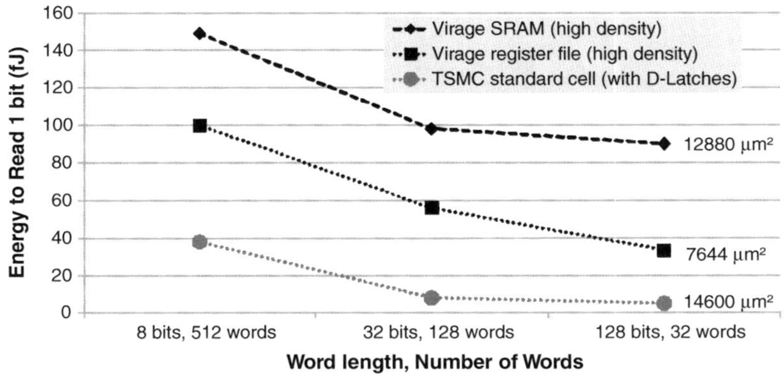

Fig. 3.11 Read energy of different storage types with increasing word lengths (4 kbits, single port, TSMC 40 nm GP technology)

Depending on the employed DMEM size, the addresses are computed at run-time, i.e. with an address generation unit, or at compile time, i.e. stored as instructions.

For the implementation of the DMEMs/VWRs, currently Static Random-Access Memories (SRAM)s and D-Latches/D-FFs (for small sizes) are most commonly used. However, the implementation of the DMEMs/VWRs with emerging embedded Non-Volatile Memories (NVM)s [11, 60], which are more energy efficient than conventional SRAMs, is increasingly becoming attractive. Especially Resistive-RAMs (RRAM)s with 1T1R cells [55] seem to be a good candidate for this application domain. The increased write cycle time of RRAMs, currently at least 10× higher than for SRAMs, has an influence on the scheduling aspect. However, because of the deterministic data flow, the software-controlled memory access and the wide memory interfaces (less accesses), the increased write cycle time can most-likely be hidden. The influence on latency can probably be kept low (e.g. by adapting the data processing order accordingly). Also Sing Torque Transfer Magnetic RAMs (STT-MRAM)s seem to be promising [5, 60].

3.4.6 Instruction Control

A processor, which supports ILP, consists of multiple components that need to be supplied with different instructions. In a conventional instruction hierarchy, at every cycle, always all components are fed with new instructions from a big L1 Instruction MEMory (IMEM). However, especially in heterogeneous architectures, typically not all components require a new instruction at every cycle. In addition, the energy cost for accessing the L1 IMEM is typically rather high [14]. Therefore, when employing a conventional instruction hierarchy, storage area due to the increased code size and energy due to the frequent expensive L1 IMEM accesses is wasted.

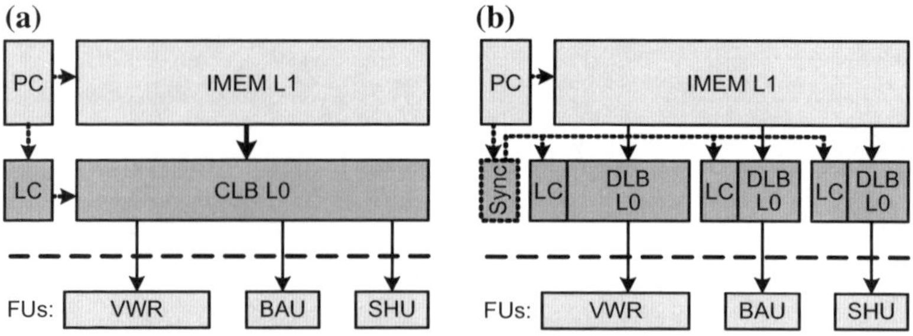

Fig. 3.12 **a** Centralized and **b** distributed loop buffer architectures in comparison

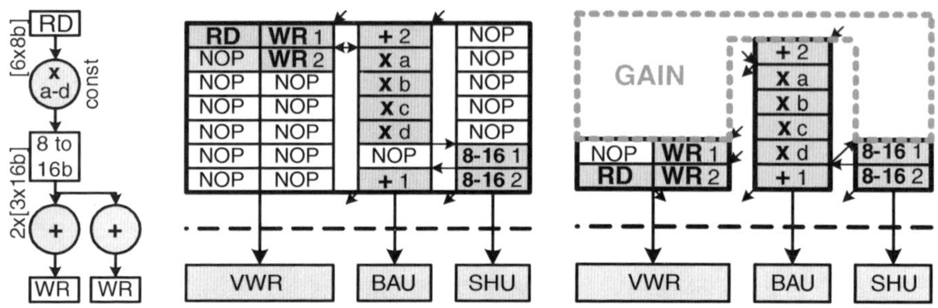

(**a**) DFG. (**b**) Centralized Loop Buffer. (**c**) Distributed Loop Buffer.

Fig. 3.13 **b** Centralized and **c** distributed loop buffer content for **a** a simplified MAC operation. The storage and the energy costs of the distributed loop buffer approach are much smaller

As mentioned in Sect. 3.3.6, the algorithms from the considered domain are loop dominated. To reduce the number of accesses to the L1 IMEM, loop caches or Centralized Loop Buffers (CLB)s can be used [23, 27, 63]. Figure 3.12a shows the instruction hierarchy when a CLB is employed. When leveraging on this technique, the instructions for the loop body are fetched from the small CLB rather than from the big L1 IMEM, therefore the energy consumption is reduced. Figure 3.13b shows the CLB content for the MAC-operation example of Fig. 3.13a. As it can be seen, not all 3 heterogeneous units are active at the same time, therefore many No OPeration (NOP) instructions are present in the CLB. Nevertheless, these NOP instructions consume storage size and contribute to the energy consumption.

An effective way to reduce the storage size and to reduce the energy consumption is the employment of Distributed Loop Buffers (DLB)s [49]. As shown in Fig. 3.12b, each unit is supplied with a small dedicated DLB. Each DLB has an individual Loop Controller (LC), which enables to operate independently. Hence, across multiple DLBs, incompatible loop-nests can be executed. Figure 3.13c shows the content of the DLBs for the fore-mentioned MAC-operation example. As it can be seen, in contrast to the CLB approach, almost no NOP instructions are present anymore. Hence, the use of DLBs can significantly reduce the cost of the instruction hierarchy. Therefore we chose this approach for the proposed architecture template. Note, currently no suitable compiler for DLBs exists. However, as mentioned in Sect. 8.2, current effort is ongoing to enable a compiler support (but this is outside the scope of this book). Because the by a compiler required information can be extracted from the source code (loops, dependencies) and from the architecture description (which code has to go to which DLB), a compiler support for DLBs seems feasible.

Like for the data storage hierarchy, also here the employment of embedded NVMs (for instance RRAMs [28] or STT-MRAMs [60]) instead of SRAMs/D-Latches/D-FFs can be attractive.

3.5 Proposed Architecture Template

This section provides an overview of the proposed architecture template. The architecture template includes architectural concepts, such as SoftSIMD, VWR and DLB, which have previously been mentioned in [10]. However, on top of this, it includes new concepts, which are mainly related to scalability. During the design of the proposed architecture template, the delay on components and wires have been estimated and considered. This information was used to ensure, that the critical path goes via the BAU, which is the most utilized unit. As mentioned in Sect. 3.2.2, we assumed that the logical relative representation of the architecture components represents also the physical one (components of the instruction hierarchy may be placed differently).

3.5.1 Top Level

Figure 3.14 shows the top level view of the proposed architecture template. For an architecture instance, the parameters of the template, e.g. *I*/*J*/*K* in Fig. 3.14, are fixed. As mentioned in Sect. 2.2, a DSIP platform consists of several DSIP architecture instances. The DSIP architecture instances are connected with wide data-buses. On the left side of Fig. 3.14 the input interface and on the right side of Fig. 3.14 the output interface of these wide data-buses are shown. The top level of the architecture template consists of one or multiple clusters and the Shared Top Level Control (STLC).

3.5.1.1 Clusters

Multiple clusters are mainly employed to perform coarse-grain parallelization in form of DLP and ILP amongst each other. However, if also TLP should be exploited within a DSIP architecture instance, then it should preferably also be done at this level. For coarse-grain parallelization, the data communication requirements are rather limited. A communication path, which enables to connect two neighboring clusters, is shown in Fig. 3.14. As it can be seen, the path starts at the VWR (T1) and ends at the input multiplexer of the DMEM. The path has the same width as the VWR and the DMEM. The physical length of this connection can be long. To avoid an unnecessary toggling on the path, a switch is present at the source of the path. To avoid that this path becomes the critical path of the design, the path can be defined as multi-clock cycle path. In case the fore-mentioned connection is not sufficient to fulfill the needs of the application, a more traditional Network on Chip (NoC) [56] could be employed.

3.5.1.2 Shared Top Level Control

The STLC is the main control of the processor. It consists of the master control logic with the corresponding L1 IMEM Direct Memory Access (DMA), the L1 IMEM and the DMEMs for the Look-Up Table Units (LUTU)s. The master control is mainly responsible for the instruction transfers between the L1 IMEM and the distributed L0 loop buffers, for the synchronization of L0 loop buffers and for condition handling. Depending on the required control-flow handling, the complexity of the top-level control unit can range from a simple control logic with program counter to a full control processor with a dedicated scalar datapath. The program of the controller is stored in the L1 IMEM DMA. The L1 IMEM stores the instructions for the individual distributed L0 controls. The instructions are either executed from the L0 loop buffers or directly from the L1 IMEM. Often the employed clusters are homogeneous and execute the same set of instructions. In this case, the same L1 IMEM content can be shared. If several LUTUs require the same data, also the DMEM LUTU content can be shared. The STLC is typically connected to the L2 control, to the platform controller and to the STLCs of other DSIP architecture instances.

3.5.2 Cluster Level

A cluster consists of a DMEM, one or multiple Very Wide Registers (VWR)s [50], one r multiple engines, a VWR for the output (VWRo) and the luster Level Control (CLC).

3.5.2.1 Data Memory

In the DMEM the input and the temporal data is stored. The DMEM has a wide interface and typically a single read and write port.

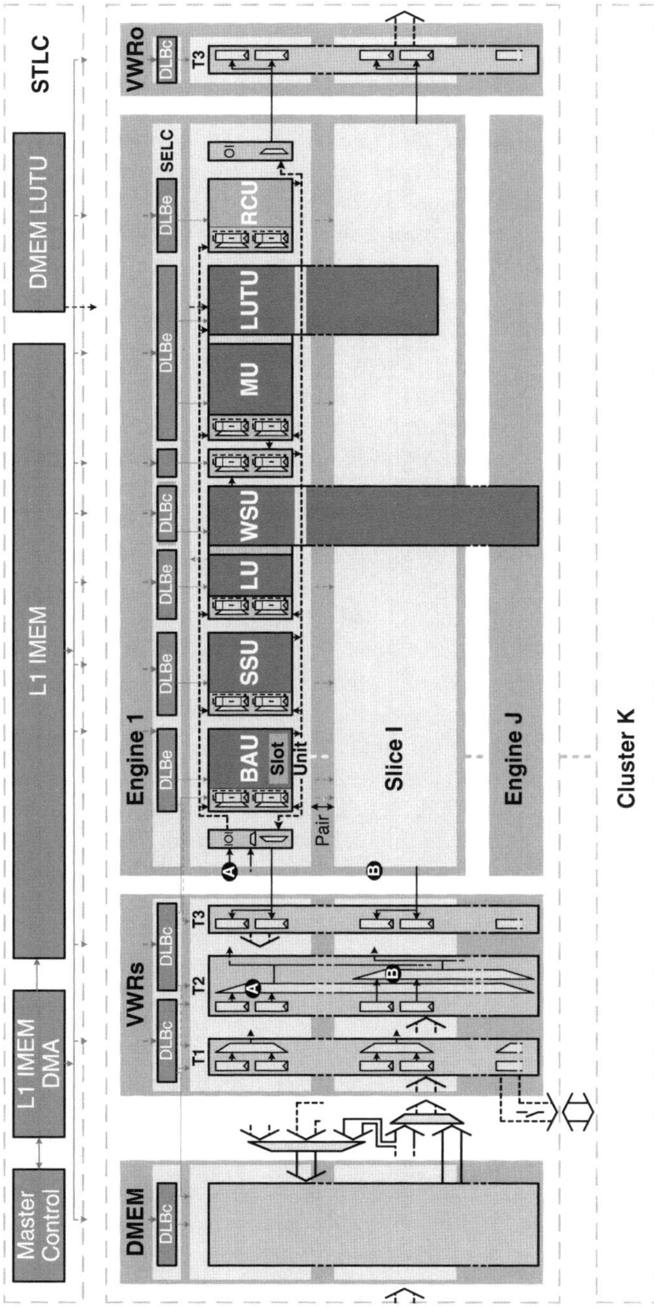

Fig. 3.14 *Top* level view of the proposed DSIP architecture template. The architecture is build in a hierarchical and scalable way. The relative order of architectural components depicts also the physical one Depending on the template parameters and the resulting delays, certain parts of the instruction hierarchy may be placed differentially. The parameters of the template, e.g. *I/J/K*, are fixed for an instance

3.5.2.2 Very Wide Registers

In this architecture, the VWRs replace the conventional register file. The VWRs have a wide interface towards the DMEM and a narrow interface towards the engines/slices. Within a cluster, one or multiple VWRs can be present. The employed VWRs are typically of different types. A VWR type is characterized by two properties: (1) the enabled data transfer operations and (2) the enabled data sharing. Figure 3.14 shows 3 different examples, denoted as T1 to T3. T1 is used for the data transfers from the DMEM to the slices. The registers of T1 are directly assigned to a specific slice and are therefore not shared. T2 is also employed for the data transfers from the DMEM to the slices, but contrary to T1, the data in the registers of T2 can be accessed by multiple slices. Hence, T2 enables to share or distribute data amongst slices. In Fig. 3.14, T2 has 2 narrow read ports, but each port can only access a limited set of registers. This structure could be used to distribute complex data (real part on port A and imaginary on port B) to multiple slices. T3 has a wide read interface, but a narrow write interface. T3 is used to write data from the slices to the DMEM. Like for T1, the registers in T3 are assigned to slices. In total, 21 different VWR types have been considered. However, 11 of them were pruned because of one or several reasons: (1) too high load on port, (2) enabled data transfers little useful and (3) additional multiplexers in the critical path of the processor required. As illustrated in the following example, the concept of employing multiple VWRs of different types offers a good combination of flexibility and energy efficiency: Let us assume the VWR configuration as shown in Fig. 3.14 and an algorithm mapping with little data sharing amongst slices. In this case, most data between DMEM and slices (and vice versa) is transferred via T1 and T3 respectively. Because T1 and T3 include little multiplexing and short wires, the energy consumption of these most frequent operations is low. However, at the same time T2 enables a sharing of data, which again reduces the storage requirements of the DMEM. A read operation of T2 is more costly than of T1, but since it occurs less frequently, the overall energy contribution is small. As for the DMEM, an individual control for the VWRs is employed.

3.5.2.3 Engines

The engines contain the arithmetic and logic datapaths that perform the actual data computations. Each engine has its individual control and can therefore execute individual instructions. Hence, like for clusters, multiple engines enable to exploit DLP and ILP. The degree of possible parallelization amongst multiple engines ranges from coarse to fine grain. Depending on the targeted algorithms, the engines are homogeneous or heterogeneous. For heterogeneous engines, the word lengths at the interfaces towards the VWRs/VWRo need to be matched. The exploitation of TLP within a cluster is also possible, but because of the fore-mentioned constraint, it may not be an attractive option.

3.5.2.4 Very Wide Register Output

The VWR for the output (VWRo) can be seen as the output interface of the processor. The processor can only write to the VWRo, but not read from it. For the output, typically only one VWR of type T3 is employed. A DSIP processor platform, which exploits the streaming nature of the wireless domain, consists of several DSIP processor instances. Thereby the output of the VWRo is typically connected to the DMEMs and to the VWRs of the subsequent processor instance(s).

3.5.2.5 Cluster Level Control

The control for the DMEM, the VWRs, the VWRo and the Word Shuffler Unit (WSU) is denoted as cluster level control. All these components are shared amongst the engines. As it can be seen in Fig. 3.14, each of these components has either one or several individual controls (DLBc)s. For instance, the VWRs could have two DLBcs, one that controls the writing and one that controls the reading. Because of the asymmetric interfaces of the VWRs, the activities of these two tasks can be very different. Thus, a split in two individual controls can make sense. A DLBc can be a implemented as a DLB, as a conventional Address Generation Unit (AGU) or as a mixture of both.

3.5.3 Engine Level

An engine consists of one or of multiple homogeneous datapath slices and a shared engine level control (DLBe). The datapath slices perform DLP amongst each other. However, slice internally, amongst multiple units, also ILP is exploited.

3.5.3.1 Slices

Because of the following reasons, it can make sense to employ multiple narrow slices instead of one wide slice:

1. *The word-length scaling of a slice increases the critical path, which is not acceptable:* An example for this is a 48-bit SoftSIMD slice, which supports 8, 12, 16 and 24-bit subwords. To enable a higher DLP while still supporting a full datapath utilization with the fore-mentioned subword sizes, the word length of the slice could be doubled to 96 bits. However, in case the full shuffling freedom of subwords should be maintained, the Subword Shuffler Unit (SSU) needs to support a shuffling over longer distances and becomes therefore much more complex. Thus, the critical path could move from the BAU to the SSU,

which is clearly not intended. A solution for this is to employ two separated 48-bit slices instead of one 96-bit slice.

2. *To enable a grouping, which can be beneficial for saving data storage and/or for enabling a more efficient data communication:* For instance, if the targeted algorithms operate on complex numbers, slices could be separated into two groups: A and B. Slices of group A operate only on the real part whereas slices of group B operate only on the imaginary part. For basic complex operations, both groups can operate on independent data and therefore the storage and data access can be orthogonalized. However, for complex multiplications, a data exchange between real and imaginary part is required. This exchange could be enabled via the VWRs or via the WSU. However, both options introduce a latency, which may lead to an inefficient scheduling. In this case, a better option is to establish a direct data connection between the two slices, which operate on the same complex data. As shown in Fig. 3.14, slices, which are directly connected, are denoted as pair.

3.5.3.2 Shared Engine Level Control

The Shared Engine Level Control (SELC) consists of several distributed loop buffers (DLBe)s, which control the units of the datapath slices. The SELC interacts with the L1 control.

Because multiple slices exploit DLP, they are generally controlled by the same control signals. However, it can be beneficial to enable a slice-individual control for the components that communicate with a shared VWR interface. Assume that the VWR T2 in Fig. 3.14 has only one narrow read port towards the slices. In case of common control signals and a data transfer from the VWR T2 to the slices of engine 1, the same data would always be written to both slices. This is clearly not intended, because the slices should perform computations on different data. For this reason, in this case, slice-individual write enable signals for the local registers (which can receive data from a shared VWR read port) of the slices are needed.

3.5.4 Slice Level

The clusters and engines of a template instance are typically rather homogeneous (topology) and perform potentially the same computations. Within a slice, the units are heterogeneous and have different functions. In the standard configuration, a slice consists of the slice interface towards the VWR (VWR_SI), the Basic Arithmetic Unit (BAU), the Subword Shuffler Unit (SSU) and the slice interface towards the VWRo (VWRo_SI). These components are sufficient to perform basic arithmetic operations and constant multiplications on different subword sizes. As mentioned in Sect. 3.3.2, by employing the CORDIC algorithm, also other arithmetic operations can be computed. We denote the BAU and the SSU as primary units. To support more fundamental operations, the following secondary units can additionally be instantiated: Logic Unit (LU), Word Shuffler Unit (WSU), Multiplier Unit (MU) and Look-Up Table Unit (LUTU). Contrary to the approach of designing ASIs that are added to FUs, we decide on the support of fundamental operations from a very limited predefined set.

For the sake of completeness, it has to be mentioned that in very exceptional cases, we consider the use of an additional more customized unit type, which we denote as Reusable Custom Unit (RCU). The reason for this extension will be explained in an individual paragraph further below this section.

As mentioned in Sect. 3.4.5, units can be grouped to reduce the number of local registers. In addition, grouping reduces the interconnect requirements because units of a group share the same communication path. Candidates for grouping are units, which, when jointly considered, do not have a high demand on storage and on data transfer operations. To evaluate this, the operation breakdown and cycle counts need to be examined. In this work, we group the LU and the WSU together. Both perform one operation on the datapath word in a single clock cycle and both units are typically not very frequently used. Furthermore, we group the MU and the LUTU together. Both require multiple clock cycles to compute the output of a datapath word. If both units are not used by the same algorithm (which is rather likely), the MU can potentially leverage on more local registers as in the case of a separation.

As it is shown in Fig. 3.15, the employed interconnect can generally be divided in an upper and in a lower interconnection network. The upper network is dedicated for the reading from the VWRs, the lower network for the communication between units and for the writing to the VWRs and to the VWRo. The upper/lower network is connected to the upper/lower local registers within a unit. In general, low duty cycle paths are time-multiplexed and for high duty cycle paths direct connections are used.

3.5.4.1 Slice Interfaces

A slice has an interface towards the VWRs, denoted as VWRs_SI., and an interface towards the VWRo, denoted as VWRo_SI. The interfaces are depicted in Fig. 3.15a and g respectively. The VWRs_SI enables to re-route data from the lower to the upper interconnection network.

3.5.4.2 Basic Arithmetic Unit

The BAU is the main and the most utilized unit of the architecture. It supports BAOs and MCOs. As mentioned in Sect. 3.4.2 and shown in Fig. 3.15b, it consists of only basic operators, such as shifters, adders/subtracters and a circuitry for the

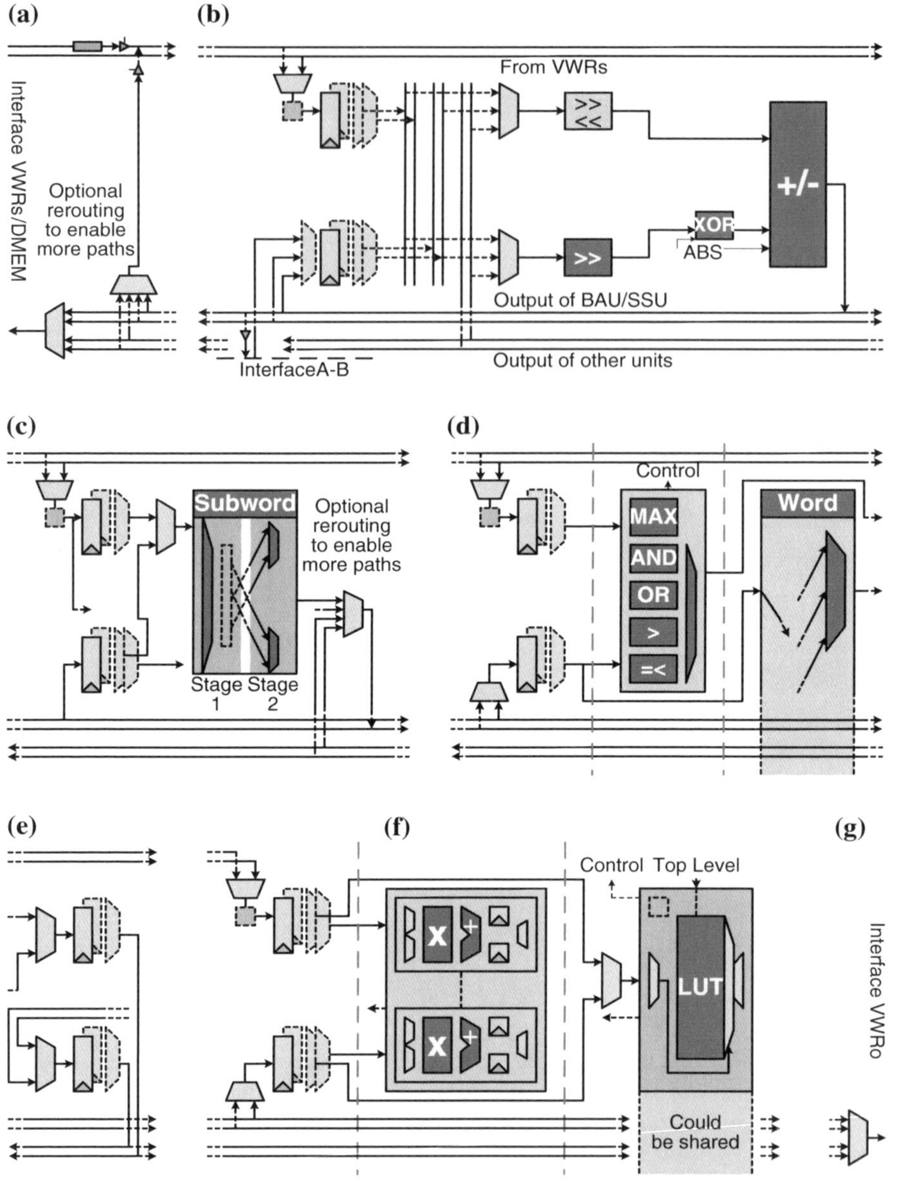

Fig. 3.15 Interfaces, units and shared registers of a datapath slice. The upper interconnection network is shown with 2 lines, because 2 VWR read register ports are considered. **a** VWRs slice interface. **b** Basic arithmetic unit (BAU). **c** Subword shuffler unit (SSU). **d** Logic unit (LU), word shuffler unit (WSU). **e** Shared output register. **f** Multiplier unit (MU), Look-up table unit (LUTU). **g** VWR output interface

computation of abs. The concept of employing a unit that contains only few simple operators has two major advantages: (1) The small area, the limited multiplexing and the short wires enable to operate at a high clock frequency and therefore to achieve an overall high area/energy efficiency. (2) Since complex operators can be decomposed to simple ones, the unit remains rather generic and can therefore be reused across different designs (domain specific vs. application specific). The BAU exploits HardSIMD or SoftSIMD (or a hybrid). A typical word length for SoftSIMD is 48 bits, since it is the greatest common divisor of 6, 8, 12, 16 and 24 bits. The number of instantiated local registers will depend on the complexity of the targeted algorithms. It will typically be in the order of 1–3 per upper and lower interconnection network. Since the BAU is the most important unit, it has a direct data access to all other units.

3.5.4.3 Subword Shuffler Unit

The SSU is schematically shown in Fig. 3.15c. This unit is mainly responsible for packing/unpacking and shuffling of subwords, which are essential operations for SIMD. The datapath can be extend to support certain bit-level transfer operations, e.g. required by Viterbi algorithm, and/or to provide data re-routing functionalities. To enable high scalability and high reusability, the architecture is composed of multiple stages. The SSU is directly connected to the BAU, because both units operate closely together.

3.5.4.4 Logic Unit

As mentioned in Sect. 3.4.2, the LU is used to perform boolean operations, comparisons and max-computations of a vector. It is coupled to the L1 master control for condition handling. As shown in Fig. 3.15d, it shares the local registers with the WSU.

3.5.4.5 Word Shuffler Unit

If a frequent data exchange between multiple engines/slices is required and the communication band-width via the VWR is not sufficient, a WSU can be instantiated. To reduce multiplexer complexity, i.e. to reduce the amount of required shuffling patterns, a grouping of slices can be considered.

3.5.4.6 Multiplier Unit

The multi-precision Multiplier Unit (MU) is employed to perform variable multiplications (MVO)s. As mentioned in Sect. 3.4.2, it will typically be implemented with several multi-precision multipliers which calculate the output of a datapath word in a sequential way. Compared to a conventional full-precision multiplier solution, this approach enables a more efficient hardware utilization. Besides, the layouts of small multipliers fit better in the general layout structure of a slice. For the packing of the subwords, the SSU will be used. Figure 3.15f shows the grouping of the MU with the LUTU.

3.5.4.7 Look-Up Table Unit

The LUTU can be used for storing constants that are needed for different purposes, e.g. constants for the approximation of sin/cos and correction values, constants for comparisons and control constants. Since the subwords of a datapath word will be processed sequentially, it is a multi-cycle unit. The LUTU can optionally be shared across multiple engines/slices.

3.5.4.8 Reusable Custom Unit (Employed in Exceptional Cases)

It can occur that the decomposition of a certain complex operation (sub-DFG) into primitive operations, which are then mapped onto different units, will lead to a very inefficient implementation. An example is the max*-operation of LDPC decoders which would have to be decomposed to BAOs and BLCOs [38]. This operation is the main operation of the decoder and the output of the previous computation is required as input for the subsequent computation. Thus, a direct feedback with strong data dependency exists. A decomposing and mapping of this *max**-operation onto two units would mean, that intermediate data is stored at least in one local register. Because of the feedback dependency, the units cannot be fully utilized and therefore the efficiency is reduced by 50 %. Thus, for such exceptional cases, which will mostly occur in the outer modem, the design of a Reusable Custom Unit (RCU) is recommended. We want to stress that this unit should only perform an operation, which is reusable across different algorithms. Contrary to the traditional approach of adding ASIs to the datapath of existing FUs, we add an additional unit to the slice, which basically does not affect the datapaths of existing units. Because of the reusability of this unit and the limited consequences, the general proposed DSIP architecture template concept is still maintained. We want to emphasize that such cases, in which the use of a RCU is crucial, will occur very seldom and typically transformations can be applied to avoid them. For this reason, we will not but much stress on this unit in the rest of this book.

3.6 Scalability of the Architecture Template

In this section we summarize the scalability aspect of the proposed architecture template for different aspects. A distinction between the technology scalability, the template hardware/instance scalability and the software/run-time scalability can be made. For the latter two, the provided flexibility across different target applications and even target application domains can be analyzed.

3.6.1 Technology Scalability

The technology scalability of an architecture can be seen as a qualitative metric that indicates, how well the architecture can cope with the requirements of future technology nodes. If the architecture cannot cope with the requirements of a new technology node in an effective manner, then the migration to a new technology node does basically not bring a benefit/gain in terms of clock frequency, area and energy.

To ensure that the proposed DSIP architecture template can efficiently be implemented in future DDSM technologies, i.e. the benefit/gain of scaling remains high, we have taken the following technology-related trends into account:

- *Wires will dominate the delay (and design metrics) even more:* We have explicitly considered the influence of wires during the architecture design. As a consequence, the most important wires (highly active, critical path) are short, wire congestions are avoided and less routing metal layers are required. Other architecture proposals partly avoid the presence of long wires in their processing cores, but consider this often only at the floorplanning phase. We go a significant step further to ensure that the most important wires are short: We transform nearly all highly occurring arithmetic operations into shift-add/sub based primitives. In this way we also avoid nearly all variable multiplications and we can therefore avoid the (frequent) use of a complex and large hardware multiplier. This approach allows us to keep the relatively small shift-add/sub stages in the BAU as the only ones that are truly active all the time. By applying the floorplan/layout scheme proposed in Sect. 3.2.2, the wire lengths in these highly active BAU stages can be significantly limited. We also ensure that the rather active wires, which are connected to the BAU, are kept short. For instance, the registers of the VWRs are directly placed next to the BAU and connected with short horizontal wires. Because of the employed DLB concept, also the active control wires, which lead to the BAU, are kept short. Especially for functional blocks with frequent DMEM access, e.g. FFT, also short wires between DMEM and VWR are important. In case the DMEM is implemented with latches, which could be an attractive option for small depths, a physical abutment/alignment with the VWR is inherently possible (follow same layout scheme as for slice/VWR). As mentioned in [50], SRAMs (and NVMs) with an adapted memory interface can also be physically aligned to the VWRs. By employing potentially multiple parallel memories, and by choosing their physical shape accordingly (memory compiler), conventional SRAMs will also fit in the defined floorplan/ layout scheme. Nevertheless, the other options are clearly preferable.

- *More regularity for high manufacturability/low variability needed:* The proposed architecture is designed for a bit-sliced layout structure, which has inherently a regular placement and routing. To ensure that the targeted layout is obtained, the semi-custom design flow, which is proposed in the appendix of this chapter, can be applied. In addition to the regular layout structure, the high reusability of the template for many different markets is very beneficial in this regard: The huge expected volumes motivate the preparation of several highly optimized macro-style IP blocks that can be reused in all instances derived from the template. This is not feasible in ASIP-style templates because the accelerators are not sufficiently reusable across such huge market volumes. The forementioned two technology-related trends and issues are the topic of several ongoing and new Master student theses at IMEC.
- *2.5D/3D integration to enable shorter signal distances and to reduce packaging cost:* Because of the hierarchical nature and the distributed control, which enable a physical separation of all major architectural components (e.g. at cluster level), the proposed DSIP architecture template enables to efficiently exploit this multi-chip single-package technology option.
- *Smaller margins for voltage scaling (and higher leakage):* For this reason, we focus on (fine to medium-grain) power gating instead of on fully dynamic DVFS, which is becoming too costly in overhead.
- *New memory technologies are appearing:* We consider the employment of new embedded NVM technologies, such as RRAM and Spin Torque Transfer Magnetic RAM, to reduce dynamic/leakage energy and area in the memory organization. Because of the software-controlled memory access and the wide memory interfaces of the proposed DSIP architecture template, the negative characteristics of new memory technologies (e.g. increased write cycle time), can be compensated. That is true for the data and instruction memory organization. Several Master students and PhDs at IMEC are investigating in this topic.
- *Decreasing reliability:* The considered measures to cope with the increasing wire influence are also beneficial to cope with the decreasing reliability, both for functional and parametric aspects. For instance, because the highly-active wires are short, the probability of a wire failure is lower. The high control over the layout, i.e. enabled by the semi-custom design flow proposed in the appendix of this chapter, enables to influence the routing and the dimensions of individual wires. The regular placement and routing has generally also a positive influence on reliability. Because of the distributed control and the ability to establish clear power islands, the employment of fine/medium-grain run-time countermeasures is simplified. The mitigation, based on such run-time countermeasures, is the topic of several Master students and PhDs at IMEC.

Since the proposed DSIP architecture template has incorporated the fore-mentioned measures, it is well-suited for future DDSM technologies. Note, the actual countermeasures for technology related issues, such as increasing variability, decreasing reliability and NVM organizations, are outside the scope of this book.

3.6.2 Hardware/Instance Scalability: Design Space and Architecture Instantiation Design Flow

The hardware/instance scalability refers to the hardware design space of architecture instances (template flexibility). As previously mentioned, compared to traditional DSIP architectures, we enable a medium instead of a coarse grain customization. However, all customizations are performed within the pre-defined parameter space. Figure 3.16 summarizes the possible set of template parameters which have been grouped in 6 categories. Note, the same categories have also been used in Sect. 3.3 and in Sect. 3.4. For the definition of the parameter range, the following two criteria need to be considered:

1. *Domain-specific system/algorithm requirements:* For instance, how many parallel SIMD slots would min./max. be required (i.e. how much DLP can min./max. be exploited)? What is the min./max. required DMEM size and bandwidth (i.e. how much background memory storage and ports are min./max. needed)? A code profiling is required to determine the feasible parameter ranges. This criterion is system/algorithm dependent, but technology independent.
2. *Scalability of the parameters of the architecture template:* Each template parameter has an influence on the area/energy cost of the architecture. Typically within a certain range, the parameter values scale proportional (e.g. linear) with the cost. However, from a certain value on the cost can increase enormously, e.g. because much more buffering is required to still maintain the clock frequency (see typical delay-area/energy curve). Thereby the critical path could move from one architecture part to an other, e.g. from BAU-VWR to SSU-VWR. Such effects have to be captured in cost models. In order to understand such effects and to enable high reproducibility/reusability, a semi-custom design flow needs to be applied (i.e. the layout represents the architecture view, see Sect. 3.2.2). This criterion is technology dependent, but system/algorithm independent.

As previously mentioned, future work is required to accurately define the supported parameter ranges of the template and to obtain the corresponding cost models.

Because of the high reuse of architecture elements in the proposed design approach, it is feasible to implement elements, such as BAU, SSU, VWR, as physical IP blocks and to fully optimize them. As a consequence, for each new platform/chip design, these IP blocks can be reused. Thus, importantly, the design

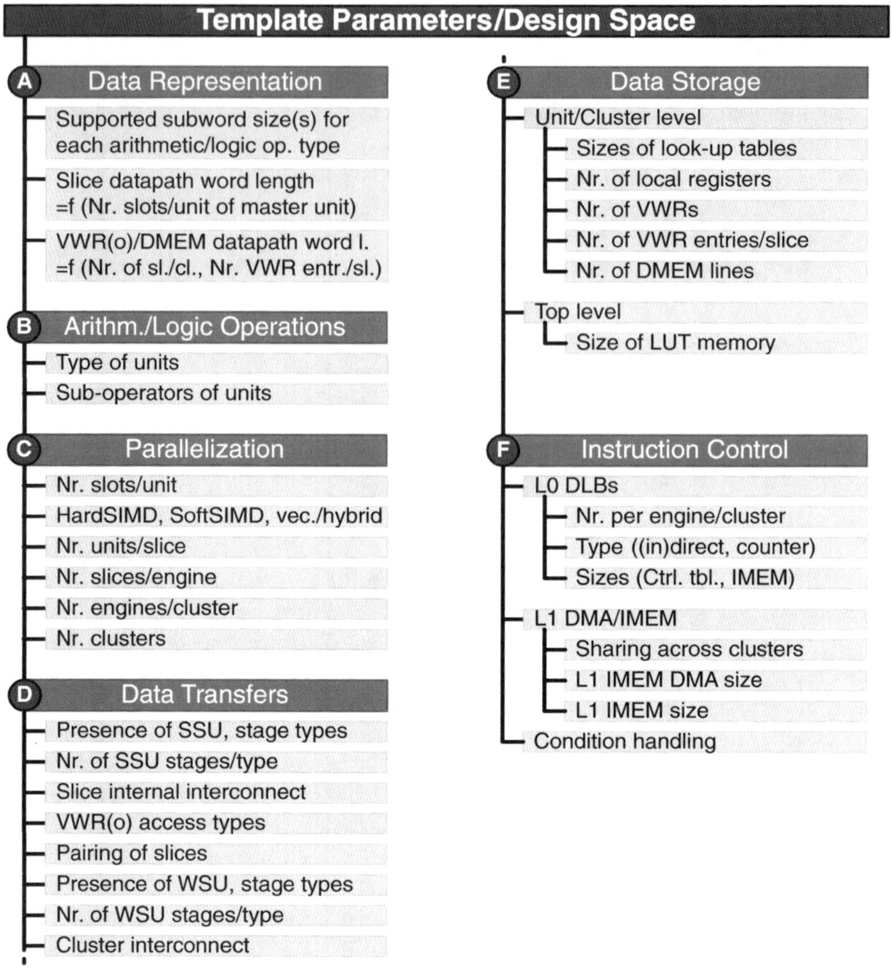

Fig. 3.16 Possible set of DSIP template parameters. The parameter values are chosen during the architecture template instantiation design flow

and verification related NRE costs, which dominate the high total NRE cost in DDSM technologies, are significantly reduced.

In Chap. 7 an instantiation design flow for the architecture template is proposed. During the application of this flow, the parameters are defined in the same category order as shown in Fig. 3.16. Conceptually, the proposed instantiation design flow and the design flow of the quantitative flexibility model, which will be presented in Sect. 3.6.3, are very similar. The instantiation design flow takes care that the instance is properly dimensioned in terms of all parameter ranges that are listed in the model (so that the instance provides sufficient flexibility).

3.6.3 Hardware/Instance Scalability: Model to Define Flexibility in a Quantitative Manner

Design metrics, such as area, energy, throughput and latency, and even yield or expected life-time, are typically directly and rather easily quantifiable. Therefore these metrics can be used and are commonly used for the quantitative comparison of different architectures and design choices. However, because no generic model exists in literature, a quantitative comparison of architecture flexibility (corresponds also to reusability in this context) is currently not possible. For this reason, flexibility is mostly described and compared in a more qualitatively manner [26, 39, 47, 48]. Nevertheless, to evaluate the flexibility of an architecture, e.g. does it fulfill the requirements sufficiently?, and preferably also to compare the degree of enabled flexibility between different architecture styles, a (generic) quantitative model is highly desirable.

Figure 3.17 provides an initial idea on how such a model could look like and how it would be used for benchmarks. The proposed design flow, which consists of 2 main steps, is illustrated in Fig. 3.17a. The first step, the profiling of the considered domain-specific code and the computation of the values for the benchmark parameters of the table (thereby system requirements, such as max. latency and min. throughput, are considered) are architecture independent. Thus, this effort can be shared amongst all architectures that should be compared against each other. The second step is an architecture-specific processing, i.e. architecture characteristics are considered, which includes code transformations and a rough code mapping. For instance, multiplications may need to be converted to shift-add/sub operations (to be supported), but because of this, the architecture needs to support more DLP than estimated in the first step.

Figure 3.17b illustrates the quantitative flexibility comparison of different architecture templates/styles and/or instances. Each benchmark parameter of the table is of type boolean (flag indicating the presence of a functional capability) or of type integer (range that is required for a parameter to be usable). If an architecture can support the required functional capability (boolean flag=true), and if it can be customized within the required parameter range (integer range), then it provides optimal flexibility. The distance between the required and the supported (capability) value quantifies the flexibility. Because the area/energy efficiency of an architecture template/instance is characterized by a separate analysis (see for instance the case studies of Chaps. 4, 5 and 6 respectively), the main goal of the model is to evaluate, whether an architecture provides sufficient flexibility to support the min. requirements (=max. value of range) for all benchmark parameters in a true/false manner.

In the flexibility evaluation of Sect. 3.6.4 we make a link to the fore-mentioned model by (1) considering all benchmark criteria that would be included in the table and by (2) indicating the type of benchmark parameter for each sub-evaluation. However, we consider the further definition, implementation and application of the proposed quantitative flexibility model as future work.

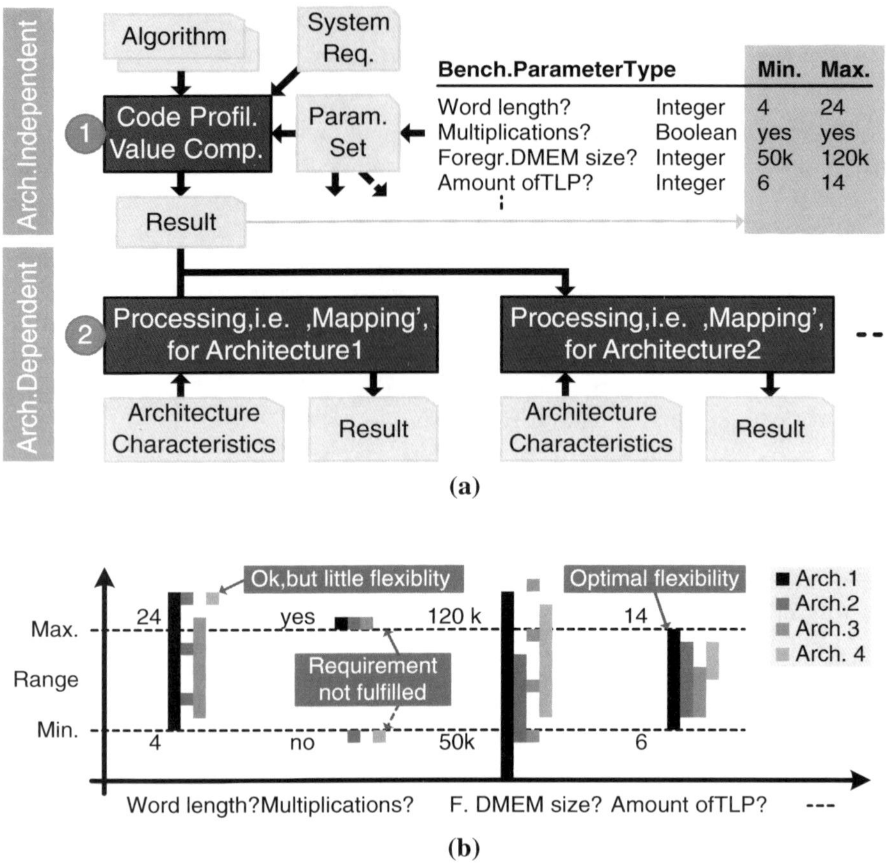

Fig. 3.17 Idea of a quantitative model for the evaluation/comparison of flexibility. **a** Design flow to obtain quantitative comparable requirements for one or a set of given application domains. Based on the obtained requirements, different architecture templates/styles and/or instances can be compared. **b** Quantitative comparison of capabilities of different architecture templates/styles and/or instances. The main goal of the model is to evaluate, whether an architecture provides sufficient flexibility to support the min. requirements (=max. value of range) for all benchmark parameters in a true/false manner

3.6.4 Hardware/Instance Scalability: Flexibility Evaluation of the Proposed Architecture Template

In the following we first summarize the system/algorithm-related requirements of the targeted application domain. Then, we describe and motivate how the proposed architecture template is capable of supporting these requirements. This can be seen as a partially quantitative and a partly qualitative evaluation of the hardware scalability, i.e. flexibility of the proposal. As mentioned above, we consider all benchmark criteria that would also be evaluated with the quantitative flexibility model and we indicate the type of benchmark parameter, i.e. boolean (true/false flag) or integer (parameter range), for each evaluation. In general, we assume that low-level code transformations, which consider the characteristics of the architecture template, are applied before the mapping. Like previously, we distinguish between 6 categories.

(1) **Requirements on data representation**

As mentioned in Sect. 3.3.1, different fixed-point word lengths as well as real and complex data needs to be supported.

- *Support of different word-lengths (range)?* BAOs, MCOs, MVONs and TSOs are performed on the BAU. Since the BAU can be implemented as a HardSIMD unit, as a SoftSIMD unit or as a hybrid unit, sufficient freedom for the support of different word lengths exists. MVOWs are computed on the MU. Since it is based on multi-precision multipliers, multiple word lengths are supported. That is valid for both of the previous categories across the entire range that is required for our wireless target domain (4 to 32 bits). Larger dynamic ranges are not needed. BLCOs are basically linear bit-level operations. Thus, also the support of multiple word lengths in the LU can cheaply be enabled (e.g. separation of subwords with AND gates). For the organization of subwords (shuffling, resize), the SSU is used. An architecture instance can include multiple units of the same type. Each unit could support different word lengths.
- *Support of real and complex data (flag)?* Real data is inherently supported. The support for complex data is enabled in several ways: (1) The real and the imaginary parts can be computed on the same hardware sequentially. For that, sufficient data storage needs to be available. (2) Within a unit, the real and the imaginary parts are computed in parallel. The SSU is used to pack the two parts in datapath words and to exchange data between the two parts. (3) The complex data is computed in parallel with 2 slices. To exchange data between the 2 slices without latency, the "pair"-connection can be used. For all 3 cases, the VWR, the SSU and/or the WSU enable the required data access/data transfers patterns.

(2) **Requirements on arithmetic and logic operations**

As mentioned in Sect. 3.3.2, all operations of the considered domain can be categorized into the following groups: BAOs, MCOs, MVOWs, MVONs, TSOs and BLCOs. BAOs and MCOs are clearly computational dominant. TSOs and BLCOs are only part of few algorithms/functional blocks and have generally a relatively low duty cycle.

- *Support of BAOs (flag)?* BAOs are efficiently supported with the BAU.
- *Support of MCOs, MVOWs and MVONs (flag)?* MCOs, i.e. multiplications with a constant, can efficiently be mapped onto the BAU. The shift-add/sub sequence is generated at compile time and stored in the IMEM/DLB. MVOWs, i.e. variable multiplications with a wide value range, are supported with the MU. An alternative mapping, where the actual computations are performed on the BAU, is mentioned in [17]. MVONs, i.e. variable multiplications with a narrow value range, can also be computed with the MU. Alternative mappings, which

typically lead to higher energy efficiency, are mentioned in Sect. 4.4.4.2 and in [17] respectively. Our analysis of low-level optimized domain-specific algorithms has shown that most multiplications have at least one of the following properties [16]: (A) one operand is constant (MCO), (B) one operand value remains the same for multiple multiplications and (C) one operand can only take a very limited number of values (MVON). Thus, the energy-efficient methods proposed in Sect. 4.4.4.2 and in [17] can be applied for almost every multiplication.

- *Support of TSOs (flag)?* Because of the relatively low duty cycle, a direct implementation of TSOs in hardware (i.e. as DSIs) would lead to a low area efficiency, and/or in case of an operator sharing across instructions, potentially to a low energy efficiency. For this reason, as mentioned in Sects. 3.3.2 and 3.4.2 respectively, we decompose TSOs to a sequence of simple/basic operations and we utilize the primitive hardware units BAU, MU, LU and LUTU to perform the computations.
- *Support of BLCOs(flag)?* BLCOs are computed with the LU. To provide sufficient flexibility for the domain, a LU, which supports the basic logic operations AND, NAND, OR, NOR, XOR and the comparison operations $>$, $\geq$, $=$, max(vector), is proposed.

(3) **Requirements on parallelization**

As shown in Fig. 1.3, for LTE-Advanced, the baseband implementation needs to offer a very high computational power of 1000 GOPS to fulfill the high throughput constraints. In the same time, the baseband implementation must cope with the low latency constraints of demanding standards, such as IEEE 802.11ac and IEEE 802.15.3c. As mentioned in Sect. 3.3.3, because of the limitation regarding clock frequency, the high throughput and low latency constraints can only be satisfied with a high degree of parallelization at all distinct levels, namely instruction-, data- and thread(/task)-level parallelism. Thus, the architecture needs to support ILP, DLP, and TLP efficiently. In addition, to fulfill the low latency requirements, an efficient data storage hierarchy/interconnect and a high clock frequency are essential.

- *Scaling of the template to support all required degrees/levels of parallelization (flags)?* As shown in Fig. 3.9 and explained in Sect. 3.4.3, the proposed architecture template is highly scalable and supports DLP, ILP and TLP at different granularities (fine and coarse grain). Thus, it can be optimally customized for the requirements (high flexibility). To demonstrate the flexibility of the template, in the following the dimensioning of a possible platform configuration, which offers a computational power of 1000 GOPS (real data), is illustrated: For efficiency reasons, 5 customized architecture instances are designed/employed, i.e. a DFE, a FFT, a MIMO, a DEMAP and a FEC instance. Since we assume that the MIMO detection instance needs to offer relatively the highest computational power, i.e. 400 GOPS, we explain the configuration on this instance: Because operations are performed on 16-bit data, the datapath of the MIMO detector instance needs to be at least 400×16 bits=6400 bits wide.

Since a DLP of 8 is well exploitable in the considered MIMO detection algorithms (here assumed), a slice with a datapath word length of 8×16=128 bits is designed. Within a slice, a BAU, a MU and a LU will work in parallel (to support all arithmetic/logic operations) To enable sufficient run-time flexibility/programmability, 4 engines, each with one slice, are combined to one cluster. Since a cluster has a datapath width of 4×128=512 bits, about 6400/512=13 clusters are needed/employed.

- *Fine-grain ILP/DLP and how it is supported (range)?* Fine-grain parallelization refers to the parallel computation of operations of a rather small DFG, where the operations have rather strong dependencies. In the proposed architecture template, fine-grain parallelization could be performed within a slice and/or across slices. The following examples illustrate this: (1) Butterfly structure with the real operations $y0 = x0 + x1 \times w$ and $y1 = x0 - x1 \times w$: The addition and subtraction will be performed on the BAU. The multiplication could also be performed on the BAU (e.g. in case w is a constant) or on the MU. In case of the latter, the add/sub operations on the BAU and the multiplication on the MU can be performed in parallel (pipelined). (2) DFG with the real arithmetic and comparison operations $z = max(a0 + b0, .., a7 + b7)$: The 8 additions could be performed in parallel with 8 slots on the BAU. The max-function will be executed on the LU. In case a pipelining is possible, the BAU and the LU could work in parallel. (3) Complex multiplication with the real operations $(a \times c - b \times d) + (a \times d + b \times c)i$: The real and the imaginary output could be computed in parallel with 2 slices, e.g. $a \times c$, $b \times d$, $-$ on slice 1 and $a \times d$, $b \times c$, $+$ on slice 2. The real multiplications are either performed on the BAU (e.g. in case c/d are constant) or on the MU. In summary, the proposed template can be scaled for the required parameter ranges of the targeted domain. A computational power of tens or even hundreds of GOPS can be achieved by adding a sufficient number of slices and clusters. This is illustrated in the cases studies of Chaps. 4, 5 and 6 respectively.
- *Coarse-grain TLP and how it is supported (range)?* Coarse-grain parallelization refers to the parallel computation of operations of a rather big DFG, where the operations have no or rather limited dependencies. Coarse-grain parallelization is typically performed across architecture instances, across clusters and/or across engines. For instance, the DFE and FEC functionalities could be computed in parallel with two architecture instances. The computations for different antennas could be computed with multiple parallel clusters. Sub-functionalities could be computed in parallel with multiple engines. In summary, again, by adding a sufficient number of engines and clusters, the proposed template can be scaled for the required parameter ranges of the targeted domain. As illustrated in the cases studies, combined with ILP/DLP, tens or even hundreds of GOPS can be achieved. A further scaling is enabled by adding multiple architecture instances.
- *Degree/type of parallelization and influence on data storage and data transfers (range)?* To avoid scheduling inefficiencies, the amount of available data storage (incl. ports for the access bandwidth) and the amount of supported data transfers (communication bandwidth) need to be scaled with the enabled degree/

type of parallelization. For instance, the amount of required local registers, local interconnects and VWRs/ports needs to increase (1) with an increasing number of units within a slice and (2) with an increasing number of subwords within a slice datapath word (e.g. see repacking of subwords with a SoftSIMD BAU/SSU). Nevertheless, since the proposed architecture template enables to configure the related parameters, sufficient flexibility, i.e. to support the entire required range of the targeted domain, is supported (compatible with tens or even hundreds of GOPS).

- *Degree/type of parallelization, energy/area efficiency and influence on minimally required clock frequency (flag)?* The chosen degree/type of parallelization has an influence on wire lengths and thus, on the achievable clock frequency of the architecture instance (in case no additional pipeline registers are inserted). For instance, the length of the control wire, which originates from the BAU DLB and ends at the lowest bit in the BAU datapath, increases with the slice datapath word length (semi-custom layout assumed). This effect, which can be captured in cost models (see definition of template parameter ranges further above), needs to be considered during the design of architecture instances. As indicated above, because of the high scalability of the template in regard of parallelization, even with a relatively low clock frequency, the required computational power can be offered. However, the low latency constraint could be an issue (the huge amount of parallelization may require a huge amount of buffering). Besides, for optimal energy/area efficiency, the architecture has to operate at a high-enough clock frequency (see optimal point in a typical delay-area trade-off curve). As it can be seen in the case studies of Chaps. 4 and 5 respectively, for TSMC 40 nm standard-cell CMOS technology, the targeted/obtained clock frequency is about 1 GHz. For future technology nodes, it will be scaled-up accordingly.
- *Why can the low latency requirements be fulfilled (flag)?* For the targeted domain, the latency requirements are very tight, e.g. about 0.15 μs for the 512-point FFT in IEEE 802.15.3c (see Sect. 6.2.1). However, because of the following main reasons, this tight requirements can be fulfilled: (1) High degree of exploitable parallelization. (2) Efficient data storage hierarchy/interconnect with sufficient bandwidth. (3) Because of the simple datapath, the architecture can operate at a high clock frequency (at about 1 GHz in 40 nm technology). The results of the case studies demonstrate this capability of the template.

(4) **Requirements on data transfers**

As mentioned in Sect. 3.3.4, data transfers are generally required because of (1) operation dependencies that are inherently present in the algorithm (e.g. shuffling for LDPC), (2) data storage hierarchies, which lead to data copies and data reuse trees (between levels and within level; e.g. from DMEM to VWR and from DMEM to DMEM), and data storage related transfers (data layout transformations that require transfers for the conversion), (3) data parallelization (e.g. transfer data from one slice to another) and (4) data representation related, e.g. fixed-point implementation (change of word length).

- *What are required data transfers for (1) and how are they supported (flag)?* In general, the following 3 types of data transfers need to be supported: (1) Shuffling of vector elements (e.g. in FFT, LDPC). Depending on the shuffling distance/element size, a shuffling can be performed (a) with the DMEM (read/write address), (b) with the VWR which enables broadcasting, (c) with the WSU and (d) with the SSU. (2) Bit-level shuffling (e.g. in Viterbi). This type of data transfer operation can be supported with the SSU. Nevertheless, in case bit-level shuffling needs to be supported, it can make sense to employ 2 SSUs, one for the subword shuffling and one for the bit-level shuffling. Importantly, both SSUs are designed based on a common/predefined set of SSU stages (reusability). (3) Selection and distribution of the highest value of a vector (e.g. in sensing or MIMO detection algorithms). This functionality is enabled with the LU in combination with the SSU.
- *What are required data transfers for (2) and how are they supported (flag)?* With the combination of direct connections, different VWR types, shuffler units and the slice interconnection network, the architecture template provides the required connectivity to transfer data throughout the complete storage hierarchy and within one level of the hierarchy. In general, the data transfers follow the streaming nature of the domain. Block-based algorithms, such as FFT and LDPC, read/write more often from/to the DMEM than streaming-based algorithms, such as FIR and offset compensation.
- *What are required data transfers for (3) and how are they supported (flag)?* The architecture needs to support the data exchange and the data distribution across the parallel hardware. Within a slice, the SSU and the interconnection network is used. Across slices, the VWR, the WSU and the "pair"-connection enable the required connectivity. To enable data transfers across clusters, connections between VWRs are made.
- *What are typical data transfers for (4) and how are they supported (flag)?* For the support of multiple word lengths, the repacking functionality is required. The repacking of subwords is supported with the SSU.

For all of these 4 data transfer categories, also sufficient data transfer bandwidth has to be available (parameter range). Nevertheless, this aspect has already been addressed in the paragraph about parallelization further above.

(5) **Requirements on data storage**

As mentioned in Sect. 3.3.5, in general, array and scalar data of different sizes, with different life-times, with different access orders and with different access frequencies need to be supported.

- *Data storage hierarchy in the template (flag)* The data storage hierarchy in the template consists of DMEMs, VWRs and local registers. For constants, also LUTs are used. As it will be motivated further below, these heterogeneous storage elements together with the interconnects between them provide sufficient flexibility to support the above mention requirement.

- *Requirements on DMEMs (flag) and support (range)?* The DMEMs, which are part of clusters, are used to store input data and intermediate data. Depending on the functional blocks/algorithms, different DMEM sizes are required. Especially for block-based algorithms, such as FFT and LDPC, relatively big DMEMs are needed. As mentioned in Sect. 3.5, the word lengths of the DMEMs scale with the amount of employed parallelization/slices within the clusters. Since multiple physical DMEMs can be used, the word-length scaling is generally not a bottleneck. Nevertheless, the depth of the DMEM, which can have a negative influence on the clock frequency of the processor, is an important parameter. To get an idea on the maximal required DMEM depth, we estimate the requirements for a 2 k FFT (LTE standard), which is very demanding in that regard: Assuming complex data, 16-bit word length and the fact that 2 full FFT blocks need to be stored (input, intermediate), a DMEM size of $2048 \times 2 \times 16 \times 2 = 131$ kbits is needed. Considering 4 clock cycles per butterfly (3 of them for the shift-add/sub multiplication) and 11 stages, 44 clock cycles for the computation of an element output and 90112 clock cycles for the whole FFT output are required. Thus, the computation of the whole FFT with 1 complex slot (=2 real slots) at 1 GHz results in a latency of 90 μs. This high latency is acceptable for LTE, which has very relaxed requirements regarding latency. To fulfill the throughput requirement of 30.72 Msps (FFT outputs), 2 complex slots are needed. Hence, for the 64-bit datapath (4 real slots), the required DMEM depth is 131 kbits/64 bits=2 k. If we consider a strict latency requirement of 1 μs for the same FFT size (realistic for future standards), the required degree of parallelization changes to 90 complex slots and thus, to a DMEM depth of 45. For streaming-based algorithms only a few DMEM lines are required. From the above examples it can be concluded that DMEMs with a depth of 1-2048 need to be supported. Since the DMEM I/O is basically directly connected to registers in the template, the maximal required clock frequency of the DMEM corresponds to the clock frequency of the processor. Single-ported SRAMs with a depth of 2 k and an operation frequency >1 GHz in 40 nm G CMOS technology are very common and therefore commercially available (e.g. from Synopsys). For small DMEMs, e.g. up to a depth of 16, the implementation with standard-cell latches can be considered. Because of the wide access, for most instances one single-ported physical memory/cluster will suffice the bandwidth requirements. However, to provide also sufficient bandwidth for the most demanding cases, the DMEM can optionally be implemented with several physical memories that are connected with additional multiplexers to the VWRs. Importantly, to enable an efficient use of the DMEM (required storage size, number of accesses), a good data layout is crucial. Thus, the data, which needs to be written to the DMEM, should be organized up-front (e.g. by using the WSU).
- *Requirements on VWRs (flag) and support (range)?* As explained in Sect. 3.5.2.2, the VWRs, which provide an intermediate storage between DMEM and local registers, replace the conventional register file. To enable sufficient flexibility (data access) and intermediate storage (hide latency, avoid a spilling to the DMEM), an architecture instance will typically include 1–4

VWRs of different types. Note, each type of VWR enables different data transfers and each VWR can store 2 words per slice. To avoid bandwidth bottlenecks, the number of VWRs/ports needs to increase with the number of units/slices.

- *Requirements on local registers (flag) and support (range)?* The local registers are used to store data within a slice, i.e. data with a short life time. To ensure that critical parts of DFGs can be scheduled efficiently, i.e. intermediate data does not need to be spilled to the VWRs/DMEM, sufficient local registers (and access flexibilities) need to be present. A rough scheduling of common DFGs, e.g. FFT butterfly, MAC operation, shuffling for SoftSIMD, has shown that, to provide sufficient flexibility, maximal 4 registers per unit are needed in our target domain.
- *Requirements on LUTs (flag) and support (range)?* In the LUTs, which are part of the LUTUs, typically constants for TSOs and BLCOs are stored. A reasonable size of LUTs, which are typically implemented with latches, is in the order of a few hundreds of bits. This amount of storage is sufficient for enabling basic TSOs. Since multiple LUT contents can be load from the L1 DMEM LUTU to the LUTs in the slices, the effective storage size can also be higher. For the LUT-based implementation of TSOs, which enable an approximation of the computation with very high precision, a huge amount of LUT storage size can be necessary. However, in this case it is typically more efficient to perform the computations online, e.g. with the CORDIC algorithm.

(6) **Requirements on instruction control**

As mentioned in Sect. 3.3.6, the instruction control needs to support the following features efficiently: (1) multiple incompatible loop nests, which can be different depending on controlled architecture element, (2) deterministic code, except for few variables/parameters and (3) change of algorithm parameters and algorithms at run-time.

- *Support of incompatible loops (flag) and loop buffer sizes (range)?* To support multiple parallel loops, which can have different loop sizes and execution start times, in a efficient manner, multiple parallel DLBs are employed. Because each DLB can control the execution of the loop independently, the merging of loop nests is not required. This has 2 very crucial advantages: (1) Applications, which would obtain a too complex irregular loop code when they are fully merged, do not have to be discarded. (2) The area (memory footprint) and energy overhead, which would be caused by the merging, can be fully avoided. Since the individual loop sizes for the control of the main architecture elements, such as BAU and SSU, are rather small, also the required loop buffer sizes are rather small. We have seen that in practice less than 32 words are required. For the control of the DMEM, which can potentially involve bigger loops, an "AGU-like" DLB can be used. Instructions, that are outside of loops, can directly be executed from the L1 IMEM. The L1 loop/control, which synchronizes the individual DLBs (e.g. knows the iteration information of for-loops), is controlled via the L1 DMA IMEM. A 3rd nested loop is supported by adding the L2 control.

- *Support of zero-overhead conditions in the inner loop (flag)?* Since algorithms of the considered domain are inherently deterministic or can be transformed to be deterministic, the employed instruction hierarchy is explicitly designed for the efficient execution of deterministic code. However, to support a broader range of algorithms and to avoid the execution of code that is not required (e.g. because less than the maximal number of loop iterations need to be performed), a limited amount of condition handling is still enabled. A conventional if/case-condition, where the branches can have different code lengths, is supported via the L1 IMEM DMA. The condition checking of the variables, which are unknown at compile time, is done in the LU. The start addresses of the different branches in the L1 IMEM DMA, which are known at design time, are stored in the L1 jump registers. At every jump, the L1 IMEM DMA will control/prepare the L0 DLBs so that the required code is executed. Currently only 1 level of this type of condition handling is supported. This is normally sufficient, e.g. in can be used for the early termination of loops. Note, platform level decisions, e.g. change of standard/mode, are handled by the platform level controller (middleware) and are propagated "top-down". In addition to the fore-mentioned if/case-condition handling via L1, a code structure of the form $result = max(a, .., z)$ is directly supported via the LU (comparison, subword routing) and does not require a change in control flow. Both methods can be nested. As mentioned in Sect. 3.3.6, to handle conditions, the predication technique can be used. However, the use of predication induces typically overhead/inefficiencies. Because of the presence of DLBs in the proposed architecture template, this overhead/inefficiency can be avoided (also for inner loops).
- *Support of multiple algorithm parameters and algorithms (flag)?* In the proposed architecture, multiple algorithms and algorithm parameters can be stored in the L1 IMEM. At run-time, the L2 (platform) controller can communicate a required change to the L1 DMA controller. The L1 DMA controller can load the new instructions to the DLBs and then start the new execution. In order to keep the L1 IMEM access cost low (which is important, since the L1 IMEM is still accessed for code outside loops), the size/depth of the L1 IMEM has to be kept limited. A reasonable maximal L1 IMEM depth is in the order of 64 words, which is typically sufficient to store several algorithms. Note, because of the presence of DLBs, the total instruction word is very wide. In case even more algorithms need to be executed on the same architecture instance, they can be provided from the L2 IMEM and loaded to the L1 IMEM on demand.

3.6.4.1 Summary

In conclusion, the proposed architecture template provides clearly sufficient and all necessary flexibility, because it can effectively support the entire set of requirements from the targeted wireless application domain. As future work, we plan to perform a more detailed quantitative flexibility evaluation and comparison with other architecture styles that target the same application domain. Note, the area/energy efficiency evaluation/comparison is done separately and can be seen in the case studies of Chaps. 4, 5 and 6.

3.6.5 Software/Run-Time Scalability

The software/run-time scalability refers to the flexibility of an architecture instance. Architecture instances need to provide flexibility in order to support different standards/modes and dynamic BER performance/energy trade-offs. During the design of an architecture instance, the enabled flexibility is decided. An architecture instance could include many units and many local registers, so it would be very flexible. However, because of the high degree of enabled flexibility, the energy efficiency is low. Thus, a careful consideration of this trade-off is essential. In general, it is also important to optimize targeted algorithms in order to reduce the flexibility requirements. For instance, algorithm transformations may eliminate the need for certain units and data-flow patterns.

Because of the following reasons, the proposed architecture template concept can cope with the flexibility requirements of the domain: (1) employment of several architecture instances to implement the functionally of the heterogeneous blocks of the platform, (2) the architecture is programmable and the employed units are rather generic and (3) the template provides a medium-grain customization space.

Since the architecture template is strongly linked to technology, circuit-level techniques, which increase the benefit of run-time scalability, can rather easily be included and efficiently be exploited. For instance, because of the clearly defined layout structure, even fine-grain power islands are already pre-defined. Thus, in combination with the distributed control, the capability for power gating is enabled. Power gating at unit level, engine level and/or cluster level could be considered. In addition, once technology scaling offers sufficient support for this, coarse-grain DVFS at cluster level could be an interesting option.

The high flexibility of the proposed DSIP architecture template enables to effectively exploit the concept of system scenarios, i.e. to deal with dynamic application conditions [8, 33, 40]. This is especially the case, because the employed DLBs support incompatible loop nests and zero-overhead conditional execution in the inner loops. That is a strong enabler for the run-time scalability.

3.7 Summary of Combined Innovative Concepts

To cope with the combined requirements of high performance, high energy efficiency, high flexibility, high reusability and technology scaling-friendliness, see Sect. 1.1, several innovative concepts have been included in the proposed DSIP architecture template design approach. In the following the main innovative concepts are summarized. Thereby the main differences compared to traditional DSIP and ASIP solutions are mentioned.

3.7.1 Main Template-Specific Concepts

The proposal combines the following main template-specific concepts:

- *DSIP architecture template with customization:* Like for conventional DSIPs, the architecture template is domain-specific, which enables high reusability. However, the main difference is that the platform is implemented with several DSIP architecture instances which are customized for each functionality/sub-domain. The customization, which is only done within a pre-defined space, is medium instead of coarse grain. Contrary to traditional ASIP platforms, all instantiated processors are based on the same architecture template and do not include ASIs/HAs.
- *Strong link of architecture and algorithm:* The following concepts are combined to establish this link: (1) The characteristics of domain-specific algorithms have been explicitly considered during the architecture template design. (2) The proposed architecture instantiation design flow of Chap. 7 is based on the algorithm-architecture co-design approach. (3) The template-specific cost models can guide algorithm selection and optimization steps. This combination offers a good-trade in between traditional approaches.
- *Strong link of architecture and technology:* To enable an efficient scalable implementation in advanced DDSM technologies, the influence of wires is explicitly considered in the architecture template. Because of the layout-aware architecture, fine-grain power gating can easily be exploited. The semi-custom design flow proposed in the appendix of this chapter ensures that the targeted layout is obtained.

3.7.1.1 Main Architectural Concepts

In the following the main innovative architectural concepts, which have been included in the proposed DSIP architecture template, are listed. For a more specific comparison with traditional DSIP and ASIP (BP) solutions, we refer to Table 2.1.

- ***Multiple small generic units*** *instead of a single big customized FU:* Employing several small generic units, such as shift-add/sub unit and multiplier unit, enables high reusability, high hardware utilization with parallel computations and template scalability. Amongst others, conventional bigger FUs suffer from long wires and from a high amount of multiplexers.

- ***Software SIMD*** in the considered domain, in which a wide range of different subword sizes are required, a software SIMD datapath can enable a better hardware utilization of arithmetic operators and thus, a higher efficiency.
- ***Scalable regular multi-stage shuffler*** *instead of a irregular/synthesized single-stage shuffler:* Shufflers are strongly wire dominated. To avoid wire congestions and long wires, we implement the shuffling functionality with multiple regular and reusable stages. For each stage, the routing and the possible physical distance to the next stage, i.e. to create more space for the wires, can individually be considered.
- ***Very Wide Registers (VWR)s*** *instead of a conventional symmetrical narrow-ported register file:* The wide interface of the VWRs to the DMEM reduces the DMEM access energy and reduces the control effort. We employ several VWRs of different types, which enables to optimize for certain data-flow patterns and therefore to reduce the amount of multiplexers and wires.
- ***Local software controllable registers*** *instead of conventional pipeline registers:* This choice enables more flexibility and a better hardware utilization. For instance, data transfers, which involve components with different control loops, can more easily be hidden.
- ***Data-interconnect*** *optimized for the typical case and not for the worst case:* The data-interconnect within a slice, but also the data-interconnect of the VWRs and between clusters, is optimized for the typical, i.e. for the most frequently occurring, data-transfer patterns. A communication between components, which need to communicate, but very seldom, is still enabled. However, the hardware that enables this "slow" transfers does not decrease the energy efficiency of typical transfers.
- ***Distributed Loop Buffers (DLB)s*** *instead of conventional centralized loop caches:* The separation of a big control loop into several smaller ones enables a fine-grain datapath control, which typically reduces the storage requirements and the control energy significantly. Besides, because the small loop buffers can be placed close to the datapath, the energy in wires can be reduced.

3.8 Conclusions

In this chapter, the proposed DSIP architecture template and the corresponding design approach have been presented. Because of fulfilling the joint requirements of high performance, high energy efficiency, high flexibility, high reusability and technology scaling-friendliness, the proposal is suited for a wide employment in emerging wireless communication systems.

For the design of the architecture template, system requirements, common domain-specific algorithm properties as well as technology constraints, especially regarding wire scalability, have been considered. This ensures a good match between architecture and algorithm/technology.

To fulfill the requirements, existing architectural concepts where evaluated and complementing concepts were proposed. The resulting architecture template, which ensures scalability and high energy efficiency, includes innovative architectural concepts such as Software SIMD, VWR and DLB.

In the following two chapters, the proposal is benchmarked on two case studies. The first case study focuses on a MIMO detector for LTE/WLAN and the second case study focuses on a FIR filter for high-throughput 60 GHz systems. The results of these case studies indicate close to standard-cell ASIC energy efficiency, which is at least a factor 2–3 better than that of state of the art programmable solutions. Thus, the proposed design approach of this book is very promising.

A.1 Appendix: Proposed Back-End Semi-custom Design Approach

This appendix provides concepts/ideas for a back-end design approach which leads to a significant higher implementation efficiency as the conventional standard-cell flow (especially for DDSM technologies), but without sacrificing design time in a significant way. To enable this, the proposed approach is limited to architecture templates, such as the one proposed in this book. Preliminary experimental results, which are based on this proposal, will be shown in the appendix of Chap. 5. It should be emphasized that the content of this appendix is mostly theoretical at this stage and more effort is needed to develop, evaluate and refine this ongoing work further.

A.1.1 Motivation

As explained in Sect. 2.7.2, for the implementation of a hardware block, generally three different types of design flows can be applied: (1) conventional standard-cell design flow, (2) semi-custom design flow (based only on standard cells or also on custom cells) and (3) full-custom design flow. These three different types of design flows provide basically a trade-off in terms of design time and resulting implementation efficiency (i.e. area, energy of the design/layout). For both criteria, the differences amongst design flows can be in the order of several magnitudes [13]. Because of the tight time-to-market requirements and the strict limitations on development cost, today mostly the conventional standard-cell design flow is applied. In this design flow only (generic) standard cells are used and typically wires on each routing layer are either routed horizontally or vertically. Amongst further other drawbacks, the conventional standard-cell design flow cannot sufficiently cope with the increasing wire influence in DDSM technologies [22]. Thus, despite of migrating to a newer technology, the benefit, i.e. the reduction in terms of area and energy, may strongly be limited because of this reason. This is clearly creating a big dilemma. The application of conventional semi-custom design flows is basically also not a viable solution, since it usually implies a strong negative impact on design time and this because of the requirement to develop and characterize a new cell library for each design. In addition, many current semi-custom design flows focus only on the layout generation of the arithmetic datapath of the design, and neglect that the data movement from/to the peripheral (e.g. memories, arithmetic datapath of other cores) contributes significantly to the overall design efficiency. Therefore, to achieve overall high implementation efficiency, the complete floorplan needs to be taken into account. So if hard macros blocks are used, they should be present in a form which enables to change their physical shape, and that property should be exploited. Furthermore, most proposals are based on semi-custom designed hard-macro cells, which cannot "easily" be reused across different design instances. Hence, technology-dependency and reusability is a big issue. Thus, from these observations it is clear that the need for a new back-end design approach, which targets high implementation efficiency in combination with sufficient low design time, is mandatory.

3.1.1.1 Overall Proposed Back-End Design Approach

To summarize, the two main criteria that the new back-end design approach has to fulfill are: 1) acceptable design time (comparable to the conventional standard-cell design flow) and 2) highest possible implementation efficiency (comparable to existing semi-custom design flows and ideally close to the ones of full-custom design). Unfortunately, these two criteria imply contradictory constraints. The overall proposed back-end design approach offers a reasonable compromise. It is depicted in Fig. A.1. It combines the following main concepts/ideas which should ensure that both criteria can be fulfilled in the best-possible way:

- *Enable in general a big layout/design optimization space by removing restrictions of the conventional standard-cell flow:* For instance, by enabling the use of custom cells and by removing the restrictions of an equal cell height, the design can significantly be more optimized (increases efficiency). Nevertheless, as a negative effect, the degree of automation will decrease (increases design time!). Fortunately, the following concepts will largely compensate for that.
- *Customize only parts of the architecture/circuits that have a significant impact on efficiency:* To increase the efficiency of the design in a significant manner, the most impacting and critical parts of the architecture/circuit need to be optimized (e.g by using custom cells). However, for parts of the architecture/circuit that have overall little impact and are less critical (e.g. certain control logic, parts of the datapath that are seldomly activated and are not in the critical path), no customization will be applied (i.e. standard cells are used), since the resulting effect would basically only be an increase of design time. Thus, a "hybrid" design approach, which combines custom cells and standard cells, is proposed.

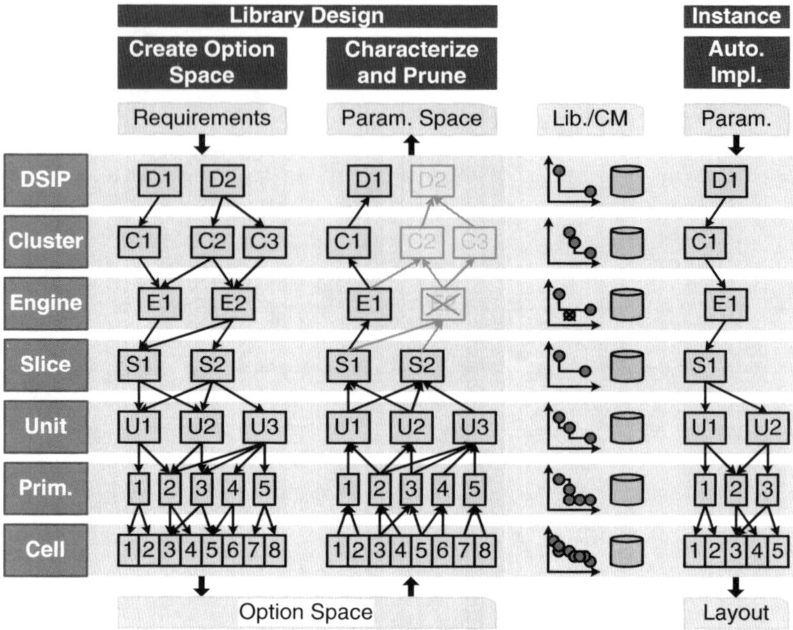

Fig. A.1 Overview of the overall proposed hierarchical back-end design approach. A clear distinction between library design and instance design is made. The library design involves two design flows: (1) top-down propagation of requirements from the architecture to the technology level. Thereby the design space is made, which means that possible design options for implementing the required architecture/component are "collected". (2) bottom-up characterization and pareto-optimal filtering from the technology to the architecture level. Thereby the previously "collected" design options are implemented, characterized e.g. in terms of delay and energy, and non-pareto optimal design options are discarded at each level. The results of this flow are then stored in libraries, which are used during the design of template instances

- *Limit the design space at the (micro-)architectural level by using an architecture template:* The conventional standard-cell design flow can be applied to obtain the layout for any type of digital circuit description/architecture. Nevertheless, to enable this generality while keeping the number of different cells in the library acceptable (important for automation), only cells with primitive functionality (=less customization/efficiency) are present. Because an architecture template limits the number of architectural options significantly, this high degree of generality offered by the conventional standard-cell design flow is clearly not needed. Thus, the proposed design approach supports only a limited design space (decreases design time). In general, to determine the actual required architecture design space, i.e. the DSIP architecture template parameter space (see Sect. 3.6.2), requirements/limitations from the algorithm side as well as technology side have to be considered.
- *Reduce the "final" design space at component/cell level by applying a combination of top-down and bottom-up flow with pareto-optimal filtering:* As

mentioned above, the DSIP architecture template defines the design space at the (micro-) architectural level. Based on this information, a top-down flow is applied and thereby all feasible/relevant design options/implementations at different levels of the design hierarchy are "collected". This basically involves two sub-steps: First, all possible design options for a particular component/cell are determined. For instance, to implement the adder that is present in the BAU, all different types of adder topologies, such as carry-ripple adder and brent-kung adder, are initially considered. Second, all design options which would clearly not fulfill the requirements are already discarded. For instance, the BAU is generally the most utilized unit and should therefore determine the maximal clock frequency, i.e. the adder that is used within this unit should have a very low delay. Thus, utilizing a "slow" carry-ripple in the BAU is not feasible and therefore this implementation option can already be filtered-out. After applying this top-down flow, which is generally technology-independent (across technology nodes with similar back-end of line options), all feasible/relevant design options at each level of the design hierarchy (component/cell levels) are known. Next, a bottom-up flow is applied. Thereby all previously determined feasible/relevant design options are implemented, characterized and again filtered at each level of the design hierarchy. As it will be mentioned further below, instead of making a manual layout for each design option, the designs are described in a more general/scalable way, which enables a high reuse across technology nodes. Nevertheless, the characterization and the filtering of each design option has to be technology specific, but it is assumed that this task can largely be automated. The filtering is done by discarding all design options which are not pareto-optimal (e.g. in terms of delay-energy trade-off). This ensures a reduction of the design space (decreases design time) in an optimal multi-objective way. Note, because the proposed DSIP architecture template is composed in a strong hierarchical way, see Fig. A.2 for an example, this concept can be well applied.

- *Describe design options in a scalable/reusable way by using (propagated) relative placement information at all design levels:* Instead of designing fixed cells and/or hard-macro blocks, the logical and physical structures of design options (circuits) are described in a parametrizable format, which enables high scalablility/reusability. A parameter of the logic description will usually be the word length, a parameter of the physical description could be the edge of the data input pins. In order to ensure that the most important wires (most actively used, wires in critical path) in the design are short, which is a necessity for DDSM technology (increases efficiency), the relative placement of components/cells in the architecture template is pre-defined at design time, and propagated between hierarchy levels. In general, the architecture template is the enabler of this approach. Note, the relative placement information is largely technology independent and therefore highly reusable. With the a priori specification of the relative placement, the general design space is reduced. Thus, automation becomes again more feasible (decreases design time). In comparison, traditional hard macros offer no or a very limited degree of flexibility and traditional soft macros do not include physical/layout information.

- *Leverage on a design library at all levels to enable a high reuse of design effort:* The DSIP architecture template ensures that the actual designs (instances) are always composed out of the same components/cells. Therefore, the components/cells do only need to be designed once and can then be reused. To leverage on this property, a clear distinction between library design and instance design is made. The effort for designing and optimizing the library can be averaged over all instances (decreases design time). By employing an individual design library at each hierarchy level, maximal re-usability is ensured.
- *Provide meaningful/rather accurate information to architecture and even algorithm designers:* As mentioned in Sect. 1.2.4, in practice the efficiency of the design implementation and the design time suffers because of the cultural gap. The former mainly because of over-design, the latter mainly because of design iterations. Both negative effects are caused due to the lack of information. By applying the proposed back-end design approach, a strong link between technology/layout-level and the architecture-level is enabled. Thus, a significant part of the cultural gap can be bridged, which results in less design iterations (decreases design time) and in designs which fulfill exactly the requirements/specifications (increases efficiency).
- *Automate as much as possible to increase productivity:* The top-down flow of the library design can probably not very-well automated and requires at least initially a manual effort. However, this initial effort is highly reusable because (1) the architecture template is domain-specific and the requirements (optimization goals) for a domain do basically not change and (2) the space of design options is also growing very slow. For instance, it is rather seldom that a new better adder topology is found. In the bottom-up flow of the library design the design options need to be described, the actual layouts need to be made and characterized, and the pareto-optimal filtering needs to be applied. Except for the first step, a high degree of automation seems to be feasible. The first step requires manual effort, but it is a task that is largely technology-independent and highly reusable. Since it is applied most often, the most crucial design flow in terms of design time is the library instantiation design flow. However, because of the existence of the design library, which captures basically all possible "layouts" within the DSIP template design space "in advance", the application of this design flow is rather straight-forward and can therefore be highly automated (decreases design time). Large parts of the proposed approach can already be automated today with existing EDA tools.

In the following we propose an initial library instantiation design flow, which is still only based on the standard-cell layout structure, i.e. it does not include custom cells.

3.1.1.2 Proposed Back-End Library Instantiation Design Flow

Figure A.3a depicts the main steps of the proposed back-end library instantiation design flow. As a reference, the conventional standard-cell design flow is shown in

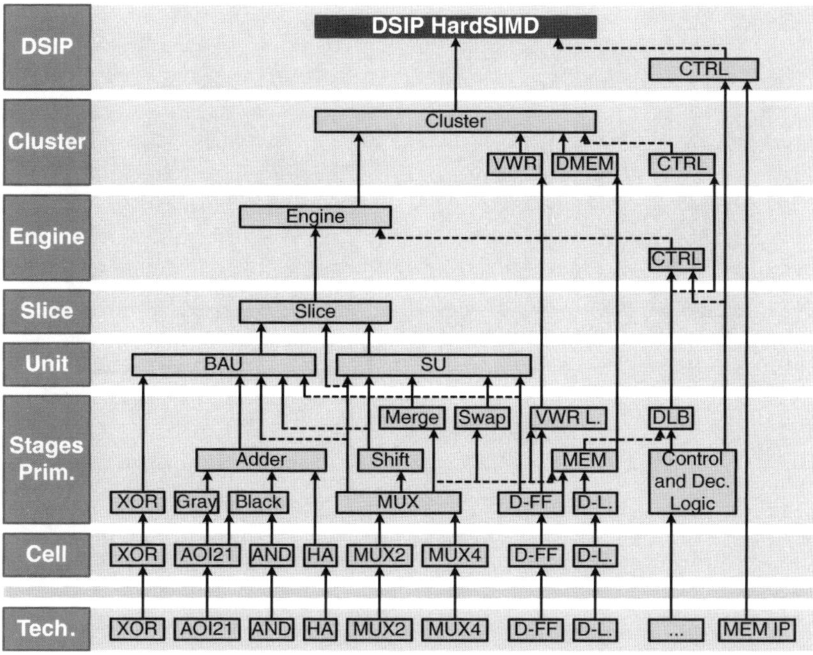

Fig. A.2 Design hierarchy present in typical DSIP template instances. Here shown in the case of the HardSIMD FIR DSIP instance which will be explained in Chap. 5. The strong hierarchy ensures high scalability and reusability

Fig. A.3b. To enable a reasonable degree of automation with existing EDA tools, the resulting layout of the proposed flow is still compliant with the conventional standard-cell layout structure. Thus, the only way to obtain higher efficiency is basically by reducing the length of important wires (i.e. wires in the crictical path and highly active wires). It is important to emphasize that this limitation will not be present in the actual targeted flow mentioned further above.

In the following the main three differences compared to the conventional standard-cell design flow are highlighted:

- *The architecture design, the logic synthesis as well as the "local" layouting (refers to the relative placement of cells) steps are combined.* Because in this combined step basically only a selection of pre-computed pareto-optimal design options takes place, this seems feasible. As mentioned earlier, design options are described in a scalable and largely technology-independent way (at least as much as possible). For instance, VHDL enables the use of generic parameters and to abstract the logic gates from the standard cell technology library, an abstraction layer can be added. Existing layout description languages, such as Cadence Structured DataPath (SDP) [7], enable already today to describe the relative placement of standard cells in a parametrizable manner. Optional pin position and routing constraints can typically be provided to layout tools in form

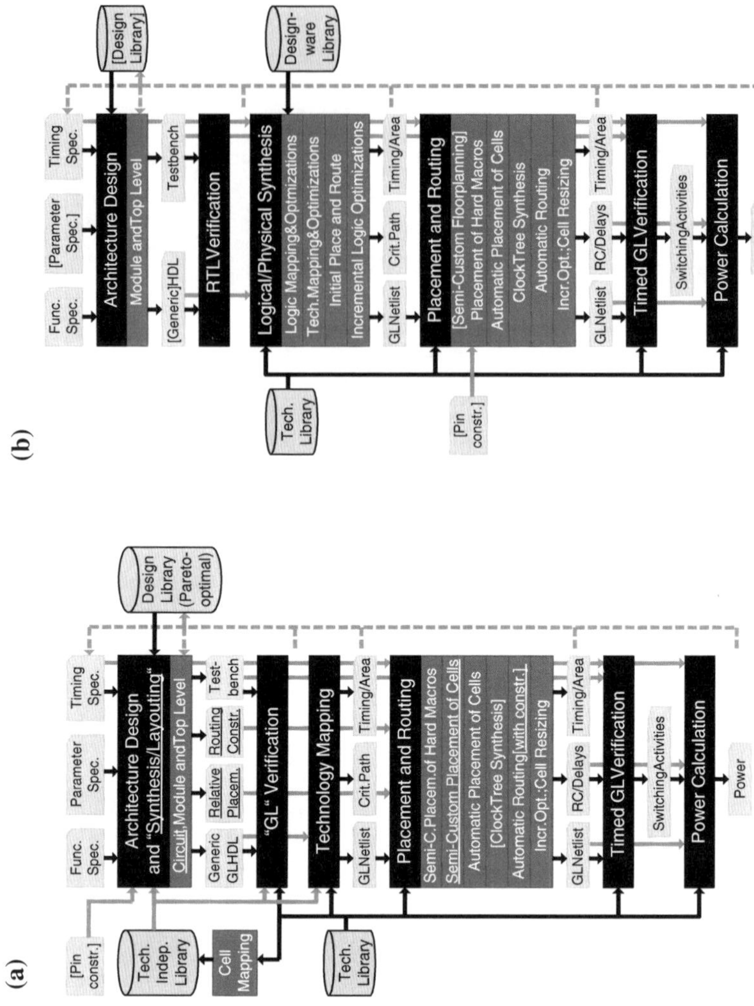

Fig. A.3 **a** Proposed back-end design flow and **b** conventional standard-cell flow, which can be used for the template instantiation, in comparison

of scripts. The selection of pareto-optimal design options, given the specific requirements for the instance, can largely be automated. After this combined step the design is represented in a generic gate-level HDL with relative placement information of cells and optional routing constraints. Generic means that certain parameters, such as drive strength of cells, can still be changed later in the design flow (e.g. for the most critical locations).

- *Semi-custom placement of cells.* All cells, for which relative placement information is present, are placed first. Note, this affects typically all the important cells from the datapath and the data memories (data plane). Only after this a priori placement has been completed, all other cells (which are typically part of

the control plane and have little overall impact) are placed automatically. An incremental optimization step can refine the positions of the cells, but the ones with relative placement information can only be moved within a certain radius.
- *Optional routing constraints are applied.* When deciding on the relative placement of cells, the relative length of the wires and potentially also the targeted routing layers are considered. The information about the targeted routing layers can be provided in this step and would ensure that the resulting routing corresponds to the targeted one.

3.1.1.3 Implementation of the Proposed Back-End Library Instantiation Design Flow and Experimental Results

A variant of the proposed back-end library instantiation design flow has been implemented and applied on the following two designs: 1) on the DSIP architecture template instance of the FIR filter (see case study 2) and on 2) a standard-cell memory. The preliminary experimental results, which are shown in the appendix of Chap. 5, are promising.

3.1.1.4 Conclusions

In this appendix concepts/ideas for a new (more holistic) back-end design approach were proposed and for the part of the template instantiation, a more concrete design flow variant was introduced. Contrary to the conventional standard-cell design flow, the proposed approach establishes a strong link between architecture and technology, which is strongly enabled because of the assumption that an architecture template is utilized. Nevertheless, this assumption applies for the targeted DSIP design approach proposed in this book. To increase the implementation efficiency, certain design/layout limitations that are present in the conventional standard-cell flow have been removed. In general this has a negative impact on automation and design time. Nevertheless, because the components of the library are often reused, the effort for designing and optimizing these components can be averaged-out over many template instances, which decreases again the average design time. The design flow which is applied most often, i.e. the template instantiation flow, can largely be automated, because most design decisions have already been considered during the library design. Thus, the proposal combines the advantages of conventional semi-custom design flows in terms of implementation efficiency with the advantages of the conventional standard-cell design flows in terms of design time.

References

1. Ahn, J.H., Erez, M., Dally, W.J.: Tradeoff between Data-, Instruction-, and thread-level parallelism in stream processors. In: ACM International Conference on Supercomputing (ICS), vol. 21 (2007)
2. Anjum, O., Ahonen, T., Garzia, F., Nurmi, J., Brunelli, C., Berg, H.: State of the art baseband DSP platforms for Software Defined Radio: A survey. EURASIP J. Wirel. Commun. Netw. **1**, 5 (2011). doi:10.1186/1687-1499-2011-5
3. Auger, F., Lou, Z., Feuvrie, B., Li, F.: Multiplier-free divide, square root, and log algorithms [DSP Tips and Tricks]. IEEE Signal Proc. Magaz. **28**(4), 122–126 (2011). doi:10.1109/MSP.2011.941101
4. Implementing SIMD in Software. BDTI (2006). http://www.bdti.com/InsideDSP/
5. Bi, X., Weldon, M.A., Li, H.: STT-RAM Designs Supporting Dual-port Accesses. Design, Automation and Test in Europe (DATE), In (2013)
6. Bougard, B., De Sutter, B., Rabou, S., Novo, D., Allam, O., Dupont, S., Van der Perre, L.: A coarse-grained array based baseband processor for 100Mbps+ software defined radio. In: Design, Automation and Test in Europe (DATE), pp. 716–721. IMEC (2008). doi:10.1109/DATE.2008.4484763
7. Encounter Digital Implementation System, Cadence. http://www.cadence.com/
8. Carvalho, E.L.d.S., Calazans, N.L., Moraes, F.G.: Dynamic Task Mapping for MPSoCs. IEEE Design Test Comput. 27(5), 26–35 (2010). doi:10.1109/MDT.2010.106
9. Catthoor, F., Danckaert, K., Wuytack, S., Dutt, N.: Code transformations for data transfer and storage exploration preprocessing in multimedia processors. IEEE Design Test Comput. **18**(3), 70–82 (2001)
10. Catthoor, F., Raghavan, P., Lambrechts, A., Jayapala, M., Kritikakou, A., Absar, J.: Ultra-Low Energy Domain-specific Instruction-set Processors, 1st edn. Springer, New York (2010)
11. Chang, M.F., Chen, P.C.: Embedded non-volatile memory circuit design technologies for mobile low-voltage SoC and 3D-IC. In: IEEE International Conference on Solid-state and Integrated Circuit Technology (ICSICT), pp. 13–16 (2010). doi:10.1109/ICSICT.2010.5667868
12. Cheng, Y.: A glance of technology efforts for design-for-manufacturing in nano-scale CMOS processes. Sci. China Ser F: Inform. Sci. **51**(6), 807–818 (2008). doi:10.1007/s11432-008-0054-9
13. Chinnery, D., Keutzer, K.: Closing the Power Gap between ASIC & Custom: Tools and Techniques for Low Power Design. Springer, New York (2007)
14. Dally, W.J., Balfour, J., Black-Shaffer, J.C., Harting, R.C., Parikh, V., Park, J., Sheffield, D.: Efficient embedded computing. Computer **41**(7), 27–32 (2008)
15. Fan, X.: A VLSI-oriented FFT algorithm and its pipelined design. In: International Conference on Signal Processing (ICSP), pp. 414–417. IEEE (2008). doi:10.1109/ICOSP.2008.4697159
16. Fasthuber, R.: Efficient Implementation of Multiplications (Slide Set). Technical Report, IMEC (2011)
17. Fasthuber, R., Raghavan, P., Catthoor, F.: An enhancement for enabling variable multiplications on a general shift-add/sub datapath. to be decided (In preparation, 2013)
18. Fort, A., Weijers, J.W., Derudder, V., Eberle, W., Bourdoux, A.: A performance and complexity comparison of auto-correlation and cross-correlation for OFDM burst synchronization. In: IEEE International Conference on Acoustics, Speech and Signal Processing (ICASSP), p. II-341-4. IMEC (2003) doi:10.1109/ICASSP.2003.1202364
19. Ghosh, A., Ratasuk, R., Mondal, B., Mangalvedhe, N., Thomas, T.: LTE-advanced: Next-generation wireless broadband technology. IEEE Wirel. Commun. **17**(3), 10–22 (2010)
20. Glossner, J., Iancu, D., Hokenek, E., Moudgill, M.: A software-defined communications baseband design. IEEE Commun. Magaz. **41**(1), 120–128 (2003). doi:10.1109/MCOM.2003.1166669

21. Gupta, R.: The Variability Expeditions: Exploring the Software Stack for Underdesigned Computing Machines. Qualcomm, UCSD (2011)
22. IRC: ITRS Roadmap on Interconnect (2009)
23. Jayapala, M., Barat, F., Vander Aa, T., Catthoor, F., Corporaal, H., Deconinck, G.: Clustered loop buffer organization for low energy VLIW embedded processors. IEEE Trans. Comput. **54**(6), 672–683 (2005)
24. Kantabutra, V.: On hardware for computing exponential and trigonometric functions. IEEE Trans. Comput. **45**(3), 328–339 (1996)
25. Karuri, K., Leupers, R., Ascheid, G., Meyr, H., Kedia, M.: Design and implementation of a modular and portable IEEE 754 compliant floating-point unit. In: Design, Automation and Test in Europe (DATE), vol. 2, RWTH Aachen (2006)
26. Kelley, B.: Software defined radio for broadband OFDM protocols. In: International Conference on Systems, Man and Cybernetics (ICSMC), pp. 2309–2314. IEEE (2009). doi:10.1109/ICSMC.2009.5345986
27. Kin, J., Gupta, M., Mangione-Smith, W.: Filtering memory references to increase energy efficiency. IEEE Trans. Comput. **49**(1), 1–15 (2000). doi:10.1109/12.822560
28. Komalan, M., Hartmann, M., Gomez Perez, J.I., Tenllado, C., Artes Garcia, A., Catthoor, F.: System level exploration of Resistive-RAM (ReRAM) based hybrid instruction memory organization. In: Memory Architecture and Organization Workshop (MeAOW) (2012)
29. Koren, I.: Computer Arithmetic Algorithms, 2nd edn. Peters/CRC Press, A.K., Natick (2002)
30. Kunze, S., Matus, E., Fettweis, G.: ASIP decoder architecture for convolutional and LDPC codes. In: IEEE International Symposium on Circuits and Systems (ISCAS), i, pp. 2457–2460 (2009) doi:10.1109/ISCAS.2009.5118298
31. Lambrechts, A., Raghavan, P., Novo, D., Ramos, E.R., Jayapala, M., Catthoor, F., Verkest, D.: Enabling wordWidth aware energy and performance optimizations for embedded processors. In: Workshop on Optimizations for DSP and Embedded Systems (FlexWare). IMEC (2007)
32. Lee, D.: Reconfigurable and area-efficient architecture for symmetric FIR filters with powers-of-Two coefficients. In: Conference on Innovations in Information Technologies (IIT), pp. 287–291. IEEE (2007). doi:10.1109/IIT.2007.4430440
33. Li, M.: Algorithm and Architecture Co-design For Software Defined Radio Baseband. Ph.D. thesis, KU Leuven (2010)
34. Li, M., Amin, A., Appeltans, R., Torrea, R., Cappelle, H., Fasthuber, R., Dejonghe, A., Van der Perre, L.: Instruction set support and algorithm-architecture for fully parallel multi-standard soft-output demapping on baseband processors. In: IEEE Workshop on Signal Processing System (SIPS), pp. 140–145. IMEC (2010). doi:10.1109/SIPS.2010.5624777
35. Li, M., Bougard, B., Xu, W., Novo, D., Van Der Perre, L., Catthoor, F.: Optimizing Near-ML MIMO detector for SDR baseband on parallel programmable architectures. In: Design, Automation and Test in Europe (DATE), pp. 444–449 (2008). doi:10.1109/DATE.2008.4484721
36. Li, M., Fasthuber, R., Novo, D., Van Der Perre, L., Catthoor, F.: Algorithm-architecture co-design of soft-output ML MIMO detector for parallel application specific instruction set processors. In: Design, Automation and Test in Europe (DATE), pp. 1608–1613. IMEC (2009)
37. Liu, D., Nilsson, A., Tell, E., Wu, D., Eilert, J.: Bridging dream and reality: Programmable baseband processors for software-defined radio. IEEE Commun. Magaz. **47**(9), 134–140 (2009). doi:10.1109/MCOM.2009.5277467
38. Mansour, M., Shanbhag, N.: High-throughput LDPC decoders. IEEE Trans. Very Large Scale Integr. Syst. 11(6), 976–996 (2003). doi:10.1109/TVLSI.2003.817545
39. Markovic, D., Brodersen, R.W.: DSP Architecture Design Essentials. Springer, New York (2012)
40. Miniskar, N.R., Hammari, E., Munaga, S., Mamagkakis, S., Kjeldsberg, P.G., Catthoor, F.: Scenario based mapping of dynamic applications on MPSoC: A 3D graphics case study. In:

International Workshop on Embedded Computer Systems: Architectures, Modeling, and Simulation (SAMOS), pp. 48–57 (2009). doi:10.1007/978-3-642-03138-0_6
41. Nahm, S., Han, K., Sung, W.: A CORDIC-based digital quadrature mixer: Comparison with a ROM-based architecture. In: IEEE International Symposium on Circuits and Systems (ISCAS), pp. 385–388 (1998). doi:10.1109/ISCAS.1998.698871
42. Nigam, T.: Scaling to the final frontier: Reliability challenges in sub 20 nm technologies. In: IEEE International Integrated Reliability Workshop (IIRW), pp. xi–xi (2011). doi:10.1109/IIRW.2011.6142574
43. Novo, D., Li, M., Fasthuber, R., Raghavan, P., Catthoor, F.: Exploiting finite precision information to guide data-flow mapping. In: Design Automation Conference (DAC), pp. 248–253 (2010)
44. Okada, K., Kousai, S.: Digitally-Assisted Analog and RF CMOS Circuit Design for Software-Defined Radio, 1st edn. Springer, New York (2011)
45. Panda, P.R., Nicolau, A., Dutt, N.: Memory Issues in Embedded Systems-on-Chip: Optimizations and Exploration. Kluwer Academic Publishers, Norwell (1998)
46. Psychou, G., Fasthuber, R., Hulzink, J., Husiken, J., Catthoor, F.: Subword handling in data-parallel mapping. In , Parallel Programming and Run-Time Management Techniques for Many-core Architectures (PARMA) (2012)
47. Rabaey, J.: Low-power silicon architectures for wireless communications. In: Design Automation Conference (DAC), pp. 377–380 (2000). doi:10.1109/ASPDAC.2000.835128
48. Rabaey, J., Abnous, A., Ichikawa, Y., Seno, K., Wan, M.: Heterogeneous reconfigurable systems. In: IEEE Workshop on Signal Processing System (SIPS), pp. 24–34 (1997). doi:10.1109/SIPS.1997.625684
49. Raghavan, P., Lambrechts, A., Jayapala, M., Catthoor, F., Verkest, D.: Distributed loop controller for multi-threading in uni-threaded ILP architectures. IEEE Trans. Comput. **58**(3), 311–321 (2009)
50. Raghavan, P., Lambrechts, A., Jayapala, M., Catthoor, F., Verkest, D., Corporaal, H.: Very Wide Register: An asymmetric register file organization for low power embedded processors. In: Design, Automation and Test in Europe (DATE). IMEC (2007)
51. Ramacher, U.: Software-defined radio prospects for multistandard mobile phones. Computer **40**(10), 62–69 (2007)
52. Salmela, P., Happonen, A., Burian, A., Takala, J.: Several approaches to fixed-point implementation of matrix inversion. International Symposium on Signals, Circuits and Systems (ISSCS), vol. 2(2), 497–500 (2005)
53. Samaras, K., Fasthuber, R., Agrawal, P., Catthoor, F.: Code Profiling for 60 GHz Baseband Processing. Technical Report, IMEC, SSET-CSI (2012)
54. Sasanka, R., Li, M.L., Adve, S.V., Chen, Y.K., Debes, E.: ALP: Efficient support for All levels of parallelism for complex media applications. ACM Trans. Architect Code Optim. 4(1) (2007)
55. Sheu, S.S., Cheng, K.H., Chang, M.F., Chiang, P.C., Lin, W.P., Lee, H.Y., Chen, P.S., Chen, Y.S., Chen, F.T., Tsai, M.J.: Fast-write resistive RAM (RRAM) for embedded applications. IEEE Design Test Comput. **28**(1), 64–71 (2011). doi:10.1109/MDT.2010.96
56. Tran, A.T., Truong, D.N., Baas, B.: A reconfigurable source-synchronous on-chip network for GALS many-core platforms. IEEE Trans. Comput. Aided Design **29**(6), 897–910 (2010). doi:10.1109/TCAD.2010.2048594
57. Woh, M., Mahlke, S., Mudge, T., Chakrabarti, C.: Mobile supercomputers for the next-generation cell phone. Computer **43**(1), 81–85 (2010). doi:10.1109/MC.2010.16
58. Woh, M., Sangwon, S., Mahlke, S., Mudge, T., Chakrabarti, C., Flautner, K.: AnySP: anytime anywhere anyway signal processing. IEEE Micro **30**(1), 81–91 (2010)
59. Wolf, M.E., Lam, M.S.: A data locality optimizing algorithm. In: ACM Conference on Programming Language Design and Implementation, pp. 30–44. Stanford University, Stanford (1991)
60. Xie, Y.: Future Memory and Interconnect Technologies. Design, Automation and Test in Europe (DATE), In (2013)

61. Xu, W., Richter, M., Sauermann, M., Capar, F., Grassmann, C.: Efficient baseband implementation on an SDR platform. In: International Conference on ITS, Telecommunications, pp. 794–799 (2011). doi:10.1109/ITST.2011.6060163
62. Yoshizawa, S., Miyanaga, Y.: Use of a variable wordlength technique in an OFDM receiver to reduce energy dissipation. IEEE Trans. Circ. Syst. (TCAS) **55**(9), 2848–2859 (2008). doi:10.1109/TCSI.2008.920098
63. Zhang, W., Hu, J.S., Degalahal, V., Kandemir, M., Vijaykrishnan, N., Irwin, M.J.: Reducing instruction cache energy consumption using a compiler-based strategy. ACM Trans. Archit. Code Optim. **1**(1), 3–33 (2004)

Chapter 4
Case Study 1: DSIP Architecture Instance for MIMO Detection

4.1 Motivation, Related Work and Contributions

This section motivates this case study, summarizes related work on MIMO detector implementations and mentions the main contributions.

4.1.1 Context and Motivation

Emerging next-generation devices, which target advanced standards, such as IEEE 802.11ac, LTE-A and IEEE 802.15.3c, have to support Multiple-Input Multiple-Output (MIMO) technology. Contrary to previous single-antenna systems, in a MIMO system, the data is transmitted and received via multiple antennas. The MIMO detector, which estimates the transmitted data, is an essential element in the receiver of such a system. Advanced MIMO detection algorithms, which enable the efficient spectrum use, have a high computational complexity. The forementioned advanced standards to not only specify MIMO technology with different antenna and modulation schemes, i.e. up to 8 antennas and 64-QAM, but also high data rates of more than 1 Gbps [20]. Considering the high computational complexity of advanced MIMO detection algorithms, the high data rates, the energy limitations, the need for high flexibility, high reusability and technology-friendliness, a clear challenge for the implementation of MIMO detectors exists.

4.1.2 Summary of Related Work

In literature, a wide variety of different MIMO detector implementations with different architecture styles have been proposed. Proposals range from ASICs [9, 13, 16, 26, 28, 35, 37, 42, 48, 50–52, 54], rASICs [5, 6, 10, 18, 27, 32, 34, 36, 38,

R. Fasthuber et al., *Energy-Efficient Communication Processors*,
DOI: 10.1007/978-1-4614-4992-8_4,

40, 47, 55, 58, 60], dedicated ASIPs [1, 4, 15, 22–24], SDR processors [3, 17, 23, 25, 56, 57], GPUs [59] to FPGA circuits [21, 39] and FPGA processor implementations [11, 19, 33]. Besides the data rates, the energy efficiency and the enabled flexibility/reusability, the implementations differ mainly in employed algorithm, hard and/or soft-output capability, Bit-Error-Rate (BER) performance, supported antenna and modulation schemes. The proposed ASIPs and SDR processors can provide sufficient flexibility, therefore they are as such attractive to cope with the huge NRE cost of DDSM technology designs. However, as mentioned in Sect. 1.2, especially the limited energy efficiency of these platforms is still a major concern. Comparison results in [17, 23] and Sect. 4.6.3 show that, for the same data rate and BER performance, the programmable processors consume at least a factor 4 more energy than ASICs. Thus, the need for more efficient programmable solutions, which can close the energy gap almost fully, is still prevailing (all 5 requirements need to be fulfilled).

4.1.3 Main Contributions

In this chapter, we propose a MIMO detector design that targets high data rates at high BER performance, close to ASIC energy efficiency, high flexibility, high reusability and technology scaling-friendliness. As motivated in the introduction of this book, maximizing these metrics is crucial for the considered domain. We mainly leverage on the proposed DSIP architecture template of Chap. 3 and on the optimized MIMO detector algorithm of our previous work to achieve this goal.

The contribution of this chapter can be summarized as follows: (1) A highly efficient, scalable and programmable MIMO detector implementation is proposed. Contrary to other implementations, the proposal is based on the proposed DSIP architecture template and leverages on the Selective Spanning with Fast Enumeration (SSFE) algorithm [17]. This algorithm, which was heavily optimized for the type of considered architecture, offers high BER performance. The comparison with an ASIC equivalent [16] and with other programmable architectures proves high energy efficiency and motivates therefore the proposed DSIP architecture template design approach. (2) The scalability of the architecture instance is demonstrated. Thereby, to the best of our knowledge, we show the first programmable MIMO detector that supports MIMO 8×8 (8 transmit and 8 receive antennas) 64-Quadrature Amplitude Modulation (QAM). (3) The energy consumption for different run-time modes/situations is provided. The results strongly motivate the exploitation of a run-time manager [8, 44], which enables to further reduce the average energy consumption of the proposed design. Thus, the proposed programmable architecture can potentially achieve a higher average energy efficiency than the ones of ASICs. (4) To understand the cause for the area overhead, a detailed area comparison between the proposal and the ASIC reference is provided. Thereby also the influence of algorithm optimizations on the resulting area is studied.

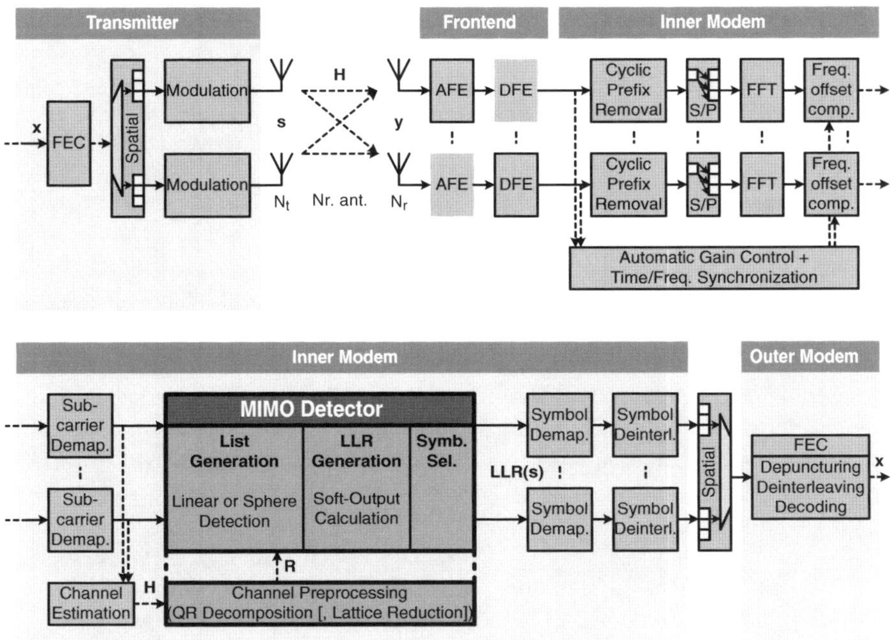

Fig. 4.1 Overview of a typical MIMO OFDM communication system. In the receiver, the front-end, the inner-modem and the outer-modem functionalities are shown. The MIMO detector estimates, based on the channel matrix **H** and on the with interference received symbol vector **y**, the most-likely transmitted symbol vector **s**

4.2 Background on Driver

This section provides background information on MIMO detection and on the flexibility requirements of this functional block.

4.2.1 MIMO Detection

Figure 4.1 shows a typical MIMO OFDM communication system. The number of transmit and receive antennas are denoted as N_t and N_r respectively. For a W-QAM modulation, a symbol represents one out of $W = 2^q$ constellation points. A 16-QAM symbol consists of 4 bits and a 64-QAM symbol of 6 bits respectively. At once, the transmitter maps one $qN_t \times 1$ binary data vector **x** to an $N_t \times 1$ symbol vector **s**. Then, the symbol vector is transmitted via multiple antennas. Because of non-idealities in the communication channel **H**, the transmitted symbol vector is received with interference. Therefore the transmitted symbols are not instantaneously known at the receiver. The MIMO detector combines the received symbols from different antennas and the information of the channel **H** to estimate the transmitted symbol vector **s**. The transmission of each vector **s** over a flat-fading MIMO channel can be modeled as $\mathbf{y} = \mathbf{Hs}+\mathbf{n}$. Thereby **y** denotes the received $N_r \times 1$ symbol vector, **H** characterizes a $N_t \times N_r$ channel matrix and **n** describes the noise vector. For the Maximum Likelihood (ML) detection scheme, the detector is designed to solve

$$\hat{\mathbf{s}} = \arg \min_{\mathbf{s} \in \Omega^{N_t}} \|\mathbf{y} - \mathbf{Hs}\|^2 \tag{4.1}$$

where Ω^{N_t} is the set containing all the possibilities of $N_t \times 1$ vector signals **s**. For all other detection schemes, not all, but only a limited number of vector signals **s** are considered for estimating the transmitted data. Hard-Output (HO) detectors provide the most-likely symbol vector **s**. Soft-Output (SO) detectors provide additionally the Log-Likelihood-Ratio (LLR) for each bit in **s**. Modern Forward-Error-Correction (FEC) decoders, such as Turbo and Low-Density Parity-Check (LDPC) decoders, require soft-input to achieve high BER performance.

As shown in Fig. 4.1, the MIMO detector functionality at the receiver can generally be decomposed into the following parts: channel preprocessing, list generation, LLR generation and symbol vector selection. The first part processes the channel matrix **H** to reduce the complexity of the actual MIMO detection. Thereby typically QR decomposition and potentially Lattice Reduction (LR) is performed [19]. The list generator computes a list of the most likely transmitted symbol vectors **s**. Popular list generators are either based on linear detection algorithms, such as Minimum Mean Square Error (MMSE) [22], or on near-ML sphere detection algorithms, such as K-Best [4]. Linear detectors and ML detectors are generally opposed in terms of complexity and BER performance. To achieve a high BER performance for a channel with high interference, either a strong detector or advanced channel preprocessing, e.g. based on LR, is needed [3]. The LLR generator computes the LLR for each bit in all likely symbol vectors **s**. Especially for complex modulation schemes, such as 64-QAM, the computational load of this task is significant. From the set of candidates, the symbol vector selector chooses the vector with the highest LLR, which is the most-likely transmitted symbol vector **s**.

4.2.2 Motivation for a Flexible Implementation

As motivated in Sect. 1.1.1.2 and Sect. 2.1.2, emerging 4G multi-standard communication devices have to support a wide variety of different modes. The main reasons are summarized as follows: (1) Mainly to enable a more flexible spectrum use, the number of modes, which have to be supported for a single standard, is increasing. (2) For high connectivity and compatibility, short range to long range communication standards across different generations need to be supported. Each standard may specify a different set of modes. Examples for standards, which specify MIMO, are IEEE 802.15.3c, IEEE 802.11n, IEEE 802.11ac/ad, IEEE 802.16e, 3GPP LTE and LTE-A [20]. (3) In addition to the modes that are required for standard compliance, modes, which enable dynamic run-time scaling, can be useful. For instance, for well-conditioned communication channels, e.g. the user is close to a base station, the detection performance and therefore the energy consumption could be reduced. To enable this dynamic adaptation, different algorithm parameters (e.g. K in K-best) and/or different algorithms must be supported. This wide variety of modes, which need to be supported by a single device, rises the need for a flexible/programmable solution. In the context of a MIMO detector, a definition of a mode could include (1) system parameters, such as (a) number of antennas (e.g. ranging from 2 to 8), (b) modulation scheme (e.g. ranging from Quadrature Phase Shift Keying (QPSK) to 256-QAM) and (c) output type (hard- or soft-output), (2) specifications on the required detection/BER performance (e.g. ranging from MMSE to ML performance) and (3) specifications on the latency/throughput (e.g. ranging from several Mbps to above 1 Gbps).

4.3 Algorithm Optimizations and Characteristics

In this section we provide background information on the selected algorithm and on its characteristics. The latter is considered for the architecture design (see Sect. 7.3).

4.3.1 Algorithm Choice and Applied Optimizations

In literature, a wide variety of different MIMO detection algorithms/schemes, such as Reconfigurable Reduced Dimension Maximum Likelihood (RRDML) [27], Minimum Mean-Squared Error (MMSE) [35, 57], Tuple Search Detector (TSD) [1], Multi-pass Trellis Traversal (MTT) [59], Layered ORthogonal Lattice Detection (LORD) [5], K-Best [4], Scalable Block-based Parallel Lattice Reduction (SBP-LR) [3] and Selective Spanning with Fast Enumeration (SSFE) [29], have been proposed. For this work, we have chosen the SSFE algorithm [29] for list generation and the multiplier-less algorithm [30] for LLR generation. The LLR generator performs the soft-output computation which enables to increase the BER performance by 2–4 dB (compared to hard-output detection only). For the symbol selection simple comparison operations are used. The channel preprocessing, e.g. QR decomposition and LR, is outside the scope of this work. We chose the above mentioned algorithm combination because of the following main reasons:

- In our previous work, this algorithm combination was heavily optimized. Amongst others, the following transformations have been applied [17]: (1)

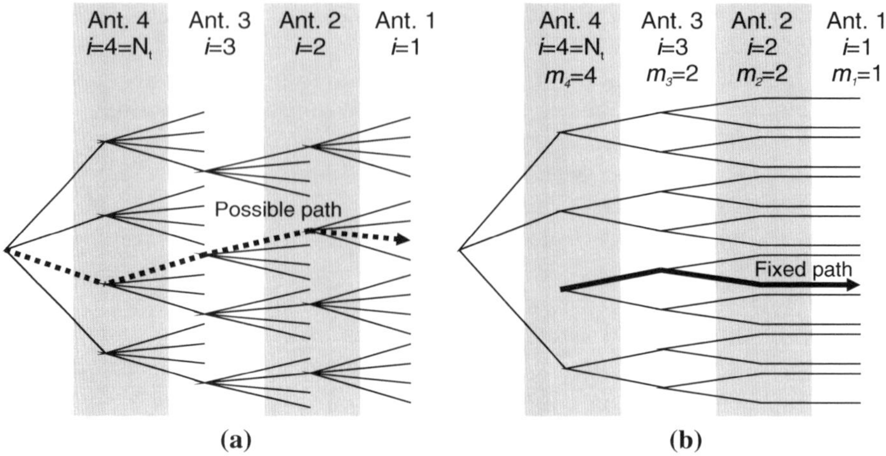

Fig. 4.2 Search tree-topologies, which correspond to data-flow structures, of the **a** traditional K-Best and the **b** selected SSFE algorithm. A 4×4 MIMO system with Quadrature Phase Shift Keying (QPSK) modulation is considered. The K-Best algorithm first spans the K nodes at level/antenna $i+1$ to $K \times W$ nodes. After spanning, the algorithm sorts the $K \times W$ nodes, the K best nodes are selected and the rest of the nodes are deleted. Then, the next level/antenna is considered. These approach results in an irregular and in a non-deterministic data-flow structure, because the sorting depends on the actual data values. In contrast, mainly because the spanned nodes are never deleted, the data-flow structure of the selected SSFE algorithm is completely regular and deterministic. Thus, the SSFE enables a more efficient implementation on data-parallel processors. **a** K-Best. **b** SSFE

Multiplications, which have a limited operator range, have been transformed to shift, addition and subtraction operations. This is crucial, since the proposed architecture concept is designed for these type of operations. (2) The data-flow structure has been transformed so that it becomes regular and deterministic. Figure 4.2 compares the data-flow structure of the optimized SSFE algorithm with the data-flow structure of the traditional K-Best algorithm. Regularity and determinism is essential for enabling an efficient implementation on parallel architectures. (3) The memory requirements have been reduced and both algorithm parts, i.e. the list generator and the LLR generator, have been co-optimized. These optimizations reduced the storage overhead and removed the data memory bottleneck.

- The algorithm scales with the number of antennas, supports 16/64-QAM as well as hard- and soft-output. Therefore this algorithm can be utilized for different modes.
- A SSFE algorithm instance is parameterized with the values of a scalar vector $\mathbf{m} = [m_1,..,m_{N_t}]$, $m_i \leq W$. The entries in this vector specify the number of scalar symbols s_i that are considered at antenna N_i. Thus, with the parameter **m**, the SSFE algorithm enables a scalability in terms of BER performance and computational complexity (see Fig. 4.3). For instance, with $\mathbf{m} = [1, 1, 1, 64]$, near-

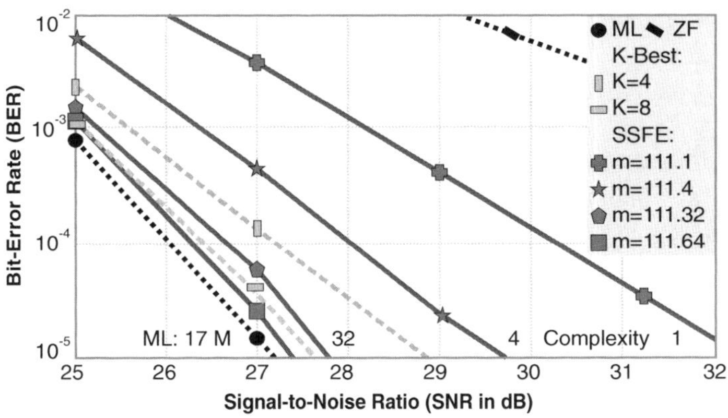

Fig. 4.3 BER performance of the ML, the Zero-Forcing (ZF), different K-Best and different SSFE algorithm instances in comparison. The trade-off between BER performance and computational complexity of the SSFE algorithm is shown. The difference between near-ML performance ($m = 111.64$) and SIC performance ($m = 111.1$) is 3–4 dB SNR and a factor 64 in computational complexity. In comparison, the ML algorithm is 17 M/64 = 265625× more complex than the near-ML SSFE algorithm instance. Simulation setup: MIMO 4 × 4, hard-output, 64-QAM, FEC code-rate 1/2, 3GPP LTE Suburban macro [31]

ML performance, with $\mathbf{m} = [1, 1, 1, 1]$, Successive Interference Cancellation (SIC) performance, can be achieved. The difference in computational complexity is a factor 64. This scalability can be exploited to decrease the energy consumption and/or to increase the throughput in well-conditioned communication channels.

- Since the SSFE algorithm can be instantiated for different BER performance requirements, it can efficiently be combined with LR algorithms. Note, the combination with a LR algorithm could further increase the available BER performance-complexity trade-off space.
- In our previous work we have implemented the selected algorithm combination on the ADRES processor and as an ASIC [16]. Especially the latter enables a quantitative and fair comparison with the proposed programmable architecture of this work.

In the rest of this book, unless otherwise indicated, we refer to the chosen algorithm combination simply as algorithm. The applied DSIP design approach enables to make a trade-off between efficiency, i.e. more customization towards a certain algorithm, and flexibility/reusability, i.e. enabling the efficient mapping of a broader range of different algorithms. In this work we design the architecture instance for the characteristics of the chosen algorithm. However, we have seen that the characteristics are also rather representative for other MIMO detector algorithms, such as [2], therefore a rather high reusability is enabled.

4.3.2 Algorithm Characteristics

In the following we provide a summary of the main characteristics of the selected algorithm. This information was considered during the design and dimensioning of the proposed architecture instance (see Sect. 7.3). Algorithms with similar characteristics can efficiently be mapped. However, if an algorithm has very different characteristics, e.g. instead of 16 bits, 20 bits word length is needed, or instead of constant multiplications, variable multiplications are computationally dominant, then the efficient mapping onto the designed instance is not possible. To enable the efficient mapping also for that case, a broader specification would have to be considered from the beginning on. We consider the design of architecture instances, which are suited for a set of more diverse algorithms, as future work.

4.3.2.1 Required Arithmetic and Logic Operations

As mentioned in Sect. 3.3, we distinguish between the following operation types: (1) Basic Arithmetic Operations (BAO)s, e.g. addition, subtraction, shift, (2) Multiplications with a Constant Operator (MCO)s, (3) Multiplications with Variable Operators and a multiplier which can adopt a Wide value range (MVOW)s, e.g. a 16×12-bit multiplication in which the multiplier can adopt all 2^{12} different values, (4) Multiplications with Variable Operators and a multiplier which can adopt only a Narrow value range (MVON)s, e.g. 16×3-bit multiplication, where the multiplier can only adopt max. 2^3=8 different values, (5) Trigonometric and Special Operations (TSO)s, e.g. cos, sin, log, and (6) Boolean Logic and Comparison Operations (BLCO)s, e.g. AND, OR, $\geq$.

The selected MIMO detection algorithm has the following operation breakdown, which is especially representative for the most complex modes: 66 % BAOs, 0 % MCOs, 1 % MVOWs, 24 % MVONs, 0 % TSOs, 9 % BCLOs. Constant multiplications and trigonometric/special operations are not present. Most operations are BAOs, which have a rather low implementation complexity.

4.3.2.2 Required Word-Length Support

The signals which represent the constellation points are represented with 3 bits, all other signals with 16 bits. The 3-bit representation is only effecting the MVON operations.

4.3.2.3 Data Dependencies and Parallelization

The following main parallelization options can be exploited: (1) parallel computation of real and imaginary part, (2) parallel MIMO symbol computation, (3) parallel evaluation of max. m_{Nt} constellation points, (4) parallel computation of list generator and LLR generator and (5) parallel computation of different operations within list/LLR generator. Considering data dependency, data exchange frequency and regularity of data flow, a parallelization of (1), (2), (3) and (5) are most promising (see Sect. 7.3). Nevertheless, the actual choice has to additionally consider the characteristic of the targeted architecture template and the throughput requirements.

4.3.2.4 Data and Control Flow

The data-flow structure of the algorithm is streaming based, completely deterministic and does not contain conditions. The algorithm is loop-dominated.

4.3.2.5 Data Storage Requirements

The actual data storage requirements depend on several aspects, such as general architecture, exploited parallelization options and chosen parameters of the algorithm. Nevertheless, because of the applied algorithm optimizations, the requirements on the data storage are basically lower than for other algorithms, such as K-Best. For instance, compared to K-Best, the access is more regular and because of the co-optimization of list generator and LLR generator, the number of temporal values is lower. Because of the streaming nature of the algorithm, big data memories, e.g. as it would be the case for the FFT, are not required.

4.4 Proposed DSIP Architecture Instance

In this section, the proposed scalable DSIP architecture instance for MIMO detection is explained. For more information on the design/dimensioning of the architecture instance, we refer to Sect. 7.3. In the following we describe the proposed hierarchical architecture in a top-down manner.

4.4.1 Top Level

Figure 4.4 shows the top-level view of the proposed scalable architecture. The top level consists of one or multiple identical clusters and the Shared Top Level Control (STLC). The number of employed clusters depends essentially on the targeted throughput requirements.

4.4.1.1 Clusters

We employ multiple clusters to perform coarse-grain parallelization across different MIMO symbols. Thereby DLP is exploited. The cluster level is explained further below.

4.4.1.2 Shared Top Level Control

The top level control is the main control of the processor. It consists of the master control logic with the corresponding L1 IMEM Direct Memory Access (DMA), the L1 IMEM and the DMEMs for the Look-Up Table Units (LUTU)s. The master control is mainly responsible for the instruction transfers between the L1 IMEM and the distributed L0 loop buffers as well as for the synchronization of L0 loop buffers. Since all clusters are homogeneous and execute the same set of instructions (exploit DLP), the L1 IMEM can be shared amongst them. The same is valid for the DMEMs that contain the data for the LUTUs.

4.4.2 Cluster Level

A cluster consists of a DMEM, 3 Very Wide Registers (VWR)s [45], 2 engines, a VWR for the output (VWRo) and the Cluster Level Control (CLC). The datapath between DMEM and VWRs is 512 bits wide, the datapath between engines and VWRs/VWRo is 128 bits wide.

4.4.2.1 Data Memory

In the DMEM the input and the temporal data is stored. The DMEM as well as the VWRs and the VWRo have a wide interface of 512 bits. A wide interface enables to reduce the number of accesses and hence, the control effort. The required depth of the DMEM depends on the maximal number of antennas that should be supported. For MIMO $2 \times 2/4 \times 4/8 \times 8$, 8/10/14 lines are required respectively.

4.4.2.2 Very Wide Registers

Contrary to a conventional register file, multiple VWRs are employed. A VWR is characterized by an asymmetric interface [45]. The interface to the DMEM is much wider than the interface to the engines. As for the DMEM, the wide interface reduces the access cost. As shown in Fig. 4.4, 3 VWRs of different types, which are denoted as T1 to T3, are employed. Each VWR type offers an individual kind of data transfers. T1 is used for the data transfers from the DMEM to the engines.

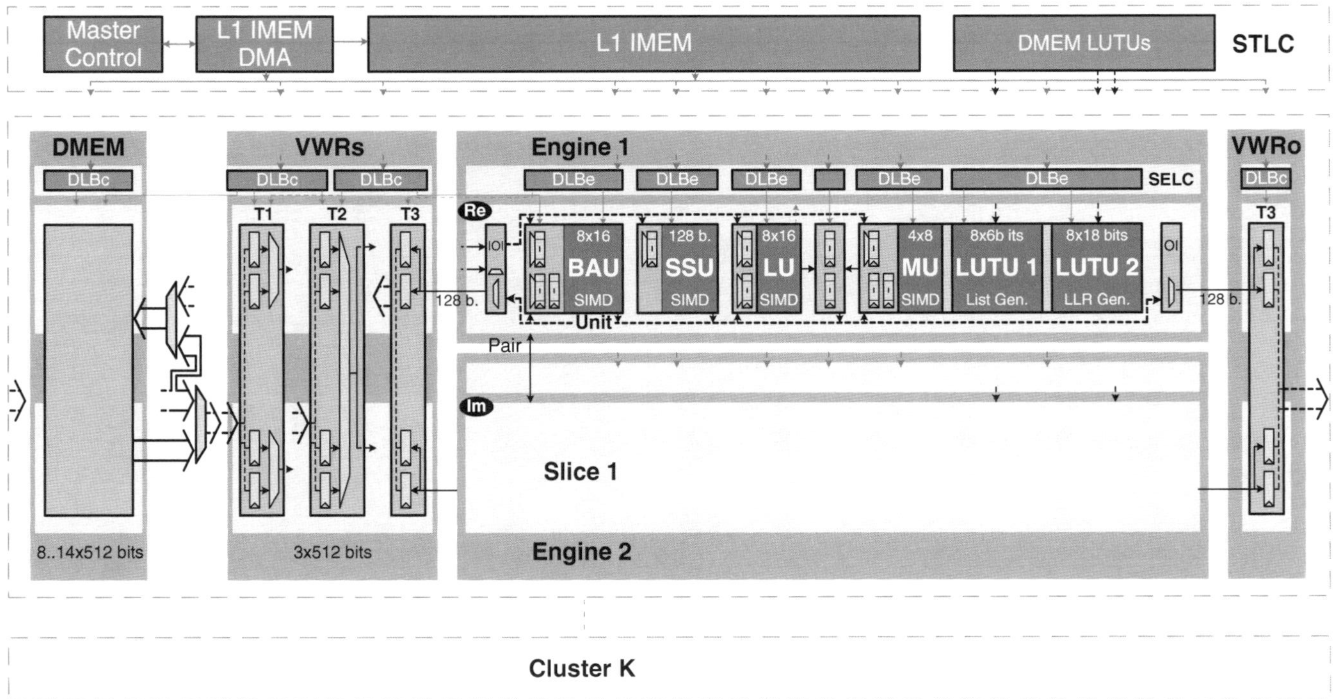

Fig. 4.4 Top level view of the proposed scalable architecture for MIMO detection. Within the BAU, SSU and LU, 8-way hardware SIMD, within the MU, 4-way hardware SIMD is performed. The units of a slice as well the DMEM/VWRs/VWRo have individual DLBs and can therefore execute control loops independently from each other. Engine 1/2 perform the computations on real/imaginary data. The direct connection between them enables to exchange data during complex multiplications. We employ multiple clusters to obtain a higher DLP

The registers of T1 are directly assigned to a specific engine. T2 is also employed for the data transfers from the DMEM to the engines, but contrary to T1, the data in the registers of T2 can be accessed by all engines. Hence, T2 enables to share or distribute data amongst engines. T3 has a wide read interface, but a narrow write interface. T3 is used to write data from the engines to the DMEM. Like for T1, the registers in T3 are assigned to engines. The concept of employing multiple VWRs, with specific data transfer options, enables to offer a wide range of data transfers while optimizing the overall energy consumption: Most of the data that has to be transferred between DMEM and engines and vice versa is engine specific. For that purpose, T1 and T3 are employed, which require little multiplexing, ensure short wires and locality. Thus, these transfers consume little energy. However, with T2, the distribution of data to engines is still enabled. A read operation of T2 is more costly than of T1, but since it occurs less frequently, the overall energy contribution is small. The total storage capacity of the 3 VWRs is $3 \times 512 = 1536$ bits.

4.4.2.3 Engines

A cluster consists of 2 homogeneous engines. Engine 1 computes the real part while engine 2 computes the imaginary part of the complex data. During a complex multiplication, the required data is exchanged. Each engine has its individual control and can therefore execute individual instructions. Hence, multiple engines enable to exploit DLP and ILP. Further below, the engine level will be explained in more detail.

4.4.2.4 Very Wide Register Output

The VWRo can be seen as the output interface of the processor. The processor can only write to the VWRo, but not read from it. The VWRo is of VWR type T3.

A DSIP processor platform, which exploits the streaming nature of the wireless domain, consists of several DSIP processor instances. Thereby, the output of the VWRo is typically connected to the DMEM and the VWRs of the subsequent processor instance.

4.4.2.5 Cluster Level Control

The control for the DMEM, the VWRs and the VWRo is denoted as cluster level control. As it can be seen in Fig. 4.4, each of these components has either one or several individual L0 distributed loop buffers (DLBc)s. For deciding on the number and sizes of DLBs, the characteristics of control signals (e.g. frequency of toggling, number of instructions within loop) for the considered algorithm were analyzed. The employed DLBs offer a storage capacity ranging from 8 to 20 instructions.

4.4.3 Engine Level

An engine consists of one datapath slice and the engine level control (DLBe).

4.4.3.1 Slice

A slice represents the arithmetic and logic datapath of the processor, which will be explained further below.

4.4.3.2 Engine Level Control

As it is shown in Fig. 4.4, the Shared Engine Level Control (SELC) consists of several distributed loop buffers (DLBe)s [46] which control the units of the datapath slice. The concept of employing separated controls for the units has 2 main advantages: (1) For inactive units, no instructions have to be fetched. (2) The individual loop nests are much smaller than the loop nest of a common (merged) loop. For this reason, the size and hence the energy consumption of the loop buffers can significantly be reduced.

4.4.4 Slice Level

As it can be seen in Figs. 4.4 and 4.5, a slice consists of several heterogeneous units and the slice interfaces. The components of a slice are connected via an upper and a lower interconnection network. Via the upper interconnection network, the input data of the VWRs is routed. The lower interconnection network is used to exchange data between units and to write data to the VWRs/VWRo. Both, the upper and the lower interconnection network are 128 bits wide. In the architecture a differentiation between primary and secondary units is made. Primary units are the main and most active units of the architecture. To ensure that scheduling inefficiencies are minimized, primary units have individual local registers and an enabled direct data communication to other units. Secondary units are mainly present to extend the set of support operations. Because secondary units are typically less frequently utilized than primary units, local registers and data connections can be shared. The reduction of local registers and data connections leads to a reduction of multiplexers, and hence, to a reduction in energy consumption.

In the proposed architecture, the primary units are the Basic Aritmethic Unit (BAU) and the Subword Shuffler Unit (SSU). The secondary units are the Logic Unit (LU), the Multiplier Unit (MU) and the two Look-Up Table Units (LUTU)s. The units have the following functionalities (a more detailed description is provided further below): On the BAU, the most computational dominant operations, i.e. BAOs, MCOs and

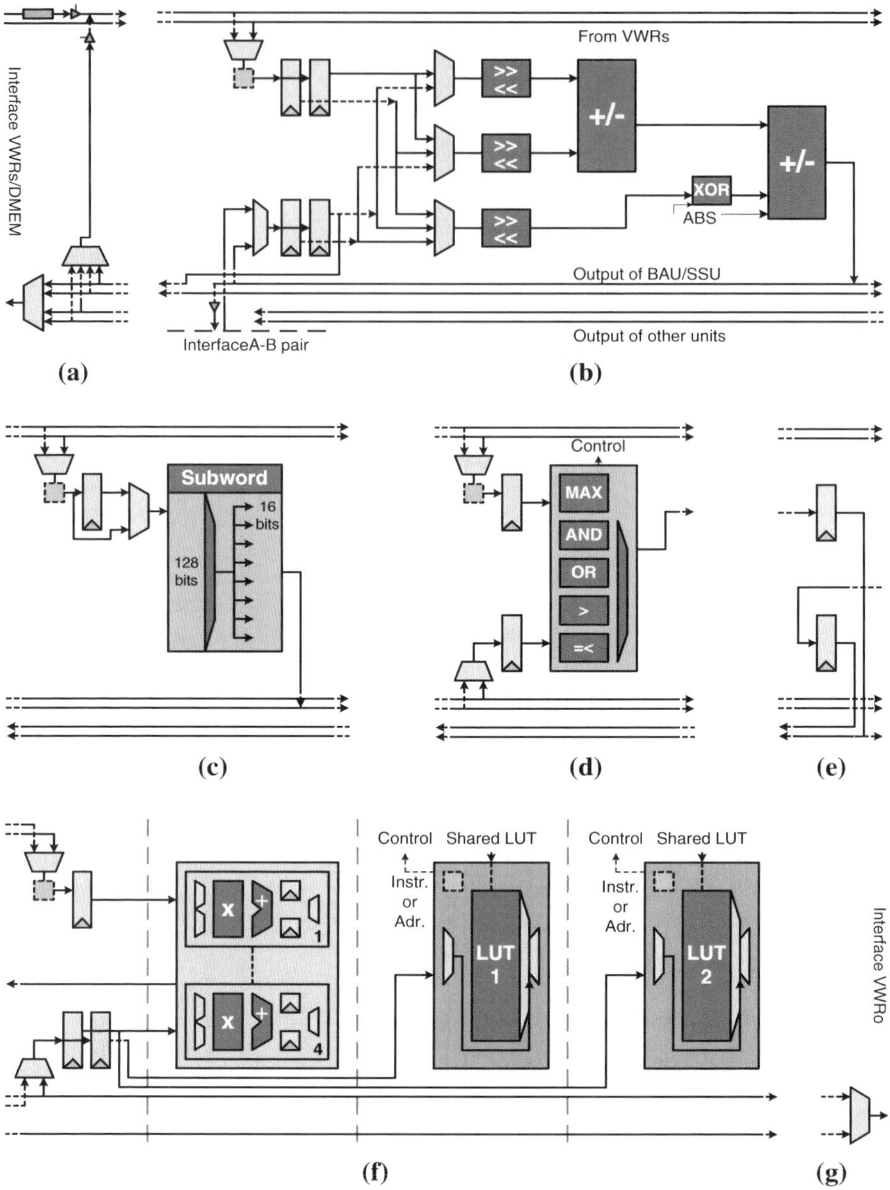

Fig. 4.5 Datapath components/units of a slice. The datapath is 128 bits wide. The upper and the lower interconnection networks are shown. **a** VWRs slice interface. **b** BAU. **c** SSU. **d** Single-cycle units: LU. **e** Shared output register. **f** Multi-cycle units: MU, LUTU1, LUTU2. **g** VWR output interface

MVONs, are computed. The SSU is employed for arranging subwords in the 128-bit wide datapath of a slice. Logic and comparison operations are performed on the LU. MVOW operations are executed on the MU. The LUTUs store the instruction sequences which are required for performing MVON operations on the BAU.

The LU computes the output of the 128-bit 8-slot datapath word input in one clock cycle. Therefore it is denoted as single-cycle unit. The MU and the LUTUs require multiple clock cycles to compute the outputs of all the 8 subwords of a 128-bit datapath word. For this reason, they are referred as multi-cycle units. Because of this property, the MU and LUTUs are grouped together and they share the same local registers and data connections.

The type of employed units and the amount of intra-unit data parallelization (i.e. the number of slots that share the same sub-datapath) considers the operation breakdown of the algorithm, which was mentioned in Sect. 4.3.2.1. By sharing hardware within an unit, the utilization of units and thus the area efficiency can be increased. Nevertheless, the sharing increases the latency of the 8-slot vector operation. This increased latency can lead to a sub-optimal schedule and/or to increased storage requirements. Hence, this trade-off has to be carefully evaluated. It can be argued that this customization will prevent the mapping of other algorithms onto the architecture. Nevertheless, we have seen that other low-level optimized MIMO detection algorithms, such as [2], have a similar operation breakdown. Hence, it seems feasible that the proposed or a slightly modified architecture instance will also be suitable for other algorithms.

4.4.4.1 Slice Interfaces

A slice has an interface to the VWRs, denoted as VWRs_SI, and an interface to the VWRo, denoted as VWRo_SI. The interfaces are depicted in Fig. 4.5a and 4.5g respectively. The VWRs_SI enables to re-route data from the lower to the upper interconnection network.

4.4.4.2 Basic Arithmetic Unit

The BAU is the main and the most utilized unit of the architecture. It supports BAO, MCO and MVON operations. As it is shown in Fig. 4.5b, it consists of only basic operators, such as shifters, adders/subtracters and a circuitry for the computation of abs(). Therefore, as explained in Sect. 3.5.4.2, the unit can operate at a high clock frequency, contributes to a overall high area/energy efficiency and is rather generic. Because 2 consecutive add/sub-operations occur very frequently in the algorithm, the datapath contains 2 adders/subtracters. For the local data 4 registers are present. The unit has a 8-way hardware SIMD/vector datapath with a subword size of 16 bits. We exploit this DLP to perform the evaluation of multiple constellations points in parallel. The factor 8 is a common divider of 16/64(-QAM).

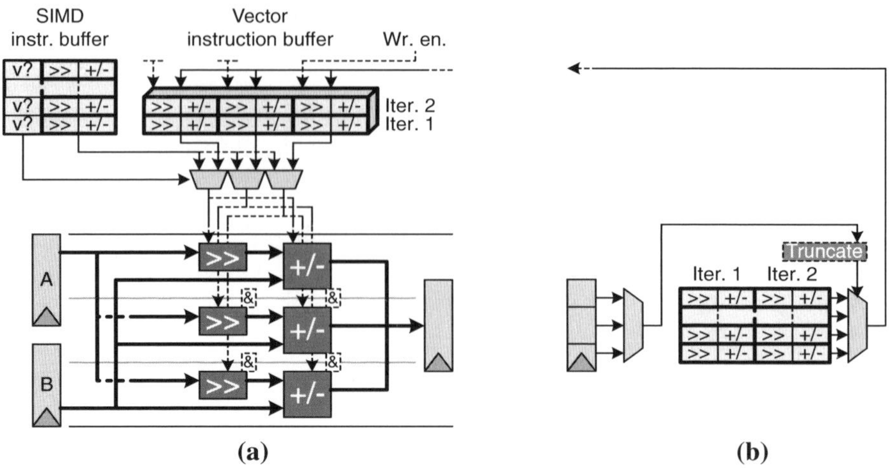

Fig. 4.6 Architecture extension to support MVON operations with the BAU. The BAU supports SIMD mode and vector mode (individual shift factors and negation controls for each slot). Nevertheless, in vector mode, alls slots must execute the same number of iterations. For all possible multiplier operands, the corresponding shift-add/sub sequences are stored in the LUTU. At run-time, based on the actual multiplier value of a slot, the corresponding instruction sequence is loaded from the LUTU to the vector instruction buffer. This table-look up can sequentially be performed for all slots. Once the instructions for all slots have been fetched, the multiplication can be executed on the BAU. Thereby the vector mode is used. The sequential loading of instructions from the LUTU to the vector instruction buffer can be done in the background, i.e. the BAU does not need to stall. **a** BAU with vector support. **b** LUTU with instructions

For constant multiplications, the decomposed sequence of shift-and-add/sub instructions is known at compile time and can therefore be stored in the loop buffer. However, for a MVON operation, the actual required shift-and-add/sub sequence depends on the multiplier value (2nd input operand of the multiplication), which is unknown at compile time. To enable MVON operations on the BAU, we propose the hardware extension as it is shown in Fig. 4.6. Each possible/valid multiplier value is first represented in Canonical Signed Digit (CSD) format [43] and then the multiplication is decomposed into a shift-and-add/sub sequence. Thereby a scaling factor can be considered. The maximal number of required iterations depends on the multiplier word length. For multiplier values which require less than the maximal number of iterations, No OPerations (NOP)s are inserted. In general, the value of the multiplier and the row read address of the LUT, where the corresponding instruction sequence is stored, are matched. After a table look-up, the instruction sequence is provided to the BAU where the actual iterative computation takes place. Since every SIMD slot may have a different multiplier value, i.e. requires to execute a different shift-and-add/sub sequence, the datapath needs to be capable to operate in a vector mode (the shift factor and the negation control signals can be different for each slot). In the vector mode, the shift-add/sub instructions are fetched from the vector instruction buffer. To avoid a duplication of LUTs, the table look-up and the writing

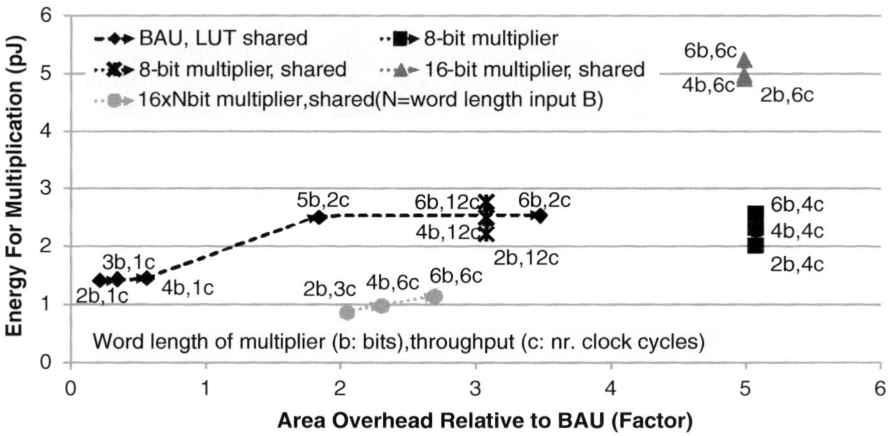

Fig. 4.7 Comparison of different options for performing MVON operations. A 48-bit datapath with 3 slots and the use of all representable multiplier values is assumed. The considered multiplier word lengths range from 2 to 6 bits. The multiplicand is 16 bits wide. Either a single hardware multiplier is shared amongst the slots or 3 parallel hardware multipliers are employed. The proposed approach achieves a high throughput and a high energy efficiency because the hardware of the BAU is simple and the multiplications are optimized at design-time (i.e. CSD coding applied). The results are based on TSMC 40 nm standard-cell implementations

to the vector instruction buffer for the vector slots are done sequentially. During the filling of the vector instruction buffer, the BAU can still operate in SIMD mode and does not need to stall.

Alternatively to the proposed approach for supporting MVON operations, a conventional hardware multiplier solution could be used. Figure 4.7 compares the proposal to a conventional solution for a 48-bit datapath with 3 slots. As it can be seen, especially when the word length of the multiplier (2nd operand) is small, the proposed solution is a better choice. This is the case for the targeted applications. For instance, the multiplier for the MVON operation in the considered algorithm has only 3 bits. A dedicated hardware multiplier, i.e. the multiplier word length matches exactly the requirements, would be more energy efficient. However, employing a customized multiplier would limit the reusability and increases the area. A conventional multiplier, i.e. both operands have the same word length, is less efficient for the targeted multiplier word length.

4.4.4.3 Subword Shuffler Unit

In the considered algorithm, all operations are performed on 16-bit data and the 8 slots of a slice, which evaluate constellation points in parallel, have little data dependency. Thus, the only required subword shuffler functionality is the broadcasting from one slot to all other slots, which is depicted in Fig. 4.5c.

4.4.4.4 Logic Unit

The LU is employed to perform BLCOs, such AND, OR, $\geq$ and maximum value of a vector. As it is shown in Fig. 4.5d, the LU has a control output which is connected to the master control. This feature enables to handle conditions. In this application, the LU is employed to select the constellation point with the highest LLR. The relative operation breakdown of BLCOs is rather low. Thus, a multi-cycle implementation approach could be considered. Nevertheless, because of the following main reasons, the outputs of all 8 SIMD slots are computed in parallel: (1) the hardware requirements to perform logic and comparisons operations is low, (2) the latency of the max() operation, which is frequently needed, is reduced, (3) no additional latency for condition handling is introduced.

4.4.4.5 Multiplier Unit

The MU, which is schematically shown in Fig. 4.5f, is used to compute MVOW operations. It is implemented with 4 parallel 8-bit multi-precision multiplier which are shared amongst the 8 slots. Because of the following main reasons, this implementation choice leads to high efficiency: (1) The area of the 4 8-bit multi-precision multipliers is much smaller compared to the area of 8 16-bit full-precision multipliers. (2) In the considered application domain, often a precision of 8 bits is sufficient and the enabled flexibility enables to exploit this characteristic. (3) In the algorithm, the relative operation breakdown of MVOW operations is very low and an increased latency does not harm the scheduling. With the implemented MU, a 8/16-bit multiplication requires 2/8 clock cycles for a 8 subword vector to complete. Because of the low duty cycle, the MU shares the local registers and the data connections with the LUTUs.

4.4.4.6 Look-Up Table Units

The LUTUs can be used to store data and control related constants. Examples for data-related constants are offset factors and threshold factors for comparisons. In this application, we use the LUTUs to store the instruction sequences that are needed for the computation of MVON operations. The required instruction sequences for the MVON operations of the list generator are different from the ones of the LLR generator. Thus, the sharing of a single LUT would require a frequent reloading of the LUT content. To avoid this, two LUTUs, which are denoted as LUTU1 and LUTU2, are employed. For the computation of the real part of a complex number a different LUTU2 content is required than for the computation of the imaginary part. Since each engine computes either part, the required storage is uniformly distributed across the engines. The LUTU1 and the

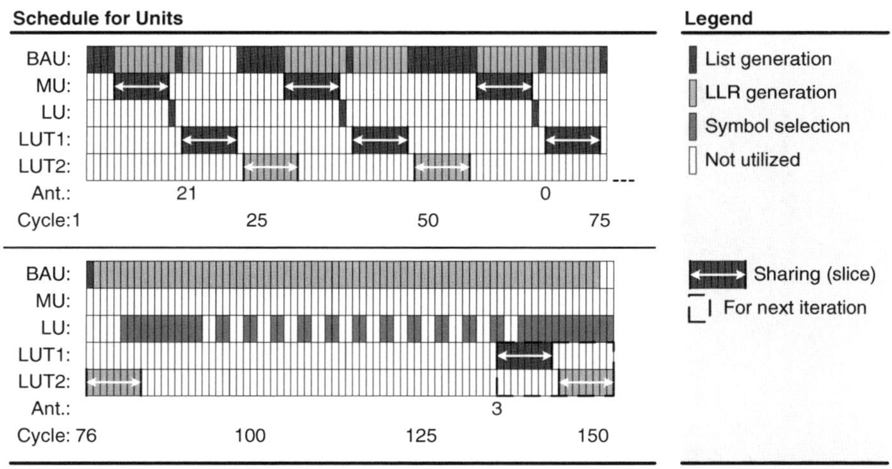

Fig. 4.8 Schedule for MIMO 4 × 4, 64-QAM, $m_{Nt} = 64$ for a slice. Only 1 iteration is shown. For a complete MIMO detection, 8 iterations are required. The BAU, which is the most important unit (it employs 8-way SIMD), is almost 100 % utilized. All slices/engines/clusters use the same schedule

LUTU2 have a size of 8 × 6 = 48 bits and 8 × 3 × 6 = 144 bits respectively (8 possibilities for a 3-bit multiplier value, 6 bits for encoding the shift-add/sub instruction). Like the MU, the LUTUs are multi-cycle units which process the subwords sequentially.

4.5 Software Mapping and Hardware Implementation

In this section the implementation results are shown and discussed.

4.5.1 Software Mapping and Scheduling

Figure 4.8 shows the schedule for 1 iteration of a MIMO 4 × 4 64-QAM detection for 1 slice. As it can be seen, the BAU is almost 100 % utilized. A few wait cycles are introduced because of the LUTU sharing. The latency of the MU, which was introduced because of sharing multipliers amongst slots, does not decrease the mapping efficiency. One iteration requires 153 clock cycles. For the complete MIMO detection, 8 iterations (64 constellation points/8 SIMD slots) are needed.

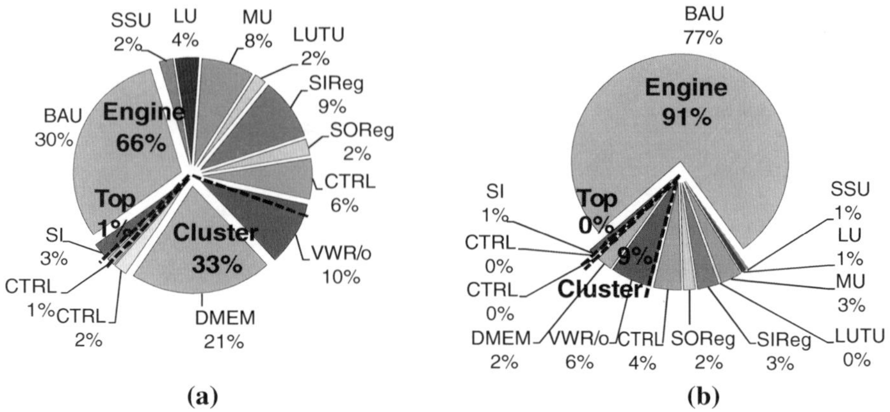

Fig. 4.9 Area and energy breakdown for an architecture instance with 13 clusters. The architecture instance is dimensioned for MIMO 4 $\times$ 4 soft-output near-ML 64-QAM detection. The top level control is shared across all clusters. SI/SOReg: Shared input/output registers of the secondary units. **a** Area breakdown. **b** Energy breakdown

4.5.2 Hardware Implementation and Results

The proposed architecture instance has been synthesized with Cadence RTL Compiler, placed and routed with Cadence Encounter. For the energy estimation, the activity of components and the extracted parasitics have been considered. The implementation results are based on TSMC 40 nm General Purpose (GP) nominal threshold-voltage standard cell technology. Because of employing only primitive operators, the architecture instance can operate at a high clock frequency of 800 MHz. Considering this clock frequency and the actual scheduling, the latency of a complete near-ML MIMO detection for 2/4/8 antennas is 0.5/1.5/5.4 μs respectively. Note, by employing more parallel slots, the latency can be reduced down to 0.07/0.2/0.7 μs. Based on 16 bit arithmetic (add, sub, mul.) and logic/comparison operations, a cluster offers a computational power of 40 Giga Operations Per Second (GOPS).

Figure 4.9a shows the area breakdown for an instance with 13 clusters, dimensioned for 200 Mbps throughput in MIMO 4×4 near-ML 64-QAM detection mode (same throughput as the ASIC reference, see Sect. 4.6.2.1). As it can be seen, the datapaths of the engines (=slices) contribute with about $66 - 6 = 60$ %, the DMEMs and the VWRs with about $21 + 10 = 31$ % and the complete control contributes with about 9 % to the total area. This breakdown indicates a high area efficiency since most of the area is consumed in the engines, which perform the actual computations. The contribution of the total control is rather low. The main reasons for this are (1) the employment of DLBs, which enable a "NOP compression", and (2) the sharing of the top level control across all clusters. Almost 100 % of the total area is consumed by the clusters. One cluster occupies an area of 0.07 mm^2, which corresponds to 98 kgates equivalent. A cluster offers a considerably high area efficiency of 571 GOPS/mm^2.

In Fig. 4.9b the corresponding energy breakdown is shown. The datapaths of the engines (BAU, MU, LU, etc.) consume more than 80 % of the total energy. The complete control contributes with less than 5 %. This results show that the energy is mostly spent on computations rather than on control or on data storage. In traditional ASIPs and DSIPs, the distribution of the energy is different. There, the control, the data storage and the interconnect consume typically more than 70 % of the total energy [7, 14]. A cluster consumes about 12 mW when fully utilized. This number translates to a high energy efficiency of 3.3 GOPS/mW.

4.5.3 Instance Scaling

As mentioned above, the achievable throughput scales with the number of instantiated clusters. Since clusters do basically not influence each other, the area and energy scales almost linearly. The DMEM size has to be dimensioned for the highest number of antennas that should be supported. The required DMEM sizes are mentioned in Sect. 4.4.1.

4.5.4 Run-Time Scaling

As mentioned in Sect. 4.2.2, a modern MIMO detector has to support a variety of different modes. In this work we consider the following run-time programmable parameters: (1) number of antennas (2, 4 or 8), (2) modulation scheme (16-QAM or 64-QAM), (3) output mode (hard or soft-output). Additionally we exploit the scalability of the algorithm which enables to trade-off BER performance and complexity [18]. The proposed architecture can be programmed for a BER performance that ranges from Successive Interference Cancellation (SIC) to near-ML.

Figure 4.10 shows the achievable throughput of a cluster for different modes. As it can be seen, almost a gap of factor 100 between the most demanding and the most relaxed mode exists. Hence, for a certain throughput, the required number of active clusters is strongly varying across modes. By enabling a power gating of clusters, inactive clusters can be turned-off and energy can be saved.

In Fig. 4.11 the dynamic power consumption for different modes is shown. The variation in dynamic power consumption is even higher than in throughput. In SIC mode, the power consumption is a factor 30 lower than in near-ML mode. Because of the following main reasons, the dynamic adaptability to the BER performance requirements seems feasible: (1) The interference level of the communication channel varies typically rather slow. Since the architecture is clocked at a high frequency, a substantial amount of computations can be completed before a

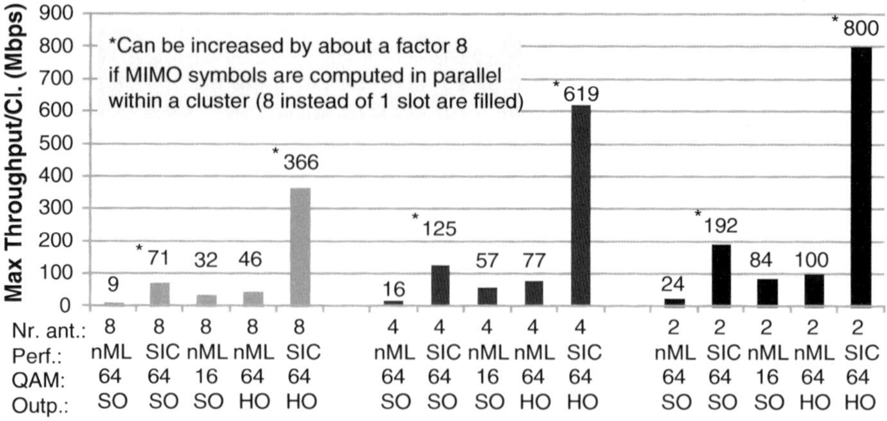

Fig. 4.10 Throughput of a cluster for different modes. In SIC mode, only 1 instead of m_{Nt} constellation points is evaluated. The Soft-Output (SO) computation contributes significantly to the total computational complexity

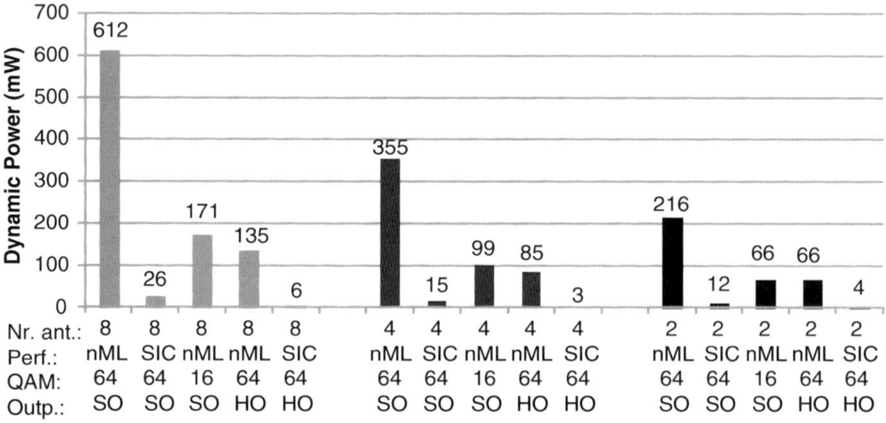

Fig. 4.11 Dynamic power consumption for different modes. All modes use only resources for achieving 600 Mbps throughput, i.e. clusters which are inactive, are powered-off. The instance was dimensioned to offer 600 Mbps in MIMO 8×8 near-ML 64-QAM soft-output mode. 68 clusters are present

potential switching would be required. (2) The architecture employs DLBs. A single DLB contains basically only the instructions for a single unit. Since the required instructions between modes, for a certain unit, differ only slightly, a fast switching is possible. Overall, this result strongly motivates the exploitation of a dynamic run-time manager [8, 44].

In the currently considered implementation a full enumeration of 16/64(-QAM) constellation points in near-ML mode is performed. For 64-QAM, this involves 8 iterations. A threshold-based stop-criterion could be introduced to further reduce the detection complexity [49]. In the proposed architecture, this feature can be supported with rather limited overhead. If this feature is exploited, additional energy savings of up to a factor 8 are possible. Since the architecture is programmable, also other algorithms can be mapped. However, because of the customization of the architecture instance, the algorithms need to have similar characteristics as the considered one.

4.6 Comparison and Discussion

In this section, we compare our proposal to an equivalent ASIC reference and to programmable processors and rASICs from literature.

4.6.1 Implemented ASIC References

In our previous work, two MIMO detector ASIC architectures, one for High-Throughput (HT) and one for Low-Area (LA), were proposed [16]. Both architectures, which can be instantiated for different antenna schemes, are shown in Fig. 4.12. The HT architecture is highly parallelized and pipelined. In contrast, the LA architecture performs iterative processing with a minimum number of ASIC datapath blocks. For throughputs above 65 Mbps, the HT architecture is more area and energy efficient than the LA architecture [18]. Since we target clearly throughputs above 65 Mbps, we select the HT architecture for a detailed comparison with the proposed DSIP architecture instance. To obtain accurate comparison results, we re-implement the ASIC by using the same technology as it is used for the DSIP architecture instance. Since the ASIC implementation and the proposed processor implementation are based on the same algorithm and on the same technology, a fair comparison is possible.

4.6.2 Comparison to ASIC References

4.6.2.1 Comparison of Different Instances

When instantiating the ASIC architecture, the supported number of antennas has to be fixed. We consider instances for 2, 4 and 8 antennas. For the proposed processor, we adjust the number of clusters so that the maximal achievable throughput corresponds to the one of the ASIC reference. Figure 4.13 shows the area and energy comparison for these 3 instances. As it can be seen, the proposed processor consumes about a factor 4 more area and a factor 2 more energy (near-ML mode). The gap between traditional programmable processors and ASICs is typically

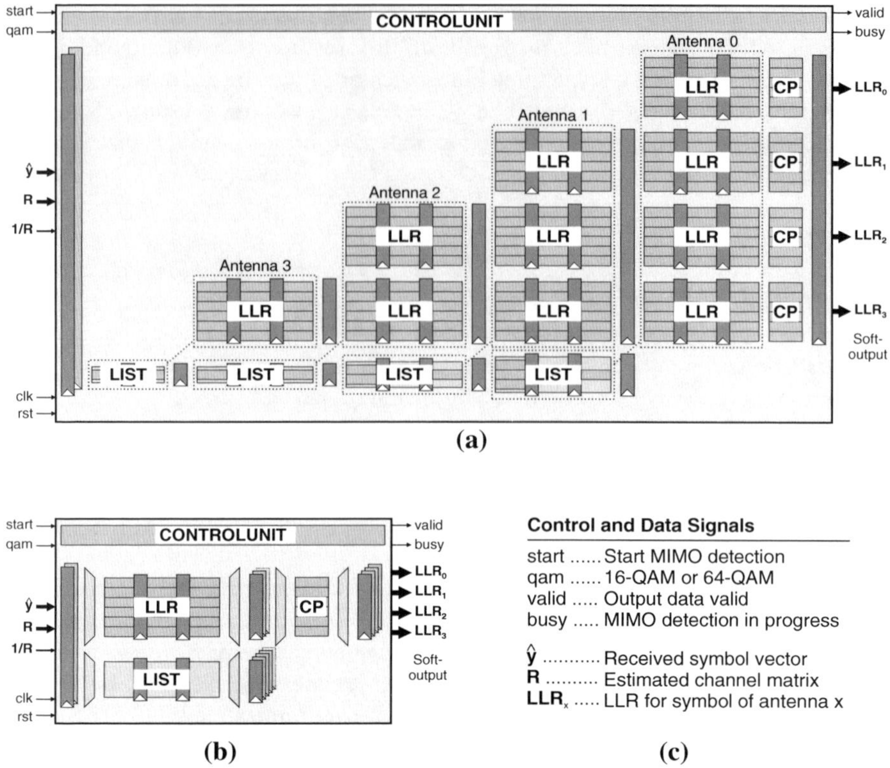

Fig. 4.12 Implemented MIMO detector ASICs from [16]: **a** high-throughput fully-parallel architecture, **b** low-area iterative architecture. Both ASIC architectures are based on the same algorithm combination as the proposed DSIP architecture. The figure shows the ASIC instances that are dimensioned for MIMO 4×4 and 16/64-QAM. The ASICs are based on list generator (LIST), LLR generator (LLR) and symbol comparison (CP) blocks. To obtain a higher throughput, the LLR and CP blocks perform a parallelization across q-bits ($q = 4/6$ for 16/64-QAM). **c** Legend

much higher [17, 23]. As shown in Sect. 4.5.4, when exploiting a run-time manager to dynamically adapt to the actual requirements, the average energy consumption of the proposal can be further reduced. For this reason, the proposed processor can potentially achieve the same or even a lower average energy consumption than the ASIC. Another interesting observation is that the processor scales similar as the ASIC. For traditional programmable architectures the scaling is often restricted by memory and/or interconnect bottlenecks.

4.6.2.2 Component Breakdown

In order to get an idea on where the difference in area originates, we show the component breakdown for the 4-antenna instances in Fig. 4.14. Interestingly, the

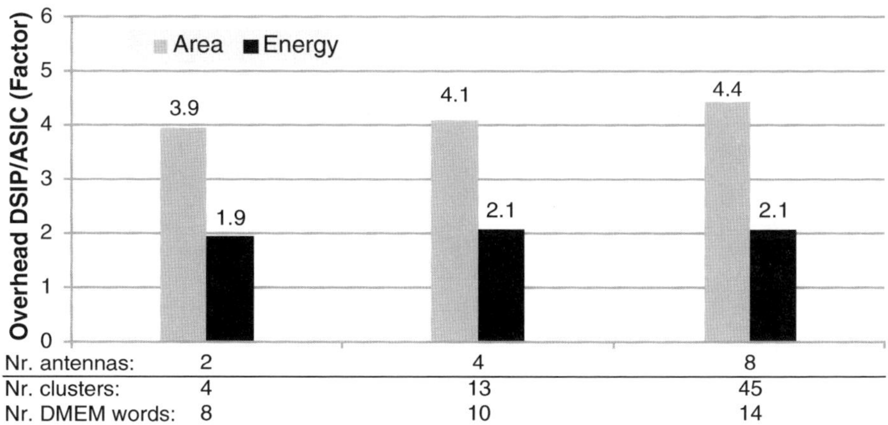

Fig. 4.13 Area and energy comparison for different ASIC/DSIP instances. The throughput and technology are normalized. Mode: near-ML, 64-QAM, soft-output

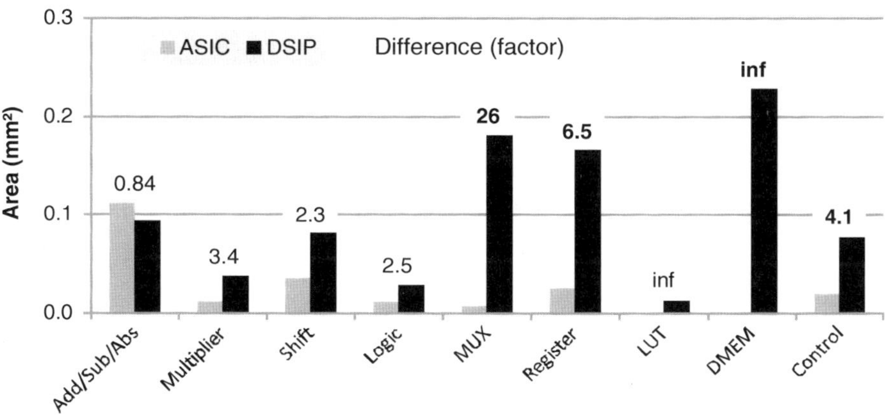

Fig. 4.14 Component breakdown of the ASIC/DSIP instances for MIMO 4 × 4, near-ML, 64-QAM, soft-output. The throughput and technology are normalized

area for the adder, subtractor and abs is a bit lower in the proposed processor. This is mostly linked to the multiplier choice and the consequence on pipelining and clock frequency. The main area difference originates from the DMEMs, the multiplexers, the registers and the control. The streaming-based ASIC architecture does not include DMEMs because (1) for the I/O-buffering, it is assumed that the adjacent hardware provides/consumes the I/O-data at the right moment in time (e.g. a new input is provided at every 64 clock cycles) and (2) the intermediate values are stored in pipeline registers instead of in DMEMs. The overhead in multiplexers is caused by the enabling of flexibility. In the ASIC, all registers, i.e. pipeline registers as well as I/O-registers, are always fully utilized. This is

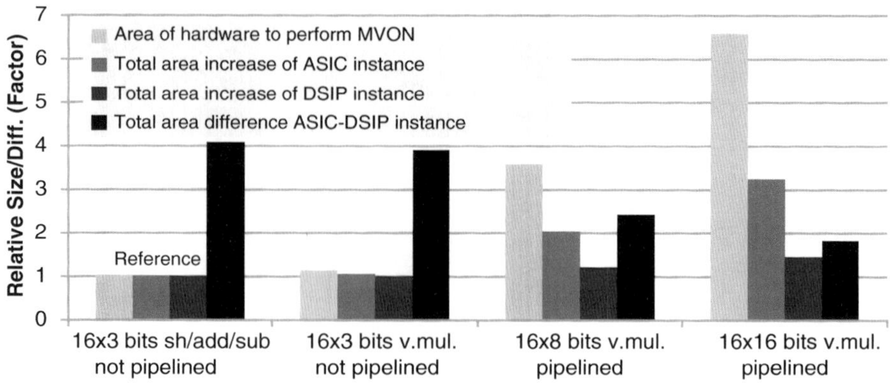

Fig. 4.15 Influence of the implementation choice for MVON on area. For all variants the same throughput is considered. In the reference, MVON contribute to 40/8 % of the total ASIC/DSIP instance area. Since the relative contribution is higher for the ASIC, the implementation choice has a higher influence. Results are based on instances for MIMO 4×4

possible, because the temporal locality of data can be fully exploited in a pipelined architecture. However, this is not feasible for programmable architectures since they require some freedom for the register allocation. For instance, in the proposed architecture, the local registers of the MU and the LU are essential but they cannot be as efficiently used as conventional pipeline registers. The difference in control is comparatively low because of 2 main reasons: (1) The ASIC implements MVON operations in a similar way as in the proposed processor and therefore a certain amount of control hardware is needed. Instead of LUTs, the ASIC uses combinatorial logic to generate the predefined instruction sequences. This logic is included in the control whereas the LUTs of the proposed processor are considered separately. (2) In the proposal, the control is efficiently implemented with DLBs.

Because the considered ASIC architecture is designed for maximal throughput in a highly parallel and pipelined way, it does not require hardware (e.g. multiplexers and storage) for iterative processing. Therefore the area efficiency is very high. Nevertheless, an ASIC, which targets lower throughputs and/or which employs an algorithm with feedback, would have a lower area efficiency because it would either be less utilized or it would have an architecture which includes iterative processing. The MIMO detector comparison of the HT architecture, i.e. fully pipelined processing, and the LA architecture, i.e. iterative processing, shows that the latter consumes, for the same throughput and for 2×2 64-QAM, a factor 2 more area [16]. For more complex configurations, e.g. 8×8 64-QAM, the difference is even a factor 4. Thus, the comparison in terms of area efficiency of our proposal with the LA architecture results in an difference of factor 2 and less.

4.6.2.3 Influence of MVON Implementation Choice

A major algorithm transformation was the replacement of multipliers with a combination of shift, add/sub operators and control logic/LUTs for the MVON operation type. This optimization was exploited in both architecture styles to increase the efficiency. Figure 4.15 shows the impact of this optimization on the total area. When conventional 16-bit multipliers are used, the total area of the ASIC is 3× higher. The total area of the ASIC is mostly occupied by arithmetic and logic operators and therefore this optimization has a big impact. In case of the proposed processor, where a large part of the area is consumed by registers, multiplexers and the DMEMs, the impact is lower. Thus, when comparing the unoptimized versions of both architecture styles, the total area gap is reduced to less than a factor 2.

4.6.3 Flexible Implementations from Literature

To further evaluate the efficiency of the proposal, we compare the proposal to programmable and configurable implementations from literature. Most implementations in literature focus on MIMO 2×2 and MIMO 4×4, therefore we use these antenna schemes for comparison. To our best knowledge, the implementations which focus on higher antenna schemes are all based on the ASIC [9, 28] or on the rASIC [32] architecture styles. Thus, we assume that this book shows the first processor implementation for antenna schemes above 4×4. In the following we provide a summary of the selected references.

4.6.3.1 Programmable Processors

We denote the MIMO detector implementation with a TI TMS320C6416 VLIW SIMD DSP processor [17] as reference DSP1. Beside the Viterbi and the Turbo decoder co-processors, the DSP includes 8 parallel FUs which are organized in 2 clusters [53]. The DSP supports 4-way SIMD and operates at 800 MHz clock frequency. In the work of [17], we have mapped the near-ML SSFE algorithm for MIMO 2×2 64-QAM onto this processor.

We refer to the MIMO detector mapping onto the Sandbridge SB3500 SDR processor [23] as SDR1. The processor includes an ARM sub-system, 3 cores, supports SIMD and operates at 600 MHz in 65 nm technology [41]. The K-Best algorithm for MIMO 2×2 has been mapped.

SDR2 refers to the ADRES CGA VLIW MIMO detector implementation of [3]. The ADRES employs 8 FUs and operates at 600 MHz in 65 nm technology. Each FU supports 8-way SIMD. The Scalable near-ML hard-output Block-based Parallel LR (SBP-LR) algorithm for MIMO 4×4 64-QAM has been mapped.

With SDR3, we refer to the ADRES-based implementation of [17]. We have evaluated different array and SIMD sizes for the MIMO 2×2 SSFE algorithm mapping. An ADRES instance with 8 CGA and 3 VLIW FUs, all with support for 16-way SIMD, offers the best utilization (85 % scheduling density). Because the throughput of the default ADRES for this application is low, 4 ASIs have been designed and added to the instruction set. The datapaths of the ASIs correspond to the datapaths of the implemented ASIC blocks. The 16-QAM and 64-QAM modulation schemes with near-ML BER performance have been considered.

ASIP1 refers to the MIMO detector implementation of [57]. The multi-core SIMD Floating Point Processor (FPP) includes MIMO pre-processing and operates at 400–500 MHz in 65 nm technology. The soft-output 4×4 MMSE algorithm with 64-QAM modulation has been mapped.

ASIP2 is a further FPP which implements near-ML hard-output MIMO detection for 4×4 [56]. In contrast to ASIP1, this implementation leverages on the LR-Aided MMSE (LRA-MMSE) algorithm with 16-QAM modulation. The ASIP operates at 400 MHz in 65 nm technology.

We refer to the Adaptive Stream Processing Engine (ASPE), which was customized for MMSE 2×2 MIMO detection in [15], as ASIP3. The ASPE is a modular coarse-grained data-processing architecture. It mainly consists of FUs and Storage Units (SU)s, which have an algorithm-specific interconnection. The architecture in [15] includes 4 FUs and 4 SUs, which all employ 2-way SIMD on 16-bit data. The processor operates at 250 MHz in 0.18 μm technology.

ASIP4-7 are based on the Transport Triggered Architecture (TTA) [12]. ASIP4 refers to the MIMO detector implementation of [23]. The near-ML K-Best algorithm, with $K = 8/16$ for MIMO 2×2 has been mapped. The TTA operates at 280 MHz.

ASIP5 implements the hard-output near-ML 2×2 K-Best algorithm, with $K = 16$ and 64-QAM, on the TTA [4]. The processor operates at 280 MHz in 0.13 μm technology.

ASIP6 implements the SSFE algorithm with algorithm-specific FUs in 0.13 μm technology [24]. The mapping onto the TTA for 2×2 16-QAM and 4×4 64-QAM are shown.

The reference ASIP7, which has been described in LisaTek, employs the hard-output MMSE algorithm [22]. MIMO 2×2 and 4×4, QPSK, 16-QAM and 64-QAM are supported.

4.6.3.2 Reconfigurable ASICs

rASIC1 refers to the implementation of [5]. It employs the near-ML soft-output Layered ORthogonal Lattice Detection (LORD) algorithm for MIMO 4×4. The implementation supports varies modulation schemes, i.e. QPSK to 64-QAM, and operates at 500 MHz in 45 nm technology.

We denote the MIMO detector implementation of [34] as rASIC2. The implementation supports 2×2 and 4×4, 16-QAM and 64-QAM, and operates at 137 MHz in 0.13 μm technology. It leverages on the near-ML hard-output K-Best algorithm with early pruning technique.

The reference rASIC3 implements the Modified Fixed-Complexity Soft-Output (MFCSO) and the MMSE algorithm [58]. The MFCSO offers approximately near-ML, the MMSE offers approximately SIC BER performance. For MIMO 2×2, the implementation supports QPSK to 64-QAM modulation schemes. In 65 nm, the rASIC can operate at 300 MHz clock frequency.

4.6.4 Comparison to Flexible Implementations

For the comparison, we scale all implementations to 40 nm technology. Thereby we use the following factors to scale between two technology nodes [18, 24]: 1.2 for clock frequency/throughput, 2 for area (if not already specified in gates equivalent), 1.5 for energy/power. For each reference, we dimension a DSIP architecture instance which offers the best match. Thereby BER performance, throughput, number of antennas, modulation scheme and output-mode are considered. In general, more pessimistic assumptions are made, e.g. the proposed DSIP offers SIC performance while the reference offers only MMSE performance. We assume that our proposal utilizes all slots in SIC mode.

4.6.4.1 Area Comparison

Table 4.1 compares the area of different MIMO 2×2 and 4×4 implementations with our proposal. Compared to other programmable processors, the proposal offers an area efficiency which is more than a factor 2 higher. Especially when compared to DSPs and SDR baseband processors, such as the ADRES or the Sandbridge SB3500, the difference is substantial. rASICs, which are fixed to a certain algorithm, are generally more area efficient than the proposal. To get an idea on the bounds, we define a pessimistic and an optimistic case. In the pessimistic case, the proposal offers near-ML BER performance while in the optimistic case it offers only SIC BER performance. As it can be seen in Table 4.1, this distinction shows a big difference.

4.6.4.2 Energy Comparison

Compared to the reference SDR3, which represents an ADRES processor with ASIs and a mapping of the SSFE algorithm [17], our proposal is about a factor $3\times$ more energy efficient. ASIP6, the 16-bit fixed-point TTA processor from [24], employs the same algorithm as in our proposal. The comparison with normalized m-parameter, i.e. BER performance/number of considered constellation points, shows, that our proposal is more energy efficient. The difference for both, 2×2 16-QAM and 4×4 64-QAM, is at least a factor 2.

Table 4.1 Coarse relative area comparison with flexible implementations from literature

Ref. 1	Ref. 2	Architecture	Algorithm	Antenna	Area Diff.
Programmable processors					
DSP1	[17]	TI TMS320C6416 VLIW	SSFE[a]	2×2	$> 20 \downarrow$
SDR1	[23]	Sandbridge SB3500	K-Best	2×2	$> 10 \downarrow$
SDR2	[3]	ADRES	SBP-LR[b]	4×4	$> 10 \downarrow$
SDR3	[17]	ADRES+ASI	SSFE[a]	2×2	$3 \downarrow$
ASIP1	[57]	FPP	MMSE[c]	4×4	$> 2 \downarrow$
ASIP2	[56]	FPP	LRA-MMSE[d]	4×4	$> 5 \downarrow$
ASIP3	[15]	ASPE	MMSE[c]	2×2	$10 \downarrow$
ASIP4	[23]	TTA	K-Best	2×2	$> 2 \downarrow$
ASIP5	[4]	TTA	K-Best	2×2	$> 2 \downarrow$
ASIP6	[24]	TTA	SSFE[a]	4×4	$> 2 \downarrow$
ASIP7	[22]	TTA	MMSE[c]	4×4	$> 3 \downarrow$
Reconfigurable ASICs					
rASIC1	[5]	-	LORD[e]	4×4	$15 \uparrow ..3 \downarrow$
rASIC2	[34]	-	K-Best	4×4	$5 \uparrow ..12 \downarrow$
rASIC3	[58]	-	MFCSO[f]	2×2	$11 \uparrow ..5 \downarrow$

Technology and throughput are normalized. For each reference, a DSIP instance, which matches in terms of BER performance, number of antennas, modulation scheme and output-mode, is considered. The proposed DSIP consumes roughly factor $\downarrow$ (less) / $\uparrow$ (more) area compared to reference. Range indicates BER performance corners (near-ML..SIC)

[a] *SSFE* Selective spanning with fast enumeration, [b] *SBP-LR* Scalable block-based parallel lattice reduction, [c] *MMSE* Minimum mean-squared error, [d] *LRA-MMSE* Lattice-reduction-aided MMSE, [e] *LORD* Layered orthogonal lattice detection, [f] *MFCSO* Modified Fixed-Complexity Soft-Output.

4.6.4.3 Reasons for Higher Efficiency

Mainly because of the following reasons, the proposal is more efficient than other programmable architectures: (1) highly-optimized algorithm (low-level operations and memory optimization, well parallelizable), (2) architecture that can exploit the low-level operations efficiently (time-multiplexing of primitive datapath operators), which results in a high hardware utilization, (3) efficient instruction control implementation (distributed local control instead of conventional IMEM), (4) register and DMEM optimizations (few local registers and VWRs instead of conventional register file), (5) implementation of multiplications (few shift-add/sub operations instead of a potentially under-utilized hardware multiplier) and (6) optimized interconnect (to reduce multiplexer hardware).

4.7 Conclusions

This chapter presents an energy-efficient programmable MIMO detector for emerging 4G communication systems. The scalable and reusable design is based on the proposed DSIP architecture template of Chap. 3. Innovative architecture concepts with algorithm-architecture co-optimizations are combined for achieving high efficiency. The implementation results show that the proposed MIMO detector is able to achieve a high area efficiency of 571 GOPS/mm^2 and a high energy efficiency of 3.3 GOPS/mW in TSMC 40 nm technology. For the most-demanding environmental conditions, the proposal consumes about a factor 2 more energy than a fully-pipelined ASIC equivalent. However, for typical environmental conditions, the proposal can potentially out-perform the ASIC in energy-efficiency. This result strongly motivates the use of programmable architectures for dynamically-varying battery-powered application domains. Mainly due to the overhead in multiplexers, registers and DMEMs, the area efficiency of the proposal is 4$\times$ lower compared to a fully-pipelined ASIC. However, compared to an ASIC with iterative architecture, the gap is significantly smaller. Compared to other programmable MIMO detector implementations, the proposal is at least a factor 2 more energy and area efficient. This result confirms the effectiveness of the applied design concepts.

Future work should further evaluate the mapping of other MIMO detection algorithms onto the proposed design and on the run-time scalability aspect.

References

1. Adeva, E.P., Shah, M.A., Mennenga, B., Fettweis, G.: VLSI architecture for soft-output tuple search sphere decoding. In: IEEE Workshop on Signal Processing Systems (SIPS), pp. 222–227 (2011). doi:10.1109/SiPS.2011.6088978
2. Ahmad, U., Li, M., Pollin, S., Fasthuber, R., Van der Perre, L., Catthoor, F.: Bounded block parallel lattice reduction algorithm for MIMO-OFDM and its application in LTE MIMO receiver. In: IEEE Workshop on Signal Processing Systems (SIPS), pp. 168–173. IMEC (2010). doi:10.1109/SIPS.2010.5624784
3. Ahmad, U., Amin, A., Li, M., Pollin, S., Van der Perre, L., Catthoor, F.: Scalable block-based parallel lattice reduction algorithm for an SDR baseband processor. In: IEEE International Conference on Communication (ICC). IMEC (2011)
4. Antikainen, J., Salmela, P., Silveny, O., Juntti, M., Takala, J., Myllyla, M.: Fine-grained application-specific instruction set processor design for the K-best list sphere detector algorithm. In: International Conference on Embedded Computer Systems (SAMOS), 1, pp. 108–115. IEEE (2008). doi:10.1109/ICSAMOS.2008.4664853
5. Bhagawat, P., Dash, R., Choi, G.: Dynamically reconfigurable soft output MIMO detector. In: International Conference on Computer Design (ICCD), pp. 68–73 (2008). doi:10.1109/ICCD.2008.4751842
6. Bhagawat, P., Dash, R., Choi, G.: Array like runtime reconfigurable MIMO detectors for 802.11n WLAN: A design case study. In: Asia and South Pacific Design Automation Conference (DAC-ASP), pp. 751–756 (2009). doi:10.1109/ASPDAC.2009.4796570
7. Bougard, B., De Sutter, B., Rabou, S., Novo, D., Allam, O., Dupont, S., Van der Perre, L.: A Coarse-Grained array based Baseband Processor for 100Mbps+ Software defined radio. In: Design, Automation and Test in Europe (DATE), pp. 716–721. IMEC (2008). doi:10.1109/DATE.2008.4484763
8. Bougard, B., Novo, D., Van der Perre, L., Catthoor, F.: Bridging the energy gap in size, weight and power constrained software defined radio: Agile baseband processing as a key

enabler. In: IEEE International Conference on Acoustics, Speech and Signal Processing (ICASSP) pp. 5384–5387 (2008). doi:10.1109/ICASSP.2008.4518877
9. Cerato, B., Viterbo, E.: Hardware implementation of a low-complexity detector for large MIMO. In: IEEE International Symposium on Circuits and Systems (ISCAS), pp. 593–596 (2009). doi:10.1109/ISCAS.2009.5117818
10. Chien-Jen, H., Chung-Wen, Y., Hsi-Pin, M.: A power-efficient configurable low-complexity MIMO detector. IEEE Trans. Circuits Syst. (TCAS) **56**(2), 485–496 (2009). doi:10.1109/TCSI.2008.2001368
11. Chu, X., McAllister, J.: FPGA based soft-core SIMD processing: A MIMO-OFDM fixed-complexity sphere decoder case study. In: International Conference on Field-Programmable Technology (FPT), pp. 479–484 (2010). doi:10.1109/FPT.2010.5681463
12. Corporaal, H.: Microprocessor Architectures: From VLIW to TTA. Wiley, Chichester (1998)
13. Cupaiuolo, T., Siti, M., Tomasoni, A.: Low-complexity high throughput VLSI architecture of soft-output ML MIMO detector. In: Design, Automation and Test in Europe (DATE), pp. 1396–1401 (2010)
14. Dally, W.J., Balfour, J., Black-Shaffer, J.C., Harting, R.C., Parikh, V., Park, J., Sheffield, D.: Efficient embedded computing. Computer **41**(7), 27–32 (2008)
15. Eberli, S., Burg, A., Fichtner, W.: Implementation of a 2x2 MIMO-OFDM receiver on an application specific processor. Microelectron J **40**(11), 1642–1649 (2009). doi:10.1016/j.mejo.2009.02.005
16. Fasthuber, R., Li, M., Novo, D., Van Der Perre, L., Catthoor, F.: Novel energy-efficient scalable soft-output SSFE MIMO detector architectures. In: International Conference on Embedded Computer Systems (IC-SAMOS). IMEC (2009). doi:10.1109/ICSAMOS.2009.5289228
17. Fasthuber, R., Li, M., Novo, D., Van Der Perre, L., Catthoor, F.: Exploration of soft-output MIMO detector implementations on massive parallel processors. J. Sig. Process. Syst. (JSPS) **64**(1), 75–92 (2010). doi:10.1007/s11265-010-0499-0
18. Fasthuber, R., Li, M., Novo, D., Raghavan, P., Van der Perre, L., Catthoor, F.: Energy-efficient run-time scalable soft-output SSFE MIMO detector architectures. Trans. High-Perform. Embed. Architect. Compilers (HiPEAC), (Special Issue SAMOS 2009) **5**(3), 1–20 (2011)
19. Gestner, B., Zhang, W., Ma, X., Anderson, D.V.: Lattice reduction for MIMO detection: from theoretical analysis to hardware realization. IEEE Trans. Circuits Syst. (TCAS) **58**(4), 813–826 (2011). doi:10.1109/TCSI.2010.2078670
20. Ghosh, A., Ratasuk, R., Mondal, B., Mangalvedhe, N., Thomas, T.: LTE-advanced: next-generation wireless broadband technology. IEEE Wireless Commun. **17**(3), 10–22 (2010)
21. Heidmann, N., Wiegand, T., Paul, S.: Architecture and FPGA-implementation of a high-throughput K+-Best detector. In: Design, Automation and Test in Europe (DATE) (2011)
22. Jafri, A., Karakolah, D., Baghdadi, A., Jezequel, M.: ASIP-based flexible MMSE-IC linear equalizer for MIMO turbo-equalization applications. In: Design, Automation and Test in Europe (DATE), pp. 1620–1625 (2009)
23. Janhunen, J., Silvén, O., Juntti, M.: Programmable processor implementations of K-best list sphere detector for MIMO receiver. Sig. Process. **90**(1), 313–323 (2010). doi:10.1016/j.sigpro.2009.06.026
24. Janhunen, J., Pitkanen, T., Silven, O., Juntti, M.: Fixed- and floating-point processor comparison for MIMO-OFDM detector. IEEE J. Sel. Top. Sig. Process. (JSTSP) **5**(8), 1588–1598 (2011). doi:10.1109/JSTSP.2011.2165830
25. Jung, Y., Kim, J.: Low-power low-complexity MIMO-OFDM baseband processor for wireless LANs. In: IEEE International Symposium on Circuits and Systems (ISCAS), pp. 601–604 (2009). doi:10.1109/ISCAS.2009.5117820
26. Kim, T.H., Park, I.C.: Small-area and low-energy K-best MIMO detector using relaxed tree expansion and early forwarding. IEEE Trans. Circuits Syst. (TCAS) **57**(10), 2753–2761 (2010). doi:10.1109/TCSI.2010.2046249

27. Kim, E.P., Shanbhag, N.R.: An energy-efficient multiple-input multiple-output (MIMO) detector architecture. In: IEEE Workshop on Signal Processing Systems (SIPS), pp. 239–244 (2011). doi:10.1109/SiPS.2011.6088981
28. Knagge, G., Bickerstaff, M., Ninness, B., Weller, S.R., Woodward, G.: A VLSI 8x8 MIMO near-ML decoder engine. In: IEEE Workshop on Signal Processing Systems (SIPS), pp. 387–392 (2006). doi:10.1109/SIPS.2006.352614
29. Li, M., Bougard, B., Xu, W., Novo, D., Van Der Perre, L., Catthoor, F.: Optimizing near-ML MIMO detector for SDR baseband on parallel programmable architectures. In: Design, Automation and Test in Europe (DATE), pp. 444–449 (2008). doi:10.1109/DATE.2008.4484721
30. Li, M., Novo, D., Bougard, B., Naessens, F., Van Der Perre, L., Catthoor, F.: An implementation friendly low complexity multiplierless LLR generator for soft MIMO sphere decoders. In: IEEE Workshop on Signal Processing Systems (SIPS), pp. 118–123 (2008). doi:10.1109/SIPS.2008.4671748
31. Li, M.: Algorithm and architecture co-design for software defined radio baseband. Ph.D. thesis, KU Leuven (2010)
32. Liao, C.H., Wang, T.P., Chiueh, T.D.: A 74.8 mW soft-output detector IC for 8x8 spatial-multiplexing MIMO communications. IEEE J. Solid-State Circuits (JSSC) **45**(2), 411–421 (2010). doi:10.1109/JSSC.2009.2037292
33. Liao, C.F., Huang, Y.H.: Power-saving 4x4 lattice-reduction processor for MIMO detection with redundancy checking. IEEE Trans. Circuits Syst. (TCAS) **58**(2), 95–99 (2011). doi:10.1109/TCSII.2010.2104013
34. Liu, L., Ye, F., Ma, X., Zhang, T., Ren, J.: A 1.1-Gb/s 115-pJ/bit configurable MIMO detector using 0.13um CMOS technology. IEEE Trans. Circuits Syst. (TCAS) **57**(9), 701–705 (2010). doi:10.1109/TCSII.2010.2058494
35. Liu, T.H., Jiang, J.Y., Chu, Y.S.: A low-cost MMSE-SIC Detector for the MIMO system: algorithm and hardware implementation. IEEE Trans. Circuits Syst. (TCAS) **58**(1), 56–61 (2011). doi:10.1109/TCSII.2010.2092819
36. Luo, D., Tsui, C.Y.: A hybrid algorithm and its re-configurable architecture for MIMO detector. In: IEEE International Midwest Symposium on Circuits and Systems (MWSCAS), pp. 297–300 (2009). doi:10.1109/MWSCAS.2009.5236095
37. Mahdavi, M., Shabany, M.: Ultra high-throughput architectures for hard-output MIMO detectors in the complex domain. In: IEEE International Midwest Symposium on Circuits and Systems (MWSCAS), pp. 1–4 (2011). doi:10.1109/MWSCAS.2011.6026425
38. Moezzi-Madani, N., Thorolfsson, T., Chiang, P., Davis, W.R.: Area-efficient antenna-scalable MIMO detector for K-best sphere decoding. J. Sig. Process. Syst. (JSPS) pp. 1–12 (2011). doi:10.1007/s11265-011-0595-9
39. Mohammed, K., Daneshrad, B.: A MIMO decoder accelerator for next generation wireless communications. IEEE Trans. Very Large Scale Integrated Systems (TVLSI) **18**(11), 1544–1555 (2010). doi:10.1109/TVLSI.2009.2025590
40. Mohammed, K., Mohamed, M.I.A., Daneshrad, B.: A parameterized programmable MIMO decoding architecture with a scalable instruction set and compiler. IEEE Trans. Very Large Scale Integrated Systems (TVLSI) **19**(8), 1485–1489 (2011). doi:10.1109/TVLSI.2010.2049592
41. Moudgill, M., Glossner, J., Agrawal, S., Nacer, G.: The sandblaster 2.0 architecture and SB3500 implementation. In: SDR Forum. Sandbridge Technologies (2008)
42. Patel, D., Smolyakov, V., Shabany, M., Gulak, P.G.: VLSI implementation of a WiMAX/LTE compliant low-complexity high-throughput soft-output K-Best MIMO detector. In: IEEE International Symposium on Circuits and Systems (ISCAS), pp. 593–596 (2010). doi:10.1109/ISCAS.2010.5537524
43. Pirsch, P.: Architectures for Digital Signal Processing. Wiley, New York (1998)
44. Portero, A., Talavera, G., Moreno, M., Carrabina, J., Catthoor, F.: Methodology for energy-flexibility space exploration and mapping of multimedia applications to single-processor

platform Styles. IEEE Trans. Circuits Syst. Video Technol. (TCSVT) **21**(8), 1027–1039 (2011). doi:10.1109/TCSVT.2011.2129750
45. Raghavan, P., Lambrechts, A., Jayapala, M., Catthoor, F., Verkest, D., Corporaal, H.: Very wide register: an asymmetric register file organization for low power embedded processors. In: Design, Automation and Test in Europe (DATE). IMEC (2007)
46. Raghavan, P., Lambrechts, A., Jayapala, M., Catthoor, F., Verkest, D.: Distributed loop controller for multi-threading in uni-threaded ILP architectures. IEEE Trans. Comput. (TC) **58**(3), 311–321 (2009)
47. Shabany, M., Gulak, P.G.: A 675 Mbps, 4x4 64-QAM K-Best MIMO Detector in 0.13m CMOS. IEEE Trans. Very Large Scale Integr. Syst. (TVLSI) **20**(1), 135–147 (2010). doi:10.1109/TVLSI.2010.2090367
48. Shiun Lin, J., Hao Fang, S., Hsuan Jen, Y., Der Shieh, M.: Design of high-throughput MIMO detectors using sort-free and early-pruned techniques. In: IEEE TENCON Conference, pp. 1513–1516 (2010). doi:10.1109/TENCON.2010.5686145
49. Studer, C., Burg, A., Bolcskei, H.: Soft-output sphere decoding: algorithms and VLSI implementation. IEEE J. Sel. Areas Commun. (JSAC) **26**(2), 290–300 (2008). doi:10.1109/JSAC.2008.080206
50. Studer, C., Fateh, S., Seethaler, D.: A 757Mb/s 1.5 mm2 90nm CMOS soft-input soft-output MIMO detector for IEEE 802.11n. In: ESSCIRC, pp. 530–533 (2010). doi:10.1109/ESSCIRC.2010.5619760
51. Sun, Y., Cavallaro, J.R.: High-throughput soft-output MIMO detector based on path-preserving trellis-search algorithm. IEEE Trans. Very Large Scale Integr. Syst. (TVLSI) **20**(7), 1235–1247 (2012). doi:10.1109/TVLSI.2011.2147811
52. Tae-Hwan, K., In-Cheol, P.: A 2.6Gb/s 1.56mm^2 near-optimal MIMO detector in 0.18m CMOS. In: IEEE Custom Integrated Circuits Conference (CICC), pp. 1–4 (2010). doi:10.1109/CICC.2010.5617463
53. Texas Instruments: TMS320C6416 Datasheet, Rev. M (2009). http://www.ti.com/
54. Tsai, P.Y., Chen, W.T., Lin, X.C., Huang, M.Y.: A 4x4 64-QAM reduced-complexity K-best MIMO detector up to 1.5Gbps. In: IEEE International Symposium on Circuits and Systems (ISCAS), pp. 3953–3956 (2010). doi:10.1109/ISCAS.2010.5537675
55. Witte, E.M., Borlenghi, F., Ascheid, G., Leupers, R., Meyr, H.: A Scalable VLSI architecture for soft-input soft-output single tree-search sphere decoding. IEEE Trans. Circuits Syst. (TCAS) **57**(9), 706–710 (2010). doi:10.1109/TCSII.2010.2056014
56. Wu, D., Eilert, J., Liu, D.: A programmable lattice-reduction aided detector for MIMO-OFDMA. In: International Conference on Circuits and Systems for Communications (ICCSC), pp. 293–297 (2008). doi:10.1109/ICCSC.2008.68
57. Wu, D., Eilert, J., Liu, D.: Implementation of a high-speed MIMO soft-output symbol detector for software defined radio. J. Sig. Process. Syst. (JSPS) (2009). doi:10.1007/s11265-009-0369-9
58. Wu, D., Eilert, J., Asghar, R., Liu, D.: VLSI Implementation of a fixed-complexity soft-output MIMO detector for high-speed wireless. J. Wireless Commun. Netw. (EURASIP) **2010**, 1–14 (2010). doi:10.1155/2010/893184
59. Wu, M., Sun, Y., Gupta, S., Cavallaro, J.R.: Implementation of a high throughput soft MIMO detector on GPU. J. Sig. Process. Syst. (JSPS) **64**(1), 123–136 (2011). doi:10.1007/s11265-010-0523-4
60. Yoshizawa, S., Ikeuchi, H., Miyanaga, Y.: Scalable pipeline architecture of MMSE MIMO detector for 4x4 MIMO-OFDM receiver. In: IEEE International Symposium on Circuits and Systems (ISCAS), pp. 2534–2537 (2010). doi:10.1109/ISCAS.2010.5537108

Chapter 5
Case Study 2: DSIP Architecture Instances for FIR Filtering

5.1 Motivation, Related Work and Contributions

This section motivates this case study, summarizes related work on digital 60 GHz baseband implementations and mentions the main contributions.

5.1.1 Context and Motivation

The 60 GHz band, ranging from 57 to 65 GHz, offers 5 GHz unlicensed bandwidth in all major markets across the globe. Consequently, this great potential for enabling very high data rate short-range wireless applications has inspired academia and industry to develop 60 GHz technology. Key applications for this technology are video and computer display streaming, e.g. uncompressed high definition content, portable point-to-point file transfers, e.g. fast video download from a data kiosk, and high-speed Wireless Personal/Local Area Networks (WPAN/WLAN)s [14]. Many 60 GHz wireless communication standards, partly differing in the targeted application, have been developed [26]. Amongst them, the IEEE 802.15.3c (WPAN) [4], the IEEE 802.11ad (WLAN) [30], the WirelessHD [41], the WiGig [14] and the ECMA-387 [40] standard.

The standards specify data rates from 1 Giga-bits per second (Gbps) up to 7 Gbps and tight latency requirements of 3–6 μs. These specifications result in very high complexity, which challenge both, the analog and the digital PHYsical layer (PHY) design. For instance, the digital baseband implementation has to offer a computational performance of more than 400 Giga Operations Per Second (GOPS), which is at least an order of magnitude higher than that of current 3G and 4G mobile receivers [8, 45].

Nevertheless, even for mobile devices that implement 60 GHz functionality, the available power budget for the complete wireless transceiver (analog and digital) is equal or less than 1 W [45]. For this reason, 60 GHz implementations need to be highly energy efficient.

R. Fasthuber et al., *Energy-Efficient Communication Processors*,
DOI: 10.1007/978-1-4614-4992-8_5,

In addition, an increasing need for more flexible and reusable implementations exists. The flexibility/reusability is mainly required (1) to support multiple standards, e.g. IEEE 802.15.3c (60 GHz) and LTE (2.4 GHz), (2) to support multiple modes, e.g. Single Carrier (SC) and Orthogonal Frequency-Division Multiplexing (OFDM), and (3) to enable high volumes, e.g. one common design is re-programmed to fulfill the local specifications (e.g. frequency band) of different markets. The latter is especially essential for advanced technologies, for which the Non-Recurring Engineering (NRE) costs, such as mask and design costs, are very high [2]. Thus, successful 60 GHz implementations have to combine high performance, high energy efficiency, high flexibility and high reusability. In addition, the designs need to be technology scaling-friendly.

5.1.2 Summary of Related Work

Previous work on 60 GHz implementations has mostly been focusing on the Analog Front-End (AFE) [3, 25, 39, 43] and on the Analog-to-Digital Converter (ADC) [47]. Several implementation results of the digital outer modem, such as [7, 23], have been reported. The existing literature on the implementation of the digital inner modem, such as [9, 28, 31], is either based on FPGAs or on ASICs. However, both architecture styles are not well-suited for the employment in multi-standard mobile devices. FPGAs can clearly not offer the required energy efficiency. ASICs are energy efficient, but the huge market volume, which would be required for the cost-effective amortization of the high NRE costs, does currently not exist.

In this case study, we focus on the efficient implementation of the matched filter, which contributes significantly (in certain modes with more than 30 %) to the total computational complexity of the digital inner modem of the receiver. As shown in Fig. 5.1, the matched filter, typically implemented as a filter with Finite Impulse Response (FIR), is the first functional block after the ADC. In literature, a wide variety of FIR filter implementations, which target mostly 2.4/5 GHz wireless systems, have been proposed. Proposals range from ASICs [22, 46], rASICs [27, 38], ASIPs [19, 36] to SDR processors [1, 21, 37, 42]. Because ASICs offer typically the highest performance and energy efficiency, commercial 60 GHz systems are traditionally implemented with this architecture style. However, ASICs are not flexible, therefore they are becoming less and less attractive. In contrast, SDR processors are flexible but state-of-the-art solutions cannot fulfill the high performance and high energy-efficiency requirements for 60 GHz systems. This is further demonstrated in Sect. 5.6.

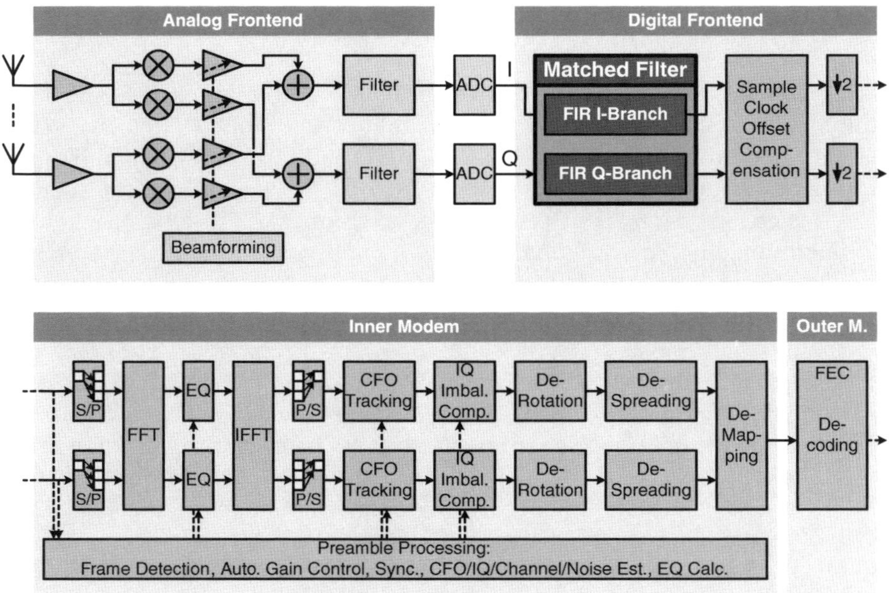

Fig. 5.1 Block diagram of the considered 60 GHz receiver. The front-end, the inner-modem and the outer-modem functionality is shown. The matched filter is situated right after the ADC and operates at twice the baseband rate. For the computation of I and Q, two identical real filters can be used

5.1.3 Main Contributions

In this case study, which is based on the proposed DSIP architecture template of Chap. 3, we propose a FIR filter design that targets high data rates, close to ASIC energy efficiency, high flexibility, high reusability and technology scaling-friendliness. As previously motivated, maximizing these metrics is crucial for the considered domain. We propose two architecture instances, each configured for a different efficiency/flexibility trade-off. The implementation results in TSMC 40 nm technology demonstrate the feasibility and high efficiency of the proposal.

The main contributions of this chapter can be summarized in the following way: (1) We propose the first instruction set processor solution, applied on the matched filter, which can fulfill the joint requirements for 60 GHz baseband computing. (2) We show the first implementation of an energy-efficient processor that exploits SoftSIMD. (3) We indicate the feasibility of a full processor solution for mobile 60 GHz baseband computing.

5.2 Background on Driver

This section provides background information on the matched filter and on the flexibility requirements of this functional block.

5.2.1 Matched Filter for the 60 GHz System

In this case study we focus on the implementation of the matched filter for the IEEE 802.15.3c standard. However, the flexibility of the proposed implementation enables a reuse for other standards, such as IEEE 802.11ad. As shown in Fig. 5.1, the matched filter, which is implemented with a FIR, is situated right after the ADC. We employ the 6-bit ADC (4× interleaved) of [47]. The filter operates at a very high sampling rate of 3.52 Gsps, which corresponds to twice the baseband rate. One channel has a bandwidth of 1.76 GHz. The purpose of the filter is to reject the adjacent channel interference, to minimize inter-symbol interference and to maximize the Signal-to-Noise Ratio (SNR) at the receiver. The filter is a Square-Root-Raised-Cosine (SRRC) filter with a roll-off factor of 0.25.

The total SIFS latency for IEEE 802.15.3c is specified with 6 μs (default), for IEEE 802.11ad with 3 μs. Considering that the Fast Fourier Transformation (FFT) and the Forward-Error-Correction (FEC) decoder will consume most of this time budget, we assume that only about 5 % are available for the filter (and the required serial to parallel conversion). This corresponds to 0.15 μs (=167 clock cycles at 900 MHz) for the most demanding standard.

5.2.2 Motivation for a Flexible Implementation

In the FPGA design of [18], the number of filter taps was dimensioned and fixed to 21. However, this dimensioning is very pessimistic. In our Bit-Error-Rate (BER) simulations we have seen, that, depending on the actual channel condition, 3–15 taps are sufficient. A typical Line Of Sight (LOS) communication requires even only a 5-tap filter. The big gap between the worst case and the typical case filter requirements clearly motivates the design of a system that can dynamically adapt to the actual requirements [5, 32]. Beside the number of taps, also the quantization scheme can be more relaxed for typical channel conditions. Figure 5.2 shows this trade-off in terms of BER performance and utilized quantization scheme. To further reduce the average computational complexity and thus, the average energy consumption, it may even be beneficial to implement different filter structures for different environmental situations. Nevertheless, to exploit this huge optimization opportunity cost effectively, the employment of a flexible/programmable FIR filter implementation is essential.

In addition, as motivated in Sect. 5.1, a flexible implementation is required (1) to support multiple standards, (2) to support multiple modes and (3) to enable high volumes.

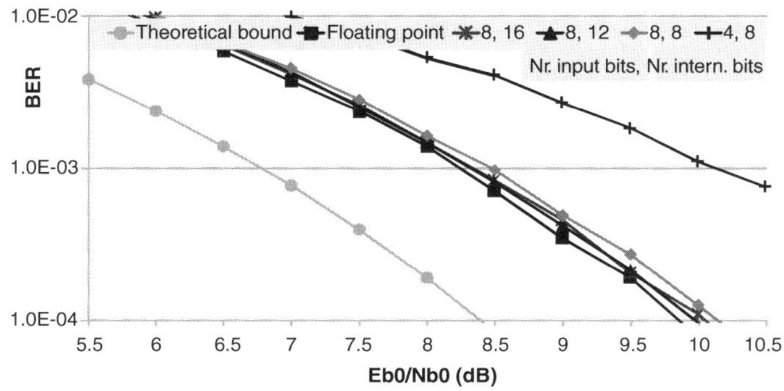

Fig. 5.2 BER performance for different signal quantization schemes of a 5-tap FIR filter. A distinction between input signals and internal/output signals is made. Compared to floating point, the 8, 12 bit-option causes only 0.07 dB degradation. Setup: Quadrature Phase-Shift Keying (QPSK), no channel coding, typical LOS channel condition

5.3 Algorithm Optimizations and Characteristics

In this section we first show the applied algorithm transformations which enable an efficient implementation on the targeted architecture style. Then, the main algorithm characteristics, which have been considered during the design of the DSIP architecture instances (see Sect. 7.4), are summarized.

5.3.1 Algorithm Choice and Applied Optimizations

In order to obtain an efficient implementation, a co-design of algorithm and architecture is essential [20]. In this section, we review the applied algorithm optimizations for the matched filter. The optimization techniques of Sects. 5.3.1.1, 5.3.1.2, 5.3.1.3 are known in literature and are more widely applicable. However, especially the optimization technique of Sect. 5.3.1.4 is novel and specific for the targeted SoftSIMD implementation.

5.3.1.1 Selection of Filter Structure

The most common FIR filter structures are the direct form and the transposed form. In the direct form, the delay line is at the input of the filter. Because the word length at the input is typically smaller than at the output, the storage requirements for the delay line are lower compared to the transposed form. In addition, because the data organization of the delay line is at the input of the filter, a more regular and efficient parallelization is possible. In contrast, the transposed form has the advantage that, by applying techniques such as Multiple Constant Multiplication

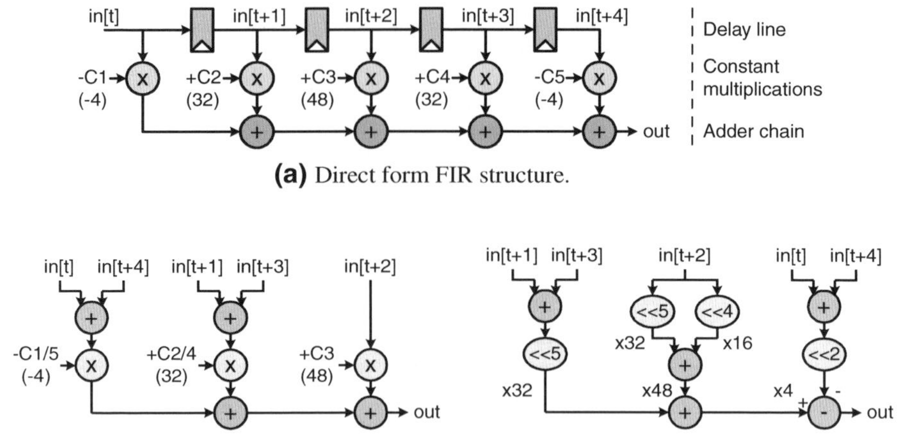

(a) Direct form FIR structure.

(b) Filter symmetry exploited.

(c) Multiplications converted.

Fig. 5.3 a–c Applied algorithm transformations to a symmetrical 5-tap direct form FIR algorithm in order to enable a more efficient mapping onto the targeted architecture style. These techniques can also be applied for a filter with more taps and typically also for a filter with different structure

(MCM) [48], multiplications can jointly be optimized. However, the application of such techniques typically results in irregular data flows, which are difficult to schedule on a parallel processor datapath. The achievable gain increases with the number of taps. Since the targeted filter of this work has only 5 taps, the benefits of the direct form are clearly prevailing. The required complex filter can be separated into two real filters, one for the real and one for the imaginary part. Both real filters are equal and can operate independently from each other. In the following, if not otherwise mentioned, we refer to a single real filter, i.e. all signals are real numbers. Figure 5.3a shows the selected direct form structure with the required coefficients.

5.3.1.2 Exploitation of Filter Symmetry

The quantized filter coefficients of the SRRC filter are symmetrical. This symmetry can be exploited to reduce the number of multiplications by almost 50 %. Figure 5.3b shows the modified Data-Flow Graph (DFG). As a consequence of this transformation, a re-ordering of input samples is required.

5.3.1.3 Transformation of Multiplications

By combining the following concepts, the cost for multiplications is significantly reduced: (1) Since the coefficients are known at design/compile time, i.e. they are constant, the multiplications can a priori be transformed to shift-add/sub sequences. (2) The cost depends on the number of required shift-add/sub iterations, i.e. the number of non-zero bits in the coefficient. By encoding the coefficient, e.g. in Canonical Signed Digit (CSD) format [13], the number of non-zero bits can be reduced. During the floating-point to fixed-point transformation, the value of the coefficient may be slightly varied to enable a more efficient implementation. (3) To avoid the computation of unnecessary bits, the right-shift instead of the left-shift algorithm can be used [6]. (4) To benefit from the fore-mentioned optimizations, an architecture, which efficiently supports shift-add/sub operations, is required. This is the case for the proposed DSIP architecture.

For the filter of this work, the multiplications with C1/5 and C3 can be implemented with a single shift operation. For C2, two shift operations and one addition are required. The negative sign of C1/5 can be propagated to the adder chain. As shown in Fig. 5.3c, the computation of C1/5 is moved towards the end of the filter structure.

5.3.1.4 Addition of an Offset for Software SIMD Mapping

In a traditional HardSIMD datapath, the sub-datapaths for the individual subwords are physically separated. However, in a SoftSIMD datapath, this is not the case. Therefore the handling of individual sign bits, which are unknown at design time, is more complex. To avoid specific hardware for the sign-bits handling, we propose to scale the input values so that no negative numbers can occur. Thus, all subwords always have the same sign. As shown in Fig. 5.4d, we first add $2^5 = 32$ (6-bit input) to obtain unsigned input values. Since the filter contains negative coefficients, an addition of a further factor is required. Considering that the multiplications with the negative coefficients (C1/C5) are performed last, this scaling factor, denoted as *scf*, is given by

$$scf = ceil\left(\frac{(in_{max} \times \sum |nc|) - (in_{min} \times \sum pc)}{\sum pc - \sum |nc|}\right) \tag{5.1}$$

whereas *nc*/*pc* denotes the negative/positive coefficient values. For this example, the resulting *scf* is 5. The input scaling needs to be compensated at the output of the filter. For that, the gain factor of the filter, which is given by

$$gain = \sum nc + \sum pc \tag{5.2}$$

needs to be considered. For this example, the gain factor is 104.

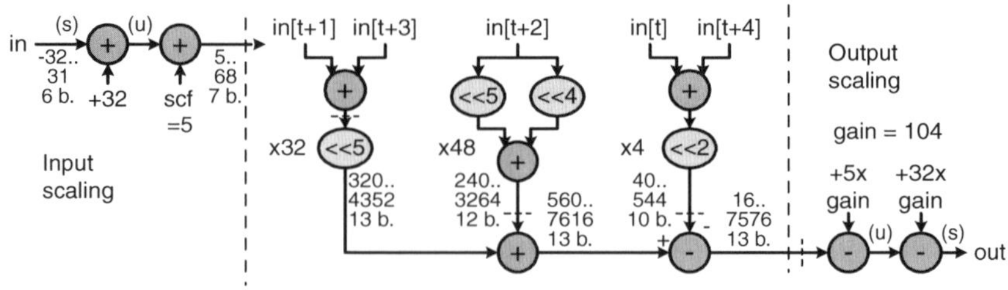

(d) Word lengths and negative numbers considered.

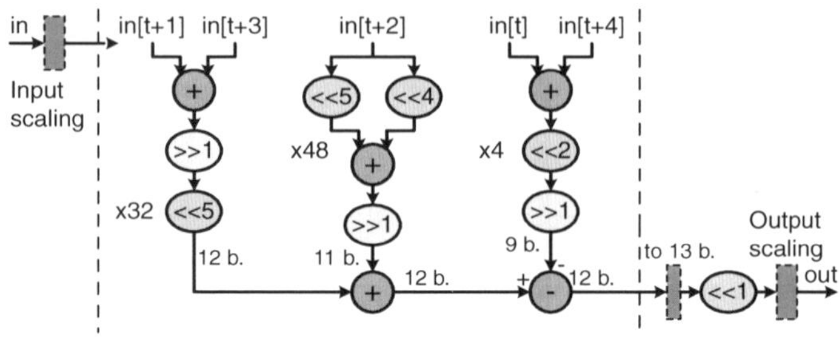

(e) Word lengths limited to 12 bits (introduces quantization noise).

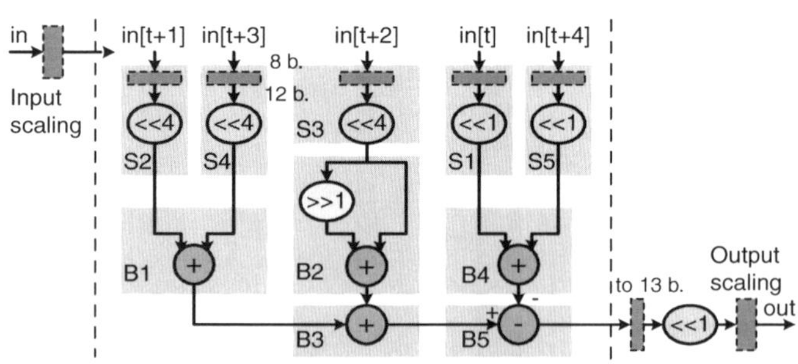

(f) Number of operations optimized.

Fig. 5.4 d–f Further applied algorithm transformations. Compared to (**a**)–(**c**), this transformations are more specific for the targeted architecture style. Also here, these techniques are applicable for a filter with more taps and typically also for a filter with different structure

5.3.1.5 Quantization Refinement

The dynamic-range analysis of Fig. 5.4d shows that for full precision, maximally 13 bits are required. This is also the case if no input scaling is applied, i.e. for HardSIMD implementation. In the considered architecture, we consider subword-lengths in a granularity of 4 bits, e.g. 8, 12, 16 bits. By choosing 16 bits, some of the Most Significant Bits (MSBs) are not utilized. Since a quantization to 12 bits causes only minor BER degradation, see Fig. 5.2, 12 bits are chosen. As shown in Fig. 5.4e, the scaling to 12 bits requires some additional shifts.

5.3.1.6 Optimization of Operation Count

In the considered architecture, the BAU can perform one right shift-add/sub operation per clock cycle. In parallel, the SSU can change the subword size in combination with a left/right shifting for one input word. We consider this characteristic to further optimize the algorithm. Figure 5.4f shows the final DFG. The operations on the BAU are denoted as B1–B5 and the operations on the SSU are denoted as S1–S5. In contrast to the conventional filter-tap computation with one Multiply-ACcumulate (MAC) operation, we compute one filter-tap with a single shift-add/sub operation. This reduces the complexity significantly. In a SoftSIMD datapath with support for 8 and 12-bit subwords, the input samples can be stored with 8 bits.

5.3.2 Algorithm Characteristics

In the following we provide a summary of the main characteristics of the selected algorithm. This information was considered during the design and dimensioning of the proposed architecture instance (see Sect. 7.4). Algorithms with similar characteristics can efficiently be mapped onto the same architecture instance. However, if an algorithm has very different characteristics, e.g. instead of 16 bits, 20 bits word length is needed, or instead of constant multiplications, variable multiplications are computationally dominant, then the efficient mapping onto the designed instance is not possible. To enable the efficient mapping also for that case, a broader specification would have to be considered from the beginning on (we see this as future work).

5.3.2.1 Required Arithmetic and Logic Operations

As mentioned in Sect. 3.3, we distinguish between the following operation types: (1) Basic Arithmetic Operations (BAO)s, (2) Multiplications with a Constant Operator (MCO)s, (3) Multiplications with Variable Operators and a multiplier which can adopt a Wide value range (MVOW)s, (4) Multiplications with Variable Operators and a multiplier which can adopt only a Narrow value range (MVON)s, (5) Trigonometric and Special Operations (TSO)s, (6) Boolean Logic and Comparison Operations (BLCO)s. The initial algorithm, see Fig. 5.3a, contains 50 % MCOs (multiplication with coefficients) and 50 % BAOs (adder chain). Because of the applied algorithm transformations, the MCOs have been converted to BAOs.

Thus, the optimized algorithm contains only BAOs, which have a rather low implementation complexity.

5.3.2.2 Required Word-Length Support

The input samples are represented with 8 bits, the internal signals for the 5-tap filter can be represented with 12 bits. However, for filters with more than 5 taps, 12-bit internal signal representations may introduce an unacceptable high quantization noise. Therefore, for this case, the support of higher word lengths will be required.

5.3.2.3 Data Dependencies and Parallelization

The following main parallelization options can be exploited: (1) parallel computation of real and imaginary part, (2) parallel computation across the complete filter DFG, (3) parallel computation across filter taps and (4) parallel computation of constant multiplications and additions (only for initial filter structure). Considering data dependency, data exchange frequency and regularity of data flow, a parallelization of (1) and (2) are most promising (see Sect. 7.4). The actual choice and degree of parallelization has to consider the characteristic of the targeted architecture template and the throughput requirements.

5.3.2.4 Data and Control Flow

The data-flow structure of the algorithm is streaming based, completely deterministic and does not contain conditions. The algorithm is loop-dominated.

5.3.2.5 Data Storage Requirements

The actual data storage requirements depend on several aspects, such as general architecture, exploited parallelization options and chosen parameters of the algorithm. Nevertheless, because the DFG is small and the considered parallelization options contain rather little dependency, the requirements are rather low. Because of the streaming nature of the algorithm, big data memories (e.g. as it would be the case for the FFT) are not required.

5.4 Proposed DSIP Architecture Instances

In this section, we propose two DSIP architecture template instances which are well-suited for the targeted filter application. Corresponding to their SIMD capabilities, we denote them as HardWare-SIMD instance (HW-SIMDi) and as

SoftWare-SIMD instance (SW-SIMDi). The HW-SIMDi is based on a 4-way SIMD datapath with 12-bit subwords and a Data MEMory (DMEM) with 768 bits storage size. The SW-SIMDi supports uniform subwords with 8, 12, 16 and 24 bits and provides a DMEM with 1.5 kbits storage size. Thus, the SW-SIMDi is more general and provides more flexibility than the HW-SIMDi. For this reason, the SW-SIMDi enables a higher reusability, which leads to larger volumes and therefore to a more cost-effective amortization of NRE costs. For more information on the design/dimensioning of the architecture instances, we refer to Sect. 7.4. In the following the hierarchical architectures are explained in a top–down manner.

5.4.1 Top Level

Figure 5.5 shows the common top level of the proposed architectures. It consists of one or multiple homogeneous clusters and the Shared Top Level Control (STLC).

5.4.1.1 Clusters

Across clusters, Data Level Parallelism (DLP) and Instruction Level Parallelism (ILP) is exploited. With the number of instantiated clusters the maximal achievable throughput can be scaled.

5.4.1.2 Shared Top Level Control

The STLC consists of the L1 master control with the corresponding L1 IMEM Direct Memory Access (DMA), the L1 IMEMa and the L1 IMEMb. The L1 master control has the full control of the processor. It initiates transfers between the L1 IMEMs and the L0 DLBs, it controls whether instructions are executed from the L1 IMEMs or the L0 DLBs and it can influence the execution of L0 control loops (start loop, restart loop, pause, jump). Like the DLBs, the L1 master control can execute loops and has an individual Program Counter (PC) and Hold Counter (HC). The L1 IMEM DMA contains the instructions for the master control. It has a size of 61 bits $\times$ 16 words and is implemented with a commercial memory macro. The L1 IMEMa contains the instructions for the L0 DLB control tables (control information for loops) and the instructions that are provided to the datapath. The sizes for the HW-SIMDi and the SW-SIMDi are 100 bits $\times$ 16 words and 156 bits $\times$ 16 words respectively. These numbers show a clear difference in control complexity. To enable a better floorplan, the L1 IMEMa is physically implemented with two commercial memory macros. The controllers of the L0 DLBs are implemented with simple logic and can therefore only handle the execution of a deterministic control flow. However, the L1 master control can

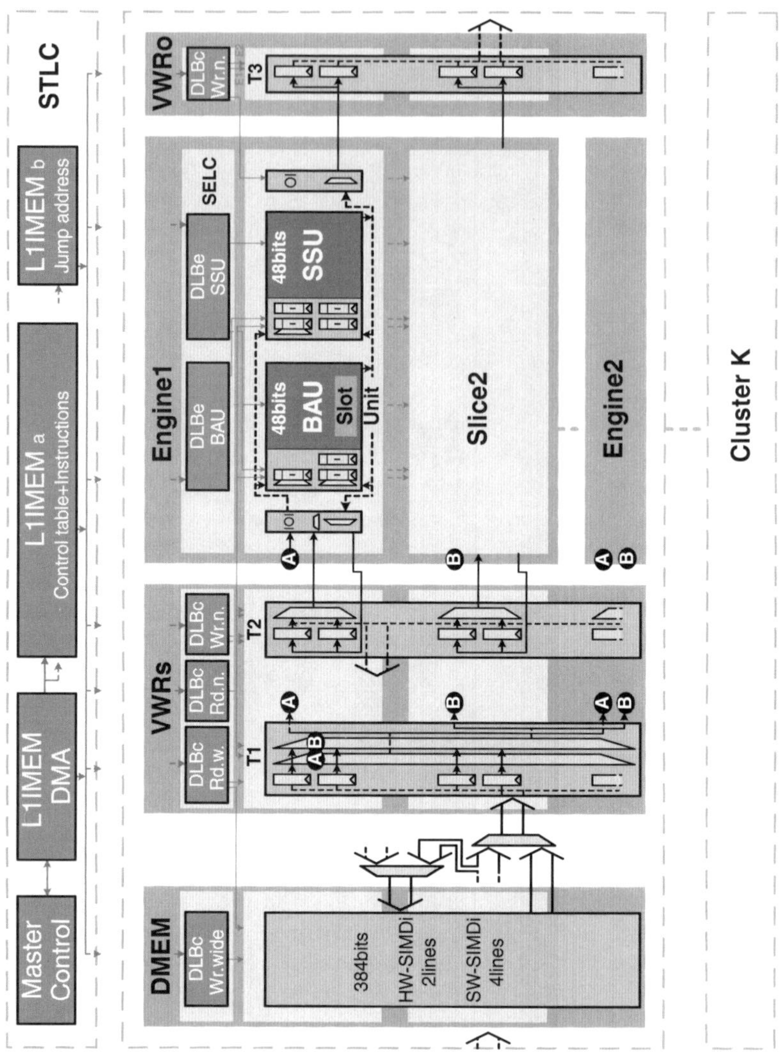

Fig. 5.5 Top level view of the proposed scalable DSIP architecture instances. The number of instantiated clusters *K* will mainly be decided based on the targeted filter algorithms (e.g. structure, number of taps) and on the throughput requirements

initiate jumps in the L0 loops so that more complicated control flows are possible. The jump addresses for the L0 controls are provided from the L1 IMEMb, which has a size of 25 bits $\times$ 4 words and which is implemented with standard-cell D-Latches. Since all clusters are homogeneous and execute the same set of instructions, the L1 IMEMs can be shared amongst them. Pipelining in the control hierarchy ensures that the critical path of the processor is in the BAU, which performs the actual computations.

5.4.2 Cluster Level

A cluster consists of a DMEM, 2 VWRs [34], 2 engines, a VWR for the output (VWRo) and the L0 Cluster Level Control (CLC). The datapath between the DMEM and the VWRs is 384 bits wide. Between the VWRs/VWRo and the engines, the datapath is 48 bits wide.

5.4.2.1 Data Memory

The DMEM as well as the VWRs and the VWRo have a wide interface of 384 bits. A wide interface enables to reduce the number of accesses and the width of the read and write address. Thus, the control energy. The DMEMs for the HW-SIMDi and the SW-SIMDi have a size of 768 bits (2 words) and 1.5 kbits (4 words) respectively.

5.4.2.2 Very Wide Registers

The VWRs, which are characterized by an asymmetrical interface [34] (384 bits vs. 48 bits), replace the conventional register file. As shown in Fig. 5.5, 2 VWRs of different types, which are denoted as T1 and T2 respectively, are employed. Each type offers an individual form of data transfers. T1 is employed for the data transfers from the DMEM to the engines. The data in the registers of T1 can be accessed by both engines. Therefore data, which is required by both engines, does not need to be duplicated. T1 has 2 narrow read ports. Each port is either connected to the slices of type A or of type B. Especially for the HW-SIMDi, where the subword sizes cannot be smaller than 12 bits, the bandwidth of 2 read ports is essential. T2 has a wide read interface to the DMEM and a narrow read and write interface to the engines. It can be used to write data from the engines to the DMEM and to store intermediate data for the engines. In contrast to T1, the registers of T2 are directly assigned to engines. The VWRs are implemented with standard-cell D-Flip-Flops (D-FF)s. It total, they have a size of $2 \times 384 = 768$ bits.

5.4.2.3 Engines

Considering the bandwidth requirements from the VWRs to the slices, the datapath word length of the VWR/DMEM interface and the scalability aspect, the use of 4 slices within a cluster is a good choice. The 4 slices could be distributed across 1–4 engines. Utilizing only 1 engine with 4 slices would minimize the area for the control, but limits the flexibility. In contrast, 4 engines would enable the highest flexibility, but also the control cost would be highest. Since it is a good-trade off, we use 2 engines, each with 2 slices. Amongst the 2 homogeneous engines and utilized slices, DLP and ILP can be exploited.

5.4.2.4 Very Wide Register Output

Once the output data is computed, it is written to the VWRo. Thus, the VWRo can be seen as the output interface of the processor. It has 4 narrow write interfaces with 48 bits and 1 wide read interface with 384 bits. Every slice has storage for 2 datapath words. The VWRo is implemented with D-FFs.

5.4.2.5 Cluster Level Control

As it can be seen in Fig. 5.5, the CLC, which controls the DMEM, the VWRs and the VWRo, consists of 5 cluster level DLBs (DLBc)s. The fine-grain separation of the control into 5 different DLBcs is based on (1) the control functionality (data flow and data word length), (2) the typical frequency of instruction fetches and (3) the time-point of instruction fetches compared to other DLBcs. This separation prevents a merging of all loops to a common loop nest, as it would be required for a centralized loop buffer [17]. Thus, the number of No OPerations (NOP)s, the storage and the energy is reduced. Table 5.1 provides an overview. The chosen depth of each DLBc considers the typical loop sizes. Because each distributed control includes a wait cycle counter, instructions, which have to be applied for multiple consecutive clock cycles, do not need to be duplicated. The DLBcs are currently implemented with D-FFs, however, they can potentially be replaced with D-Latches.

5.4.3 *Engine Level*

An engine consists of 2 homogeneous slices and the Shared Engine Level Control (SELC).

Table 5.1 Overview of control functionality, typical activity and sizes of DLBcs

Read from	Write to	Data width	Frequency	Size
In/VWR T2	DMEM	384 bits	Low	$3/5$ bits $\times 8$ words
DMEM	VWR T1	384 bits	Low	3 bits $\times 8$ words
VWR T1/T2	Slices	$1-4 \times 48$ bits	Medium	3 bits $\times 8$ words
Slices	VWR T2	$1-4 \times 48$ bits	Low-medium	24 bits $\times 16$ words
Slices	VWRo	$1-4 \times 48$ bits	Low-medium	5 bits $\times 16$ words

5.4.3.1 Slices

A slice represents the arithmetic and logic datapath of the processor. In each slice, the heterogeneous units BAU and SSU are present. The slice for the HW-SIMDi and the SW-SIMDi will be explained in Sects. 5.4.4 and 5.4.5 respectively.

5.4.3.2 Engine Level Control

As it is shown in Fig. 5.5, the Shared Engine Level Control (SELC) consists of two distributed loop buffers (DLBe)s [35], one for each unit of the datapath slice. Across slices, DLP, and across units, DLP and ILP can be exploited.

As it can be seen in Figs. 5.6 and 5.8 respectively, the implementations of the DLBes for the HW-SIMDi and the SW-SIMDi are different. Because of the rather high number of control bits, the control for the SSU of the SW-SIMDi leverages on several small L0 IMEMs. Contrary to the DLBcs and the DLBe for the BAU, the DLBe for the SSU employs a small instruction decoder between the L0 IMEM and the datapath. During the design, it was ensured that the decoder logic is not in the critical path.

5.4.4 Slice Level of the HW-SIMDi

As it is shown in Fig. 5.6, the slice consists of the BAU, the SSU, the input interface towards the VWR T1/T2 and the output interface towards the VWR T2/VWRo. The units and the interfaces of a slice are connected via an upper and via a lower interconnection network. Via the upper interconnection network the input data of the VWR T1/T2 is routed. The lower interconnection network is used to exchange data between units and to write data to the VWR T2/VWRo. For a slice, the BAU and the SSU are matched so that they operate on the same subword sizes. The slice datapaths of the HW-SIMDi and the SW-SIMDi are 48 bits wide, which is a common multiple of 8, 12, 16 and 24 bits. These subword sizes can typically be well-exploited in the considered domain. The SW-SIMDi supports all of these 4 subword sizes while the HW-SIMDi supports only one of them. For the implemented HW-SIMDi of this book, we chose a subword size of 12 bits (sufficient for

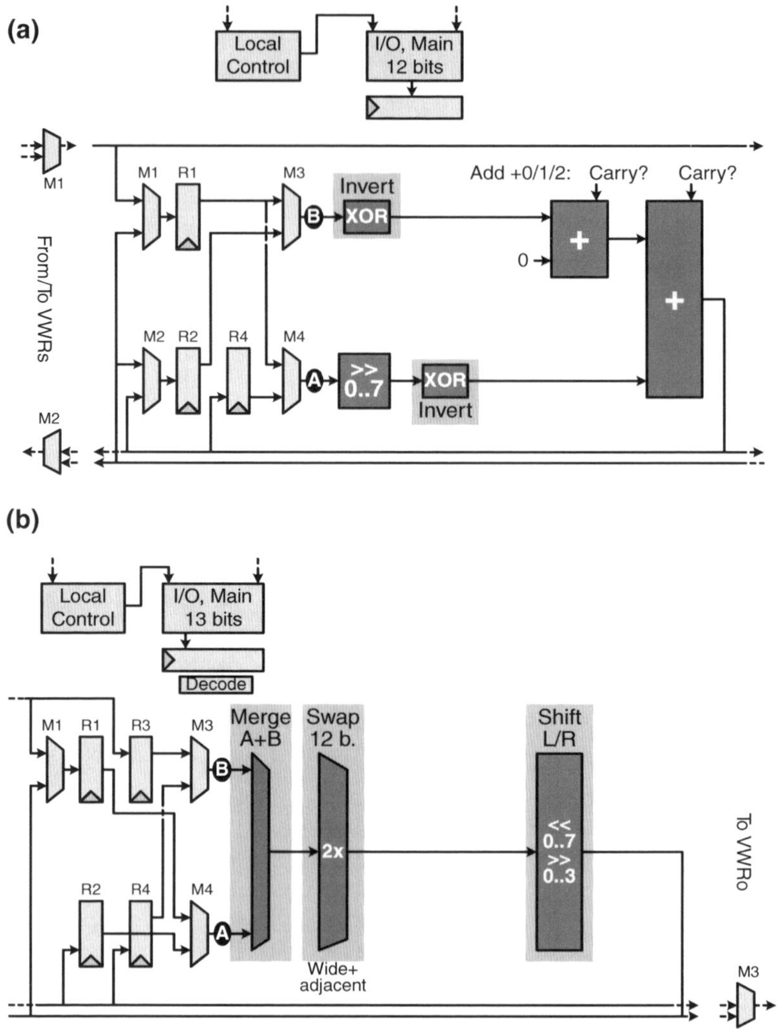

Fig. 5.6 Units, including individual engine-level DLBes, of the HW-SIMDi. The datapath is 48 bits wide. **a** BAU HardSIMD. **b** SSU HardSIMD

the considered 5-tap filter). The slice of the HW-SIMDi supports 4-way SIMD. In general, to minimize the delay of a unit, the order of the components (e.g. multiplexer, adder), the balancing of components across the input branches (A and B) and the use of inverting instead of non-inverting gates have been considered.

5.4.4.1 Basic Arithmetic Unit

The BAU, which performs the actual arithmetic computations, can be seen as the most important part of the processor. It supports the basic shift-add/sub operation $\pm(A \gg 0..7) \pm B$ in a 4-way SIMD datapath of 12-bit subwords. The input signs and the right shift factor can be programmed. Thus, basic operations and constant multiplications are supported. For constant multiplications, we generally leverage on the iterative right-shift method [6] and on the CSD coding [13] of the multiplier values.

As shown in Fig. 5.6a, the unit consists of only simple logic. Therefore, as explained in Sect. 3.5.4.2, the unit can operate at a high clock frequency, contributes to an overall high area/energy efficiency and is rather generic. The negation of a 2ers-complement number X is computed by INV(X)+1. We perform the inversion with an XOR gate and use the carry-input of the adder for +1. Since both inputs can be negated at the same operation, i.e. +2 is needed, a second adder is employed. At the input of the unit 3 local write-enable registers and 4 multiplexers are present. In a conventional register file, all storage locations are equally weighted. This is different in this design. Here, each of the local registers is customized for a different data-flow pattern. For instance, the input from the VWR T1 is routed via R1, R4 is the preferred feedback register. This approach enables to reduce multiplexers and hence, energy. Since typical data-flow patterns have been considered, the unit remains still rather generic. 12 control bits are required.

5.4.4.2 Subword Shuffler Unit

The SSU supports the following functionality: (1) combine subwords from the 2 data input words to a new datapath word, (2) change the relative position of subwords in the 4 slots and (3) apply left and right shifting. As it can be seen in Fig. 5.6b, the SSU is implemented with clearly separated stages. This approach enables high scalability and high reusability. Like the BAU, the SSU includes several local registers which enable specific data-flow patterns. For instance, R1 is used to store input data from the VWR T1 and to store the data during iterations. R2 and R4 are used to store the output data from the BAU. 13 control bits are required.

5.4.5 Slice Level of the SW-SIMDi

Like for the HW-SIMDi, the slice of the SW-SIMDi consists of 2 units/slice interfaces, includes an upper and a lower interconnection network and the datapath is 48 bits wide. The SW-SIMDi supports not only 12-bit subwords, but also 8, 16 and 24 bits. Therefore, the architectures of the 2 SW-SIMDi units are more complex (cp. Fig. 5.6 with Fig. 5.8).

5.4.5.1 Basic Arithmetic Unit

The BAU of the SW-SIMDi supports the same shift-add/sub operation as in the HW-SIMDi. However, the operation cannot only be performed on 4×12-bit subwords, but also on $6\times$, 3×16 and 2×24-bit subwords. This additional flexibility can be beneficial in two ways: (1) if the algorithms require only small subword lengths, a better datapath-word utilization can be achieved (e.g. use 8-bit instead of 12-bit subwords) [6, 12], (2) because it supports subword lengths of up to 24 bits, the unit is more generic (i.e. algorithms that require more than 12 bits can still be mapped).

Evaluation of Implementation Options for the BAU

To enable the support of multiple subword sizes in a shift-add/sub datapath, different implementation options exist. Figure 5.7 depicts 3 possible options (not exhaustive, since also hybrid solutions are possible [33]). As shown in Fig. 5.7a, the most straight-forward option is to implement multiple parallel datapaths. Each datapath supports HardSIMD for only 1 subword length. A multiplexer is employed to select the output of one of these datapaths. Operand isolation could be used to avoid the toggling of the non-utilized datapaths. However, the area of this option scales linearly with the number of supported subword sizes. In addition, the operand isolation circuit, the multiplexer and the longer wires increase the critical path significantly.

A further option is to employ only 1 datapath in which the subword separation is done in hardware. This option is depicted in Fig. 5.7b. The carry propagation in the adder and the "pollution" of the adjacent subwords in the shifter can be prevented with AND gates. Depending on the selected subword size, the decoder generates the required input bits for the AND gates. The main drawback of this option is the increased critical path, which has two causes: (1) the insertion of AND gates and (2) the separation of the wide adder into multiple narrow adders. We assume that logarithmic adders, such as Brent-Kung or Sklansky [29], are employed. For example, the adder delay for 24-bit subword size increases from $\log(24)$ to $2 \times \log(8) + 2 \times \log(4) + 3$ AND. In general, with an increasing number of supported subword sizes, this option becomes less and less attractive. Especially if also the support of non-uniform subword sizes has to be enabled, other options should be considered.

As illustrated in Fig. 5.7c, a third option for implementing the separation of subwords is the use of bit masking [6]. The adder and the shifter have the same word length as the datapath and do not include any hardware for the subword separation. Before (or after) the add/shift operation, the data bits, which cause the "pollution" of the adjacent subwords, are forced to be logical '0'. We refer to this task as masking. The masking can be done by an AND operation of the datapath word and the mask word.

The mask word can basically be provided (1) from a storage (see Fig. 5.7c-x) or (2) from a mask generator hardware (see Fig. 5.7c-y). Both choices offer different flexibility/efficiency trade-offs. The storage choice offers more flexibility, but the storage cost increases linearly with the number of required masks. All possible

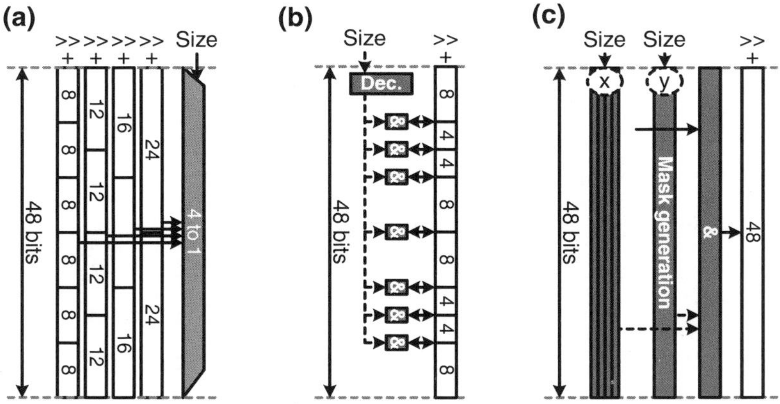

Fig. 5.7 Possible options for implementing the adder/shifter of the BAU for 8, 12, 16 and 24-bit subwords (48-bit datapath). Delays of the datapaths (assuming max. delay constraint for 24 bits) (add; shift; others): **a** log(24 bits); shift; 4-to-1 mux, **b** $2 \times \log(8 \text{ bits}) + 2 \times \log(4 \text{ bits}) + 3$ AND ; shift + AND ; 0, **c** log(24 bits); shift; AND. **a** Duplicated datapath. **b** Hardware separation. **c** With masking

masks, for the considered datapath with 4 subword sizes and a shifting of 0..7, would require a storage of $4 \times 8 \times 48 = 1.5$ kbits, which is twice the size of the combined VWRs T1/T2. Although in practice less masks will be required, this example illustrates the limitation of this approach. To enable the flexibility of supporting any software-programmed subword organization (full SoftSIMD), not only a wide set of masks need to be made available, but also the SSU must be made very flexible so that all the required data-shuffle patterns are supported. However, the implementation of a SSU, which supports all possible patterns, would be extremely costly (e.g. a 48-to-1 multiplexer for each bit in the 48-bit datapath).

In this work, we focus on uniform subword sizes and propose the use of a mask generator hardware. This combination offers a good trade-off between flexibility/efficiency. The masking could be done in the BAU, in the SSU or in both. We employ a masking hardware in both because (1) iterative operations on the BAU do not need to be interrupted and (2) the cost for the masking hardware is rather low (i.e. 11 % of the total area of the BAU). In contrast to the other two previously discussed options, this third option has little impact on the performance: The timing path for the 48-bit adder can be constraint to the width of the biggest subword length (e.g. 24 bits, t = f(log(24))). Thus, compared to HardSIMD, the critical path increases only by the delay of a single AND gate (masking).

Proposed Implementation of the BAU

The support for 24-bit subword operations may be required for the mapping of high-order filters or for the mapping of algorithms that are sensitive to quantization noise, such as matrix inversion. However, operations on 24-bit subwords are typical rare. Thus, instead of dimensioning the datapath for 24-bit subwords, we

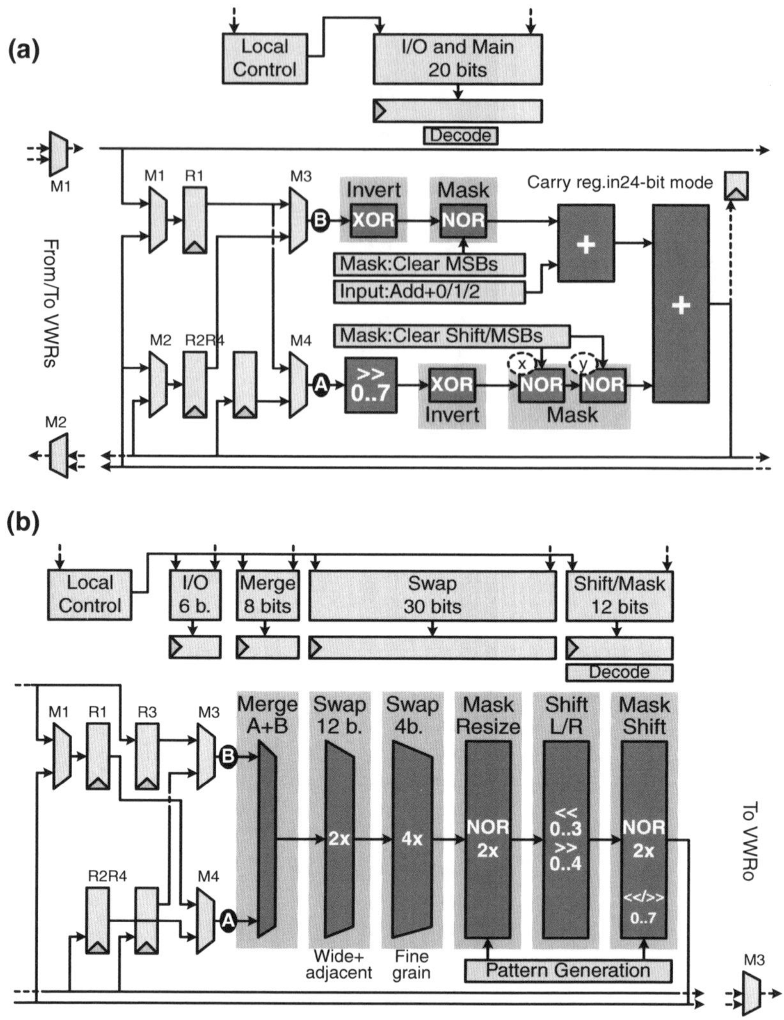

Fig. 5.8 Units, including individual engine-level DLBes, of the SW-SIMDi. The datapath is 48 bits wide. **a** BAU SoftSIMD. **b** SSU SoftSIMD

optimize the datapath for 16-bit subwords and compute 24-bit operations with 2 iterations. This approach has the advantage that the support of 24-bit operations does not decrease the performance of 8, 12 and 16-bit operations.

Figures 5.8a and 5.9 depict the proposed implementation. Like for the HW-SIMDi, the BAU consists of the local registers and the corresponding multiplexers, the shifter, the XOR gates for inversion and the two adders. The main differences are the added masking circuits, the approach of adding the value 0, 1 or 2 and the carry-bit registers for the iterative 24-bit support. Figure 5.9b illustrates the two mask generation circuits for the branch B. With the first mask, the MSBs of

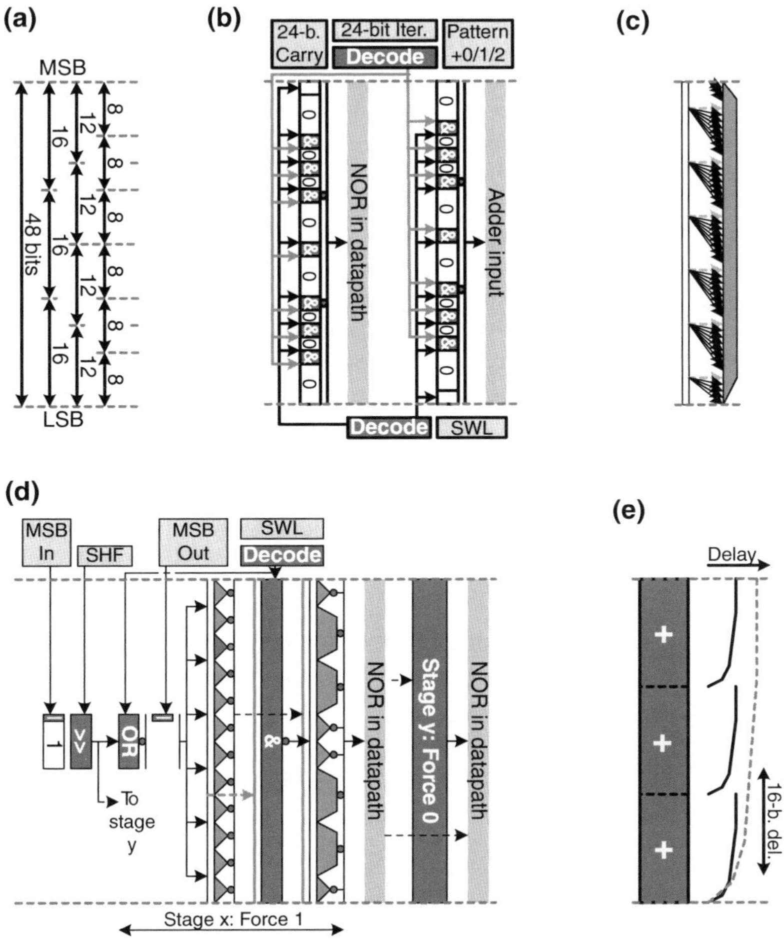

Fig. 5.9 Main components of the SoftSIMD BAU: **a** Supported subword sizes (uniform). **b** Mask generation for branch B. The left mask is used to manually set/clear MSBs. The right mask is used for adding 0, 1 or 2. **c** Right shift with range 0..7. **d** Mask generation for branch A. It is used to set/clear MSBs for shifts and to manually set/clear MSBs. **e** 3 × 16-bit adder with carry connection. Timing-path constraint to 16 bits

subwords can be set to ’0’. This ensures that no overflows can occur. The second mask is provided to the input of the first adder. It contains the information whether 0, 1 or 2 has to be added to each subword value (for 2ers complement). Since most of the 2nd adder-input bits (mask) are constantly ‘0’, the complexity of the first adder is reduced. An alternative approach would be to use the carry-in of the adder. However, to enable this approach, the 48-bit adder has to be separated and OR gates have to be inserted in the critical path, which is again not a good choice. The 8-to-1 multiplexer for the right shift is illustrated in Fig. 5.9c. In Fig. 5.9d the mask generation circuit for the branch A is depicted. It can be divided in two identical subsequent stages.

In the first stage, bits can be forced to logical '1' while in the second stage, bits can be forced to logical '0'. Both is needed, since the input can be positive or negative. For example, if a negative/positive number is shifted to the right, the MSBs of this subword must be set to '1'/'0'. The mask generation circuit can generate masks for all 4 uniform subword organizations, for the whole shift range of 0..7, for positive and negative numbers and for forcing MSBs to '0' in order to avoid overflows. This results in more than 50 different masks. As shown in Fig. 5.9d, the proposed implementation employs a hierarchical approach and exploits the uniformity of the subword organization. First, the part of the subword, which can be affected by shifting, is generated (here, the 8 MSBs). Then, by employing multiplexers, these bits are routed to the corresponding subwords of the two stages. Like for branch A, instead of an AND gate, a NOR gate for masking is used (less costly). The area of the proposed mask generation circuit is comparable to the area of 1.5 48-bit D-FFs. Thus, in a full software solution (see Fig. 5.7c-x), for the same area, only 1–2 masks can be stored. This demonstrates the efficiency of the proposed solution and motivates the choice. Note, basically for every algorithm, more than 2 masks are required. Figure 5.9e illustrates the achieved gain by limiting the timing path to 16 bits. The proposed SoftSIMD implementation of the BAU offers a good trade-off between efficiency and functionality. The overhead compared to HardSIMD is rather limited (see Sect. 5.5.2). However, beside the uniform subword organization, it has the limitation that all subwords must have the same sign. To ensure this, a dynamic range analysis and potentially a scaling must be performed. For the BAU, the number of control bits has increased from 12 to 20 bits.

5.4.5.2 Subword Shuffler Unit

In addition to the functionality of the HW-SIMDi SSU, the SW-SIMDi SSU must support the resizing of subwords (increasing and decreasing subword sizes). Resizing can be enabled by combining (1) fine-grain merging of input data, (2) fine-grain shuffling (4-bit level) and (3) masking. Figure 5.8b and Fig. 5.10b, d, e show the merging stage and the additional shuffling and masking stages. The combination of 2 × 12-bit and 4 × 4-bit shuffling stages has been shown to offer a good trade-off between flexibility and complexity. With this combination, the most important/frequent operations can be performed in one iteration. However, more complex operations, e.g. a resize from 24-bit to 8-bit subword size, requires multiple iterations. Thus, the datapath is optimized for the typical case, which is essential to achieve high energy efficiency.

The BAU performs the actual arithmetic computations, therefore it is important that the BAU and not another unit/component, such as the SSU, determines the maximal clock frequency of the processor. To ensure that the critical path of the SSU is not higher than the critical path of the BAU, the number of instantiated stages has to be limited or pipeline registers have to be inserted. In general, employing a more complex shuffling network, e.g. to support a full software solution with non-uniformity and more subword sizes, will typically degrade the overall performance/

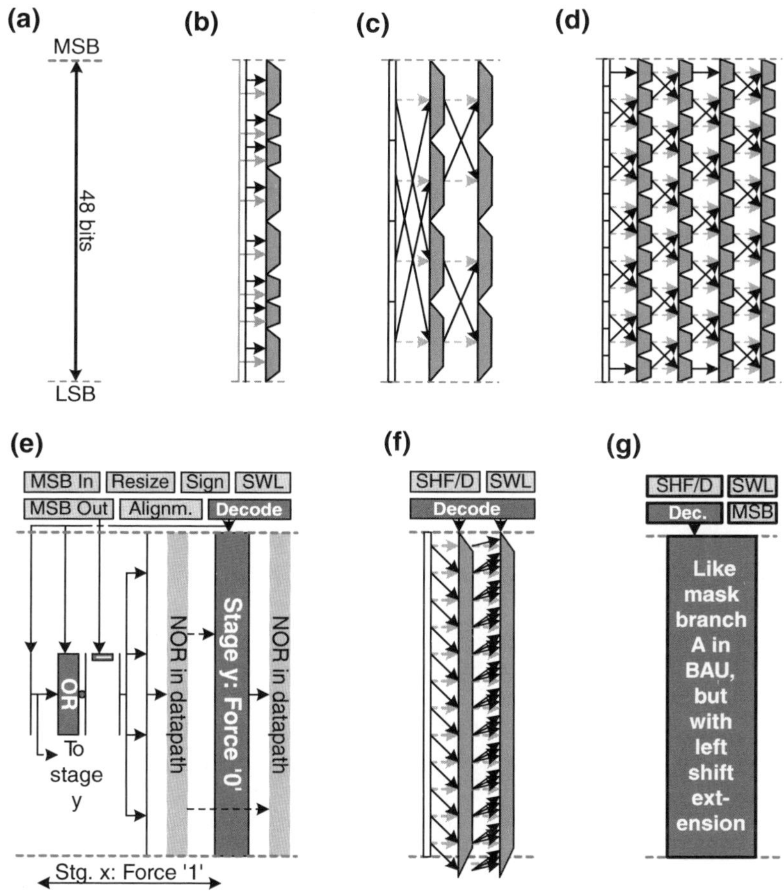

Fig. 5.10 Main components of the SoftSIMD SSU: **a** Datapath word length. **b** Merge for 8, 12, 16-bit support. **c** 12-bit swap stages: wide and adjacent. **d** 4-bit swap stages. **e** Mask for subword resize (increase/decrease word length). **f** Left/right shift. **g** Mask for shifting (similar as in BAU)

efficiency of the processor. The masking for resizing, as shown in Fig. 5.8e, exploits the same concept as in the BAU. An alignment/truncation to/of both sides (LSB and MSB) as well as the sign handling (same sign for all subwords) is supported. Beside merging, shuffling and resizing, left and right shifting is possible. For the latter, as illustrated in Fig. 5.10g, an additional masking stage was inserted.

Compared to the SSU of the HW-SIMDi, because of the additional stages, the number of control bits has significantly increased from 13 to 56 (although certain decode logic was inserted). To avoid the need for storing very similar 56-bit control words multiple times, e.g. if they differ only in 1 bit, the control word has been separated into 4 control subwords. As shown in Fig. 5.8b, each of the subwords are associated with an individual IMEM, which can be individually addressed. This approach reduces the overall storage requirements.

5.5 Software Mapping and Hardware Implementation

In this section the mapping of the 5-tap filter and the hardware implementation results are shown and discussed.

5.5.1 *Software Mapping and Scheduling*

To ensure that the targeted algorithm can be mapped efficiently, a co-design of algorithm and architecture is required [20]. For this reason, we first have applied several algorithm transformations, see Sect. 5.3, to prepare the algorithm for a mapping onto the targeted DSIP architecture template (e.g. we applied scaling for SoftSIMD). Then, while considering the mapping of the algorithm, the architecture instance has been dimensioned. For instance, the subword size of the HW-SIMDi has been chosen to be 12 bits and the VWR type T1 has been employed, since it enables to share input data and reduces storage requirements. Because the architecture instance is only based on primitive operators, i.e. no ASIs are included, it is rather generic and algorithms with similar characteristics can also efficiently be mapped.

In the following the mapping of the 5-tap FIR filter DFG of Fig. 5.4f, which uses 12-bit quantization, is explained. In the application, a filter for complex data is required. However, a separation into two real filters (I and Q) is possible. Since no data-dependency exists, each real filter is mapped onto an individual set of clusters. Both set of clusters are identical, operate in parallel and leverage on the same mapping. Thus, unless otherwise mentioned, in the rest of this section we assume the mapping of 12-bit real data. To achieve a very high throughput, we parallelize across all SIMD slots of 2 (or a multiple of 2) clusters. A slot performs all required operations of the 5-tap filter DFG, i.e. a slot computes the complete filter output. The order of the computed outputs is linear, i.e. slot 1 computes the output for $t[0]$, slot 2 computes the output for $t[1]$, and so on. All slots of an engine perform the same operation of the DFG in parallel, i.e. they operate synchronously. No data dependency between slots exists. The same input data is required by several slots, but at different clock cycles. For example, $in[0]$ is required by slot 1 at $t[0]$ and by slot 2 at $t[1]$. This means, the same subword input value is needed in multiple datapath words. We could consider this requirement outside the processor, i.e. before writing the datapath words to the DMEM/VWR T1. However, this approach requires additional hardware and moreover, it results in a high access pressure of the DMEM/VWR T1. For this reason, we store each input value only in one DMEM/VWR datapath word and perform the sharing/distribution inside the processor (locally).

The data flow, which is identical for all clusters, can be described as follows: First, the 12-bit input samples, which are linearly packed to 384-bit words (32 subwords), are written to the VWR T1. Note, we neglect the buffering in the

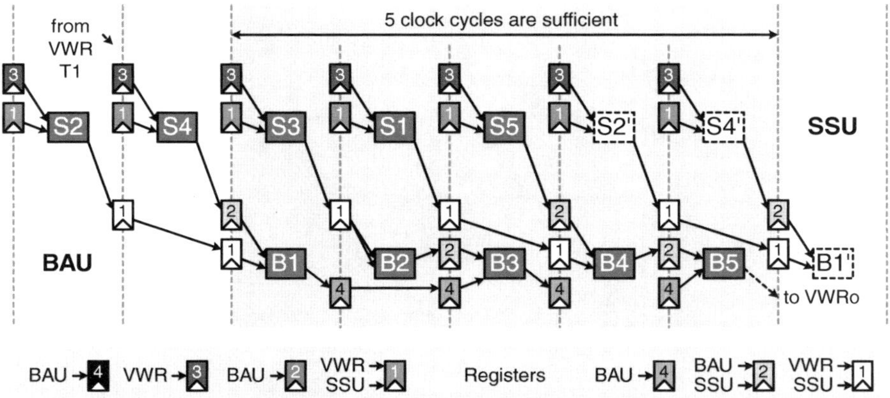

Fig. 5.11 5-tap FIR filter schedule for a slice. The operation labels S1–S5 (SSU) and B1–B5 (BAU) correspond to the ones in Fig. 5.4f. 5 clock cycles/output are sufficient

DMEM and assume that the data becomes available at the processor input at the right time. Second, the data from the VWR T1 is written to the local registers of the SSUs. Third, the SSUs generate the required datapath words and write them to the local registers of the BAUs. Fourth, the BAUs perform the computations and write the computed output values to the VWRo. From there, they can be consumed by the subsequent hardware (outside of this scope).

As shown in the schedule of Fig. 5.11, the computation of a 5-tap filter output can be done in 5 clock cycles. The same clock cycle count can also be achieved with a conventional Multiply-and-ACcumulate (MAC) processor datapath. However, the important difference is that we perform the computations on a much simpler datapath, which is clearly more energy efficient (see Sect. 5.6). The dense schedule, i.e. the computation of a filter tap requires only one shift-add/sub operation in average, was enabled by the applied algorithm transformations. In Fig. 5.11, B1–B5 denote operations on the BAU and S1–S5 denote operations on the SSU. These labels correspond to the ones that are used in the DFG of Fig. 5.4f. R1–R4 are the local registers of the units. The registers of the BAU are almost fully utilized. The SSU uses only 2 out of 4 registers. Because of the efficient schedule (manually performed), both units are always fully active and no stalls are required.

Figure 5.12 shows the schedule for the data transfers from the VWR T1 to the local registers of the SSUs. Engine 2 starts to operate 2 clock cycles later than engine 1. This delay enables to use the same VWR read port for both engines without introducing scheduling inefficiencies. The control loop of the BAU involves 5 clock cycles. Since the SSU combines subwords from 2 datapath words, i.e. 8 input subwords need to be organized, the control loop of the SSU involves 10 clock cycles. A writing to the VWR T1 occurs only every 10 clock cycles. The different loop lengths, the different number of active instructions within a loop and the asynchronous loop-starts strongly motivate the use of DLBs.

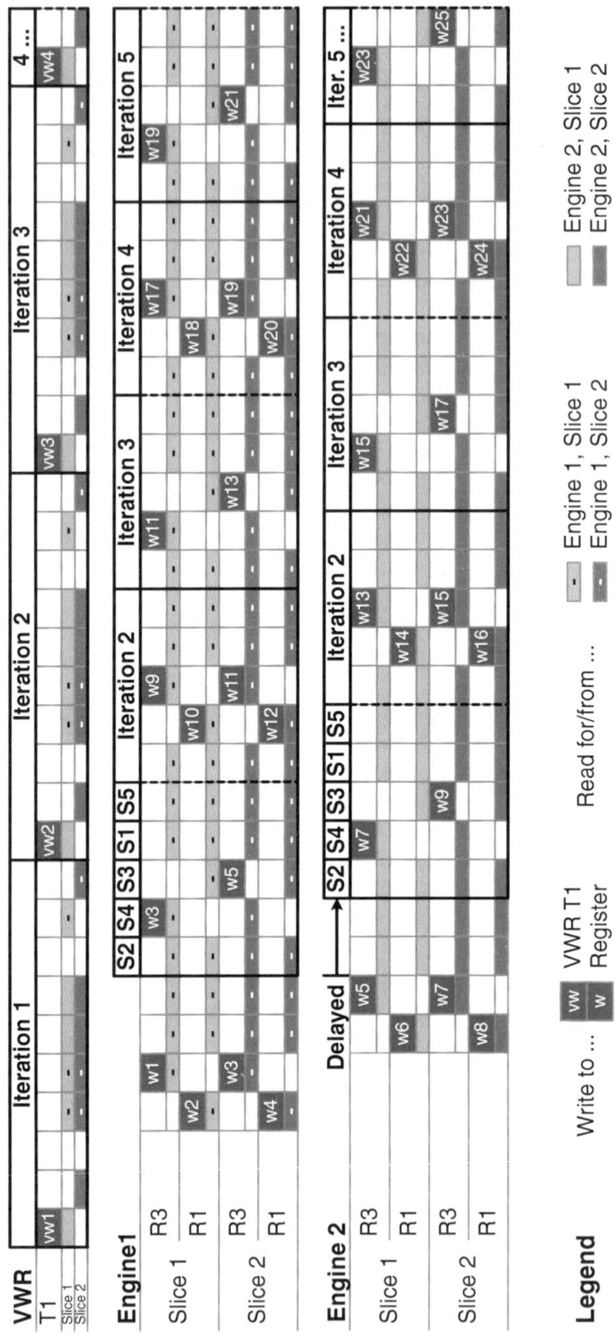

Fig. 5.12 Schedule for the accesses of the VWR T1 and the local registers in the SSUs

For 12 bits, the same schedule can be used for the HW-SIMDi and the SW-SIMDi. Nevertheless, because the input samples require less than 8-bit precision, for the SW-SIMDi, datapath words with 6 × 8-bit subwords instead of 4 × 12-bit subwords can be stored in the DMEM/VWR T1. This reduces the number of DMEM/VWR accesses, but at the cost of a longer SSU control loop.

Although we focus on the mapping of the 5-tap filter in this work (requirement for typical channel condition), it is important to mention that also filters with a higher number of taps can efficiently be mapped and similar results would be obtained. The major difference would be the increased number of VWRs/DMEM accesses, because for storing all temporal values, the number of employed local registers may not be sufficient anymore.

5.5.2 Hardware Implementation and Results

Both processor architectures, i.e. the HW-SIMDi and the SW-SIMDi, have been implemented in VHDL, synthesized with Cadence RTL Compiler Physical, automatically placed and routed with Cadence Encounter (EDI). For the energy estimation, ModelSim and Cadence Encounter Power System (EPS) have been used. The reported power/energy numbers are based on post-layout gate-level simulations (full mapping) with annotated parasitics. TSMC 40 nm General Purpose (GP) low-threshold-voltage standard-cell technology has been employed. For the implementation of L1 IMEMs, commercial memory macros have been used. During the manual floorplanning, the aspect ratio was chosen (considering the shape of the hard macros) and the hard macros where placed to the upper side of the layout. All results are based on 75 % area utilization (standard-cell vs. chip area).

Figure 5.13 shows the layout of the SW-SIMDi with 1 cluster. The placements of the main architecture components have been highlighted. The memory macros are clearly visible. Since the VWRs, the VWRo and the DMEM are based on standard cells, the placement tool had the freedom to physically intermingle them. Unfortunately, the datapath and the corresponding L0 DLBes of engine 1 have been placed far apart from each other. As a consequence, the highly active wires, which are between these architecture components, are long. In addition, it can be observed that the standard cells of the BAUs are located in areas with high routing congestions. For this reason, wires of the BAUs, which have a high toggling activity, are tentatively long and are potentially routed on the upper metal layers (especially the ones outside of the critical paths). However, for high energy efficiency, especially in DDSM technologies, the highly active wires need to be kept short (and VIAs should be avoided). This result motivates the use of a semi-custom design approach in which the relative physical placement of components/cells can be specified. Note, the placement of the components/cells has a direct impact on the routing and thus on the wire lengths.

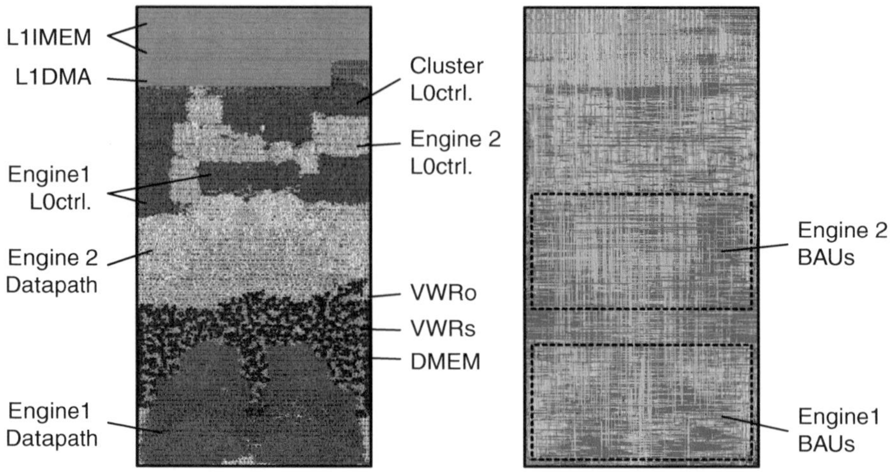

Fig. 5.13 Automatically placed and routed layout of the SW-SIMDi with 1 cluster (incl. L1 control)

Because of the simple datapath and the technology choice, the HW-SIMDi can operate at a high clock-frequency of 1.12 GHz at Worst-Case (WC) corner. The SW-SIMDi is slower, but can still operate at 900 MHz (WC). As desired, for both implementations, the critical paths are through the BAUs. By slightly varying the targeted clock frequency, it can be observed, that the timings of the BAU and the SSU are well balanced.

Based on 12-bit shift-add/sub operations, one cluster of the HW-SIMDi offers a computational power of $2 \times 2 \times 4 \times 1.12 = 17.9$ GOPS. The computational power of the SW-SIMDi varies depending on the utilized subword size. One cluster offers 3.6/10.8/14.4 and 21.6 GOPS for 24/16/12 and 8-bit subword sizes respectively. For 12 bits, the SW-SIMDi is 24 % slower than the HW-SIMDi. However, if the application suffices with 8-bit quantization, then the SW-SIMDi can be even 20 % faster.

The HW-SIMDi with one cluster consumes a total area of 0.046 mm^2, which corresponds to 64.4 k Gates Equivalents (GE)s. With 0.067 mm^2 (93.8 k GE), the SW-SIMDi consumes 45 % more area. The difference comes mainly from (1) the more complex SSU, which has a consequence on the datapath and the control storage area and (2) the instantiation of a DMEM that is twice as big. The HW-SIMDi has a high area efficiency of 389 GOPS/mm^2 (for 12 bits). With 53/161/214/322 GOPS/mm^2 for 24/16/12/8 bits respectively, the SW-SIMDi has a significantly lower area efficiency. Figure 5.14a shows the area breakdown of the SW-SIMDi with one cluster. The control contributes with 46 % and the datapath with 54 % to the total area. The significant contribution of the control has several reasons: (1) only 1 cluster is present, therefore a sharing of L1 control was not exploitable, (2) the L0 control is currently implemented with D-FFs, which consume more area than e.g. D-Latches, (3) the number of SSU control bits is rather

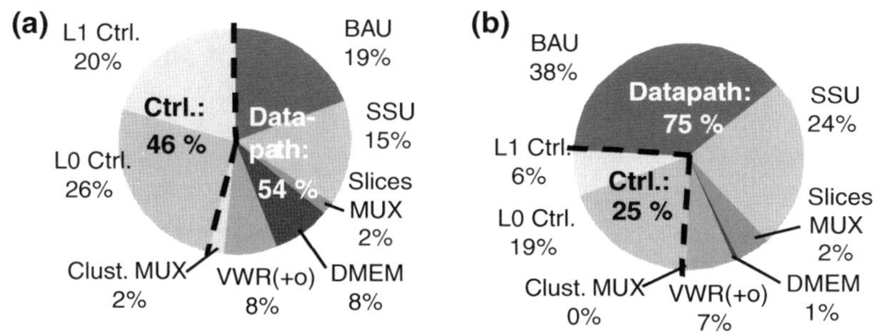

Fig. 5.14 Total area and energy breakdown for the SW-SIMDi with 1 cluster. The design consumes 67 k μm^2 total area and 24.8 mW total power. **a** Area breakdown. **b** Energy breakdown

high, (4) no code compression was applied and (5) since the complexity of the datapath is rather low, the contribution of the control appears stronger. In the datapath, the BAU and the SSU are dominant $19 + 15 = 34/54$ %. The VWRs/o and the DMEM contribute with 16 %.

Based on the mapping of the 12-bit 5-tap FIR filter from Sect. 5.5.1 (full datapath utilization), the HW-SIMDi consumes 18.5 mW and the SW-SIMDi consumes 25 mW in total (with one cluster). Thus, the HW-SIMDi offers a high energy efficiency of 970 MOPS/mW (corresponds to about 1 pJ per 12-bit shift-add/sub operation), which is close to the requirements of 4G [45]. This can also be seen in Fig. 2.8. The energy efficiency of the SW-SIMDi is 144/432/576/864 MOPS/mW for 24/16/12/8-bit subwords respectively. In Fig. 5.14b the energy breakdown for the SW-SIMDi with one cluster is shown. 25 % are consumed by the control, the rest by the datapath. The datapath energy is dominated by the BAU, which performs the actual computations. This is a clear sign for high energy efficiency. The SSU, which is fully utilized, consumes also a significant amount of energy. Because of the employed data memory hierarchy, the energy, which is consumed in the DMEM and in the VWRs/o, is comparatively low.

In this comparison for the 12-bit 5-tap FIR filter, the energy efficiency of the HW-SIMDi is clearly higher than the one of the SW-SIMDi. The main reason for this is that the 12-bit datapath of the HW-SIMDi matches exactly the word-length requirements and can therefore be fully utilized. However, if we assume that for higher filter orders up to 24 bits are required, then the BAU/SSU datapath of the HW-SIMDi has to be dimensioned for 24-bit instead of 12-bit subword sizes. Because of the wider word length of the adder, the achievable clock frequency of the 24-bit HW-SIMDi will be lower. During the execution of the considered 12-bit 5-tap filter on this 24-bit HW-SIMDi, 50 % of the BAU/SSU datapath is not utilized. This low utilization decreases the throughput and thus, the area efficiency by more than 50 % compared to the 12-bit HW-SIMDi. Nevertheless, because of the unused BAU/SSU datapath bits, the energy consumption will be lower than in the fully utilized version. Considering the energy breakdown of the SW-SIMDi and the optimistic assumption that the energy consumption of the modified BAU/SSU will be reduced by 50 %, the total energy consumption will decrease by $(38 + 24) \times 0.5 = 31$ %. However, because of the increased number of VWR accesses, i.e. only 2 subwords are packed within a 48-bit datapath word, and the fact that the BAU/SSU datapath will still consume some energy, e.g. all bits of the registers are still clocked, a realistic-optimistic total energy reduction is 20 %. Considering these assumptions, but still neglecting the reduction of clock frequency, the 24-bit HW-SIMDi will offer an energy efficiency of about 17.9×0.5 GOPS $/18.5 \times (1 - 0.2)$ mW $= 605$ MOPS/mW while executing the 12-bit 5-tap filter. In comparison, the SW-SIMDi offers 576 MOPS/mW. Considering additionally the reduction in clock frequency because of the adder, it can be estimated, that the SW-SIMDi offers the same or even a higher energy efficiency than the 24-bit HW-SIMDi for the computation of the 12-bit 5-tap filter. This example motivates the proposed SW-SIMDi. Nevertheless, when making the same assumptions for a HW-SIMDi with 16-bit subword sizes, the resulting energy efficiency of about 800 MOPS/mW is again higher than the one of the SW-SIMDi. We refer to Sect. 3.4.1 and to [33] for related information. In the following, we will again focus on the implemented 12-bit HW-SIMDi.

Note, in the appendix of this chapter implementation results of the HW-SIMDi with 1 cluster, which where obtained by applying the proposed semi-custom design approach, are shown and compared.

5.5.3 Throughput and Scalability

A cluster with 12-bit subwords offers a parallelization factor of $2 \times 2 \times 4 = 16$. For the mapping of the 5-tap filter, which requires 5 clock cycles to compute the output, the HW-SIMDi with one cluster achieves a throughput of $1.12/5 \times 16 = 3.58$ Gsps. For the same setting, the SW-SIMDi can handle $0.9/5 \times 16 = 2.88$ Gsps. The required throughput for the IEEE 802.15.3c standard is 3.52 Gsps (2× the baseband rate for one channel). Thus, for the HW-SIMDi, one cluster is sufficient. However, for the SW-SIMDi, one cluster and an additional slice would be required. The total latency, including serial to parallel conversion from 8 bits to 384 bits, is less than 70 clock cycles. Thus, the achieved latency is better than the required latency of 167 clock cycles, see Sect. 5.2.1.

Figure 5.15 shows the achievable throughput for different number of utilized slices and for different number of taps and subword quantizations. Since the filter can be well parallelized, the throughput scales linearly with the number of utilized slices. The graph shows the instance scalability and the run-time scalability of the proposed processor solution. The instance scalability, i.e. the duplicating of clusters, enables a trade-off between area consumption and achievable throughput. Often at the beginning of a new design, the exact specifications (e.g. word length, number of filter taps) are unknown (e.g. because the standard is still in draft phase and/or because not enough realistic stimuli data is available). In this case, typically

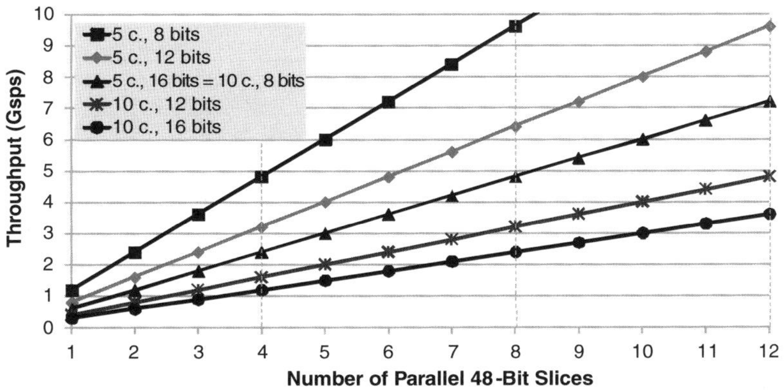

Fig. 5.15 Achievable throughput for different combinations of clock cycle requirements (to compute output) and subword sizes. The number of required clock cycles depends on the complexity of the filter (e.g. number of taps, number of non-zero bits in coefficients). 1 GHz clock frequency assumed. Computations on real data

a more conservative approach is chosen and sufficient parallel hardware is instantiated. Later in time, potentially even after the layout of the chip has already been made, more accurate specifications become available. In a traditional ASIC design, at this time, a rather limited set of options for adapting to the accurate specifications exists. Thus, the complete instantiated hardware, which is based on a conservative assumption, will typically always be active. In the proposed processor solution, we can exploit the run-time scalability to better adapt to the actual requirements. For instance, we can change the number of taps, the filter structure, the values and the quantization of the coefficients, the offered throughput and with the SW-SIMDi, even the subword sizes of the data. To lower the energy consumption, we can apply power gating to turn-off engines/clusters that are not required. In a traditional ASIC approach, power-gating of sub-filters is not possible, since it would require a reorganization of data. Because latency requirements may be violated, the application of voltage scaling and clock frequency scaling is also limited. Besides, a traditional ASIC cannot support the proposed optimization techniques, i.e. the conversion of multiplications to shift-add/sub operations and the optimization across filter taps, for multiple configurations. With the proposed processor solution, we can generally adapt to the actual requirements (1) of different standards (e.g. IEEE 802.15.3c, IEEE 802.11ad), (2) of different modes (e.g. SC/OFDM, modulation scheme), (3) of different use cases (e.g. video streaming, file transfer) [4, 30] and (4) of different channel conditions (e.g. see relationship between quantization, SNR and BER performance in Fig. 5.2) [5, 32]. Nevertheless, the latter would require a channel monitor.

Table 5.2 Overview of the implemented ASIC references to enable a fair comparison

Reference name	rASIC WC	rASIC TC	ASIC BC
Flexibility	Coefficient	Coefficient	–
Multiplication	Multiplier	Multiplier	Hardwired
Structure/Form	Direct	Direct	Transposed
Number of taps	25	5	5
Symmetry exploited	No	No	Yes
Input clock frequency (MHz)	880	880	880
Internal clock frequency (MHz)	440	440	880
Number of parallel sub-filters	8	8	4
Coefficients	16 bits	Like DSIP	Like DSIP
Input word length	8 bits	8 bits	8 bits
Word length growth (max. WL)	~WC (32)	WC (20 b.)	WC (15 b.)

5.6 Comparison and Discussion

In this section we compare the proposed processor instances to ASIC references and to programmable implementations from literature.

5.6.1 Implemented ASIC References

To enable an accurate comparison (based on same technology, technology type, number taps, quantization, coefficient values, throughput), three ASIC references have been implemented. Table 5.2 provides an overview of their characteristics. The rASIC Worst Case (WC) is the most conservative filter implementation. It enables to change the 16-bit coefficient values, employs 25 taps and the internal quantization does not add any quantization noise (WC scaling). All 8 parallel direct-form sub-filters share the same delay line and the same coefficient registers. The structure is roughly comparable with the filter of [44]. The rASIC Typical Case (TC) represents a traditional reconfigurable filter with 5 taps. Except for the number of taps and the coefficient word lengths, it is the same as the rASIC WC. The most optimistic reference is the ASIC Best Case (BC). The coefficients, which are the same as used in the proposal, are hardwired. To reduce the complexity of the multiplications, the transposed form structure was chosen and the MCM tool of Spiral was [48] used. All ASIC references employ one pipeline register per tap and have been dimensioned for the targeted throughput of 3.52 Gsps. The same hardware implementation design flow as for the proposed processor architectures has been applied.

5.6.2 Processor References from Literature

To obtain an indication on the efficiency of our proposal, we compared to programmable state-of-the-art solutions from literature. We denote the selected references as ASIP1-2 and SDR1-3. The ASIP1 of [19], named PIONEER, is based on the tool framework NoGap. It employs a 16-bit Reduced Instruction Set Computer (RISC) processor architecture which is extended with a 32-bit Digital Signal Processor (DSP) instruction set. ASIP2 is a 128-bit vector processor which includes an ASI that enables the efficient computation of FIR filters [36]. The processor, which was designed with the Target tool suite, can perform 8 parallel operations on 8-bit subwords. For reference SDR1, we have selected one Vector DSP (VDSP) of the Tomahawk SDR processor [21]. The VDSP is a 16-bit, 4-way SIMD, 5-way VLIW processor, which is based on the Synchronous Transfer Architecture (STA). Reference SDR2 is the Tensilica ConnX Baseband Engine (BBE) [37]. It is a 3-way VLIW (arithmetic, load, store) SIMD processor which can perform 16 parallel 18-bit integer MAC operations. We refer to the MT-ADRES processor of [42] as SDR3. This reference consists of 2 Coarse Grain Arrays (CGA)s with 4×4 FUs each. Each of the 32 FUs employs 4-way SIMD on 16-bit integer data.

We optimistically assume that the following number of parallel integer MAC operations per clock cycle can be computed and that all load/store operations can be hidden: ASIP1: 1 on 32 bits, ASIP2: 8 on 8 bits, SDR1: 20 on 16 bits, SDR2: 16 on 18 bits and SDR3: 128 on 16 bits. These assumptions will lead to the pessimistic lower bound, i.e. the minimum gain of our proposal compared to the references.

5.6.3 Comparison (Normalized, Pessimistic)

The number of taps and the word lengths are fixed for the ASIC references. For all processor references we assume the mapping of a real 5-tap filter (5 MAC operations), which is utilizing 100 % of all fore-mentioned parallel subword slots. We include the mapping with 8-bit and with 12-bit subwords for the SW-SIMDi.

All ASIC references have been implemented with the same technology as the proposed DSIP instances. However, this is not the case for the processor references. Thus, to enable a fair comparison, we scale all processor references to 40 nm technology. Thereby we use the following factors to scale between two technology nodes [11, 16]: 1.2 for clock freq./throughput, 2 for area (if not already specified in GE), 1.5 for energy/power. In addition to the technology, the area and energy numbers have been normalized to throughput.

Figure 5.16a shows the relative area comparison. It can be seen that there is a big difference amongst the three ASIC references (factor 73). Compared to the rASIC TC, the proposed processor architectures have an area penalty of factor 2–3.

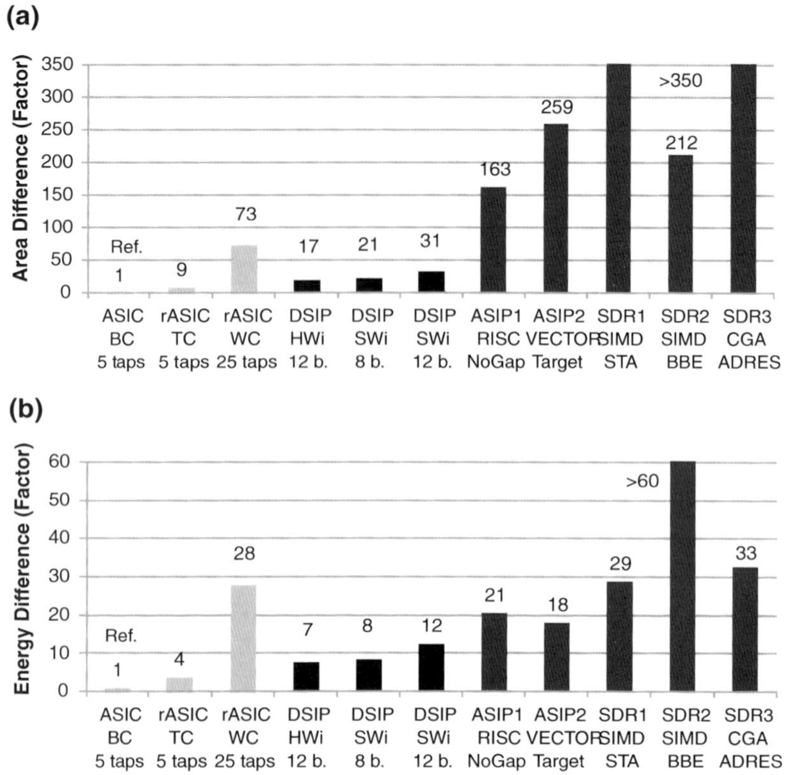

Fig. 5.16 Area and energy comparison of different filter implementations. Technology and throughput are normalized. ASIP1-2: [19, 36], SDR1-3: [21, 37, 42]. Estimations for the ASIP/SDR processors based on optimistic assumptions (i.e. in favor of the references). For a relative comparison amongst processor references from literature, more accurate schedules should be considered. **a** Area comparison. **b** Energy comparison

However, they are clearly more area efficient than the conservative rASIC WC reference. The area efficiency of the HW-SIMDi and the SW-SIMDi 8 bits is similar, but the difference of the HW-SIMDi to the SW-SIMDi 12 bits is rather big (reasons see below). The results of the different architecture instances illustrate the scalability of the proposed design approach. Compared to the processors from literature, our proposal is at least a factor 5 more area efficient.

In Fig. 5.16b the relative energy comparison is shown. Also here, the differences amongst the three ASIC references are big (factor 28). As for the area, the proposed processors consume about 2–3× more energy than the rASIC TC, but they are much more energy efficient than the rASIC WC. The differences between the proposed processors and the processor references are more than a factor 2, even for the pessimistic lower bounds. For instance, the HW-SIMDi is a factor 4 more energy efficient than the SDR processor solutions. These results demonstrate the high efficiency of the proposal.

5.6.4 Discussion

5.6.4.1 HW-SIMDi Versus SW-SIMDi

As mentioned in Sect. 5.5, the area difference between the HW-SIMDi and the SW-SIMDi is mainly caused by (1) the more complex SSU, which has a consequence on the datapath and the control storage area and by (2) the bigger DMEM. The latter could be compensated, since the DMEM is anyway not utilized. If we assume that the SW-SIMDi can operate most of the time in 8-bit mode, then basically no energy difference compared to the HW-SIMDi exists. However, if always 12 bits are required, then the SW-SIMDi has a disadvantage. In general, in the considered 5-tap filter application, the potential of SoftSIMD was not well exploited. Nevertheless, if we consider filters with more taps (where typically more different word lengths representations are required) and/or applications with higher storage requirements, e.g. FFT, then the benefits of SoftSIMD will become more visible.

5.6.4.2 SW-SIMDi Versus ASIC

The area difference between the SW-SIMDi 12 bits and the rASIC TC is about a factor 3. As shown in Fig. 5.14, 46 % of the processor area is consumed by the control, which is not present in the ASIC. The non-utilized data storage (DMEM, VWR T2, R2 and R4 of the SSUs) accounts to about 14 % of the total processor area. Considering these two differences, less than a factor 2 remains. A further reason for the difference is the number of employed registers (about 2× more storage bits than the ASIC) and the multiplexer, masking and shuffling hardware, which is not present in the ASIC. The processor operates at a higher clock frequency, does not include multipliers and employs less adders. However, the higher clock frequency has also a drawback, since it increases the cost for the adder implementations significantly. For instance, a low-performance carry-ripple adder is much cheaper than a high-performance parallel-prefix adder. As shown in Fig. 5.8a, the critical path of the processor includes the following components: 2 2-to-1 MUX, 1 8-to-1 MUX (shifter), 1 XOR and 2 NOR gates. These components enable flexibility, but their delay clearly decreases the performance and efficiency. In contrast, the ASIC includes only arithmetic operators in the critical path. The processor employs less arithmetic hardware, but it utilizes the employed hardware 5× (5 clock cycles) until the output is computed. In comparison, the ASIC reference operates fully parallel and does therefore not have the drawbacks of an iterative computation approach. If we compare to an ASIC with an iterative/serial architecture, the difference would be smaller (e.g. the comparison of a parallel and serial ASIC in [11] shows a factor 2 difference). The reasons for the area overhead are also representative for the energy overhead. Nevertheless, to enable high market volumes for ASIC designs, which is essential for the amortization of the high NRE costs [2], the designs have to be dimensioned for the WC application. The assumed WC design in this work, i.e. the rASIC WC, offers a comparatively low efficiency and has clear disadvantages compared to the proposed approach.

5.6.4.3 Proposal Versus Other Processors

The proposal is more efficient than other programmable architectures. The main reasons are the following: (1) Implementation of constant multiplications with shift-add/sub operations. Performing a very low number (CSD coding) of shift-add/sub operations is more area efficient than utilizing a hardware multiplier. In this filter application, a MAC operation is computed with a single shift-add/sub operation. (2) High hardware utilization because of basic unit. In a conventional Arithmetic Logic Unit (ALU), depending on the instruction, only a certain part of the hardware is utilized, the other part is inactive. Since we perform all arithmetic operations on the same hardware, a higher area efficiency is achieved. Besides, because of the architecture hierarchy, a better power gating is possible. (3) Optimized interconnect with reduced number of multiplexers. Because of (2), less multiplexers are required. In addition, we optimize the flexibility for typical data-flow patterns. (4) Data hierarchy. A wide DMEM and wide VWRs instead of a narrow DMEM and a conventional register file are employed. The local registers, which are individually controllable, replace the traditional pipeline registers. In general, the instantiated storage sizes are smaller than in other processors. (5) Control hierarchy. The distribution of the control enables to obtain short individual control loops which can be stored in small IMEMs. In conventional processors, typically a centralized approach, which implies bigger instruction caches and IMEMs, is employed. For the comparison of this work, a difference is also caused by the utilized subword sizes. In addition, the technology choice can have an influence.

5.7 Conclusions

This chapter introduces two scalable processors, denoted as HW-SIMDi and SW-SIMDi, which are well-suited for the computations of the matched filter functionality in 60 GHz systems. The architectures of the processors are based on the proposed DSIP architecture template of Chap. 3. Each of the processors offers a different efficiency/flexibility trade-off. The SW-SIMDi employs a SoftSIMD datapath, which enables the efficient mapping of algorithms with varying word-length requirements. To our best knowledge, this is the first work that reports an efficient implementation of SoftSIMD.

The implementation results in TSMC 40 nm technology, with the mapping of a 5-tap FIR filter, show a high energy efficiency of 970 MOPS/mW for the HW-SIMDi and of up to 860 MOPS/mW for the SW-SIMDi. Compared to state-of-the-art processor solutions, the proposed implementations are at least a factor 2 more energy efficient and a factor 5 more area efficient. An ASIC equivalent is typically still more energy efficient than the proposal. However, if the ASIC has been dimensioned conservatively, the difference is negligible. Thus, this case study proves the efficiency of the proposed design approach.

The HW-SIMDi, which performs FIR filtering with a throughput of 3.52 Gsps (=1.76 Gsps baseband rate) for complex data, consumes 37 mW. If we pessimistically assume that the FIR filter contributes only 10 % to the total computational complexity of the inner modem, we can estimate, that the processor implementation of the complete inner modem would consume about 370 mW. Considering a power budget of 500 mW for the complete digital baseband functionality and the fact, that the outer modem can be implemented with less than 130 mW in 40 nm [7, 23], we can conclude, that a complete processor solution for the inner modem seems feasible. To our best knowledge, this is the first work that indicates the feasibility of a full processor solution for mobile 60 GHz baseband computing.

In this application, with the considered 5-tap filter, the word-length requirement was basically limited to 12 bits. For this reason, the benefit of SoftSIMD could not be exploited. However, the estimations of Sect. 5.5.2 showed that the SW-SIMDi can achieve a higher energy efficiency than the HW-SIMDi for the typical operation mode. Future work should evaluate the benefit of SoftSIMD for other applications, such as FFT.

A.1 Appendix: Applied Back-End Semi-custom Design Approach

In the appendix of Chap. 3 a new back-end semi-custom design approach has been proposed. In contrast to the conventional standard-cell design flow, the proposed design approach takes explicitly the increasing influence of wires (DDSM technologies) [15] into account and targets specifically architecture templates such as the one proposed in this book. The goal of this new design approach is to obtain more efficient layouts, but without a major impact on average design time. In this appendix we show preliminary experimental results that have been obtained by applying a subset of the proposed design approach. The two designs that have been considered are (1) the FIR filter DSIP HW-SIMDi proposed in this chapter and (2) a medium-size standard-cell memory which could be used as DMEM in the architecture template. It should be emphasized that the content of this appendix is part of ongoing work and not fully mature yet, therefore only relative trends can be shown for now.

A.1.1 Implemented and Applied Semi-custom Design Flow

For the experiments, the proposed semi-custom template instantiation design flow (see appendix of Chap. 3 and Fig. 3.20a) has been implemented and applied. In the following a summary of the main steps of this design flow is provided.

1. *Architecture design, logic synthesis and layouting:* For the design and description of the logic circuits VHDL is used. VHDL supports "generics" which enable to make parameterizable designs. In order to simplify technology migration and to enable high reusability, a generic cell library is made (used as abstraction layer). This approach has strong similarities with the generic cell/logic libraries that are used in logic synthesis tools (e.g. GTECH in Synopsys). In the designed library the entity declaration of each cell is technology independent. For instance, a 2-input NAND gate has the name NAND2 and not e.g. TSMC-NAND2D1-LVTBWP. The drive strength of a generic cell is a parameter. Thus, manual sizing can be performed within the VHDL code. The implementation of this type of generic cell library is technology dependent, but since it consists basically just of "cell wrappers" (the cell content is not changed), the required effort for the library implementation is typically very low. The circuits of the proposed architecture design approach are designed by instantiating only cells from the generic library, and thus, high reusability is ensured. Since the DSIP architecture template has been designed while strongly considering the actual implementation in terms of wire-friendliness, and by ensuring the compatibility to the proposed back-end design approach (appendix of Chap. 3), the link between architecture and gate-level circuits is already largely pre-determined. In general, the circuits are composed in a fully hierarchical way and generic parameters are used wherever it is useful. An architecture instance is characterized by a certain value set of template (=generic) parameters. In parallel to the gate-level circuit design in VHDL, the layout is designed. We leverage on the Structured DataPath (SDP) feature of Cadence [10] to define the relative positions of standard cells. The relative cell placements are described with Python scripts (also in a fully hierarchical way) and then Python is used to automatically generate a SDP placement file depending on the chosen generic parameter values. Here, also generic parameters that define physical layout properties (e.g. orientation of a cell/module) are used. In accordance to the layout constraints defined in Sect. 3.2.2, typically a bit-sliced structure is targeted and the cells are abutted if possible. The relative positions of cell columns (e.g. a 48-bit 2-to-1 multiplexer and a 48-bit D-FF) are chosen in a way so that the lengths of the most important (usually horizontal) wires are minimized. In addition to the relative cell placement, optional routing constraints can be defined with TCL scripts. It should be mentioned that parts of the architecture which have little impact on the overall implementation result (e.g. certain control logic), can still be automatically synthesized and combined with the manually designed part of the design ("hybrid"). This means that the overall additional design effort should remain very limited compared to the conventional design flow, although this major step as such is very different.
2. *Gate-level verification:* After the previous step, the functional correctness of the designed circuits is verified. This type of gate-level simulation, which enables to verify different generic parameter values without involving a synthesis tool in the loop, can be done with ModelSim. In the conventional design flow a RTL simulation is made before logic synthesis and a gate-level simulation is made afterwards.

3. *Technology-mapping:* In this major step the generic VHDL gate-level netlist is translated to a fixed Verilog netlist (instance). We perform this fully automated task with Cadence RTL Compiler. Timing constraints can be specified so that an initial automatic sizing is performed. The timing report of the synthesis tool provides feedback whether the designed circuit will be able to met the targeted clock frequency. In case of a negative timing slack, a faster circuit needs to be designed. In the conventional design flow technology mapping is a automatically performed sub-step in the synthesis tool.
4. *Placement and routing:* First, all hard macros and all the standard cells, which have been considered in the SDP file, are automatically placed according to previously defined relative positions. Afterwards, all other cells, i.e. the ones that have little impact on the overall design, are automatically placed. Then, automatic clock tree synthesis is performed. Thereby the clock tree is build and the corresponding buffer cells are placed. Note, when planning the layout during the first step in this design flow, the required area and position of standard cells, which are not considered in the SDP files (incl. clock tree buffers), needs to be taken into account. Once the placement has been completed, automatic routing takes place. During this step the optionally specified routing constraints are considered. After placement and routing, an incremental optimization is performed. Thereby the position of the standard cells can be slightly moved (but the relative order remains the same), and the cells are resized so that the timing constraint is "just" met. As last step, the layout, i.e. the netlist (Verilog), the RC-parasitics (SPEF) and the actual delays of wires (SDF), is exported. We use Cadence EDI for all these steps. Compared to the conventional design flow, the layout design/generation is much more user driven by specifying relative cell positions and routing constraints.
5. *Timed gate-level verification:* In this major step the functional correctness of the design, under the consideration of the accurately estimated delays, is verified. We perform the timed gate-level simulation with ModelSim and generate thereby also a switching activity file (VCD). The timing of the design is additionally verified with Cadence Encounter Timing System (ETS). This major step is exactly the same as in the conventional design flow.
6. *Power calculation:* Typically once the layout is designed, the accurate power consumption of the design is of interest. We perform the power calculation with Cadence Encounter Power System (EPS). Thereby the RC-parasitics (SPEF) of the layout and the switching activities (VCD) are considered. Also here, this major is carried-out in the same way as in the conventional design flow.

Except for the first major step of this design flow, i.e. the architecture design, logic synthesis and layouting, all tasks are still fully automated. The manual effort performed during the first major step can largely be reused across template design instances and technologies. Additional information about the implemented design flow can be found in [24]. In the following preliminary experimental results, which have been obtained by applying the fore-mentioned design flow, are shown.

A.1.2 Experimental Results for the FIR Filter DSIP HW-SIMDi

The proposed semi-custom design flow has been applied to the FIR filter DSIP HW-SIMDi with one cluster (without L1 instruction memory hierarchy). The architecture of this design has been previously explained in Sect. 5.4. For this experiment, the TSMC 65 nm general-purpose LVT technology has been used.

To obtain a reference for the same technology (the results in Sect. 5.5.2 are based on 40 nm), the conventional design flow has been again applied. While applying the proposed semi-custom design flow, the netlist and the layout has been designed in a fully hierarchical way, which enables high scalability and reusability. To simplify the circuit design and layout planning, typically simple regular cell structures have been chosen. For instance, multiplexers have been implemented with MUX and not with AOI standard cells (the synthesis tool would typically choose the latter). Nevertheless, this trade-off between reduced cell area and improved regularity/routability should be further evaluated in future work. The adder that is present in the BAU has been implemented with a Sklansky topology, which has shown to offer a good trade-off in terms of performance/delay, cell area and wire lengths. In general, because the DSIP architecture template has been designed with the layout and circuits in mind, the actual design and implementation of the semi-custom layout has been rather straight-forward. A bit-sliced structure, where the number of standard-cell rows represents the datapath bit-width, has been chosen. The relative position of standard cells has been chosen manually, but the routing has been done fully automatically (no constraints specified).

Figure A.1 shows the semi-custom designed layout. As it can be seen, the relative physical placement of the datapath components corresponds to the architecture view of Fig. 5.5 (as desired). The L0 DLBs of the engines are closely placed to the datapath. The wires of the BAUs, which are highly active, are made very short. In the reference layout (conventional flow) 8 metal layers are used for the wires of the BAUs. In contrast, for the semi-custom designed layout not more than 6 metal layers are needed (and the upper ones are mainly used for connections with a relatively low toggling rate).

Table A.1 compares the implementation results obtained by applying the conventional and the proposed design flow. Both implementations are optimized for the same clock frequency and meet the timing with minimal slack. Interestingly, the cell area of the reference design is almost 10 % lower. This can be explained by the fact that the synthesis tool optimizes more aggressively for cell area, but at the cost of more irregular netlist structures. Despite of this, the total area of the semi-custom designed layout is 3 % lower. This gain is the result of a very high area utilization, achieved by good layout planning and cell abutment. In order to enable a full automatic placement and routing which leads to a valid design (conventional flow), empty layout area, typically in the order of 20–50 % needs to be present. The high number of Design Rule Check (DRC) violations in the automatically obtained layout shows that already a very optimistic area

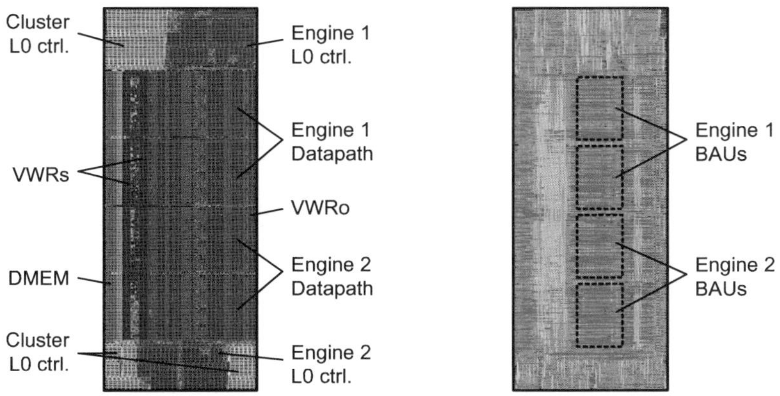

Fig. A.1 Semi-custom placed and automatically routed layout of the FIR filter DSIP HW-SIMDi with 1 cluster (without L1 control). The placement and the routing is very regular

Table A.1 Implementation results of the FIR filter DSIP HW-SIMDi with 1 cluster in comparison. The switching power, which is directly related to wires, has been reduced by 25 % in the semi-custom designed layout

	Conventional	Proposed	Difference
Timing slack (ns)	0.004	0.036	–
Std.-cell area/total area (μm^2)	76262/98709	82937/95436	+9/ −3 %
Number of DRC violations	985	166	−5.9 ×
Wire length (μm)	406000	352300	−13.2 %
Number of vias	134181	110714	−17.5 %
Internal power (mW)	26.8	26.38	−1.5 %
Switching power (mW)	14.08	10.55	−25 %
Total power (mW)	40.85	37	−9.6 %

utilization factor has been chosen. Because there is a difference of almost factor 6 in number of DRC violations between both implementations (interestingly, this even despite that already more routing layers are used in the automatically obtained layout), it can be assumed that fixing all the DRC violations will cause a higher overhead for the reference design. The total wire length and the total number of vias are 13.2 % and 17.5 % respectively lower in the semi-custom designed layout. For the dissipation of switching power mainly the wires of a design are responsible. The fact that the the total switching power of the semi-custom designed layout is even 25 % lower shows that the most active wires must have been optimized for short length (as desired). Given that both layouts have a similar standard cell area, it has been expected that also the internal power will be similar. Because the internal power dominates, the reduction of switching power has a moderate impact on the total power. The semi-custom designed layout consumes about 10 % less total power. For additional information on this experimental results we refer to [24].

A.1.3 Experimental Results for a Memory That is Based on Standard-Cells

Furthermore, the proposed semi-custom design flow has been applied to the following design: Standard-cell based memory with 128 bits and 32 words, 1 read and 1 write port (more ports could be added). This memory with medium size can be used as DMEM in the proposed DSIP architecture template. To enable a wider bit-width, the memory can simple be replicated. For the memory cells, D-latches are used. The maximal read delay has been specified to be 1 ns (=1 GHz). The experiment has been carried-out by using the TSMC 40 nm G nominal Vt technology.

The reference implementation has been obtained by applying the conventional standard-cell design flow, optimizing for area and power, given the specified timing constraint. An area utilization factor of 80 % has been chosen, which is a reasonable trade-off between area efficiency and the ability of the automatic placement/routing tool to generate a valid layout (DRC violations). While applying the proposed design flow, the gate-level netlist has been manually written. For the implementation of the read port, which is basically composed of 128 replicated 32-to-1 multiplexers, MUX 4-to-1 and MUX 2-to-1 standard cells have been used (high regularity). In the VHDL code standard cells with the lowest possible drive strength have been instantiated. However, during automatic sizing (incremental optimization) the drive strengths of the cells have been refined. To fit in the targeted layout scheme of the proposed DSIP template, 1 standard cell row for each bit has been used. The data I/O-pins are on the left/right edge of the layout and the control wires enter on the top/bottom edge. For the automatically designed layout, no constraints on layout ratio and pin positions have been specified (leads typically to best result).

Figure A.2 shows the fully automatically obtained layout and Fig. A.3 shows the layout obtained by applying the proposed designed flow. As it can be seen, the placement and the routing of the semi-custom designed layout is very regular. Table A.2 compares the two implementation results. Because the semi-custom designed layout is very regular, the standard-cell area and the core area are basically the same. When comparing the cell area after synthesis, i.e. the accurate influence of wires is not yet considered, the automatically obtained layout is about 10 % smaller (because of using AOI cells plus sharing of control gates instead of using MUX cells). Nevertheless, after placement, routing and sizing, i.e. after the actual layout is designed, the difference is negligible. Because of the chosen utilization factor, the total area of the reference implementation is 23 % higher. With the application of the proposed semi-custom design flow, the total wirelength and the total number of vias have been reduced by 82 % and 52 % respectively. This reduction has clearly a positive effect on the switching power and thus also on the total power. The total power, while performing memory read operations, has been reduced by 32.3 %. Note, in the proposed architecture template memory read operations occur generally much more frequently than write operations and most wires in the design are present in the read port (32-bit multiplexers).

Table A.2 Comparison of the implementation results obtained by applying the conventional and the proposed semi-custom design flow. With the proposed semi-custom design flow, the total area has been reduced by 23 % (area utilization factor) and the total read power by 32.3 % (mainly because of shorter wires)

	Conventional	Proposed	Difference
Timing constraint "just" met	Yes	Yes	–
Std.-cell area/total area (μm^2)	16503/20538	16379/16747	0.8/−23 %
Wire length (μm^2)	127456	70103	−82 %
Number of vias	57732	37963	−52 %
Write, internal power (mW)	12.01	12.23	+1.8 %
Write, switching power (mW)	4.61	4.47	−3.0 %
Write, total power (mW)	16.77	16.85	+0.5 %
Read, internal power (mW)	2.33	1.76	−24.5 %
Read, switching power (mW)	4.99	3.17	−36.5 %
Read, total power (mW)	7.47	5.06	−32.3 %

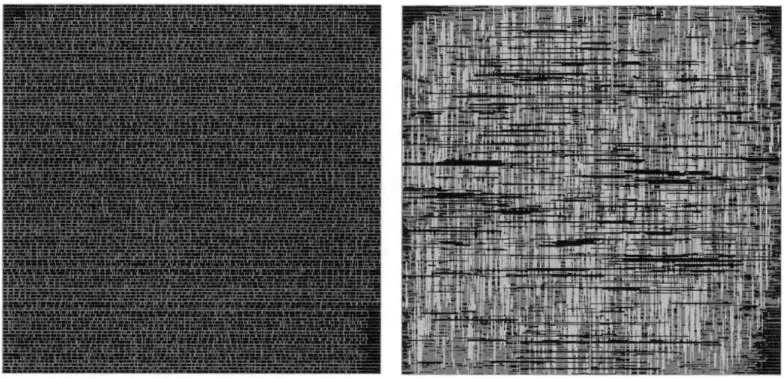

Fig. A.2 Layout of the designed 128 bits × 32 words standard-cell memory obtained by applying the conventional design flow

A.1.4 Conclusions

In this appendix the implementation of the proposed semi-custom design flow has been explained and the application results on two designs have been shown. Compared to the layouts obtained with the conventional design flow, a significant reduction of total wirelength and number of vias has been achieved. As a consequence, the switching power of the two designs has been reduced by 25 % and 36.5 % respectively. Because in the FIR filter DSIP HW-SIMDi design the internal power is dominating the total power, and the internal power has basically not been reduced, a high gain in total power has been left out. For the standard-cell memory additionally also the internal power has been reduced. The total power consumption during read operations has been lowered by 32.3 %. Although the achieved power reductions in these preliminary experiments are still rather limited,

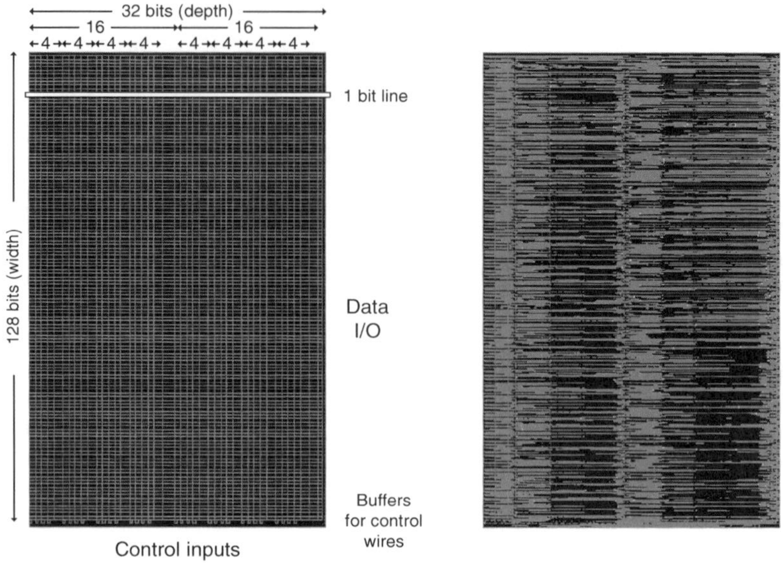

Fig. A.3 Layout of the designed 128 bits $\times$ 32 words standard-cell memory obtained by applying the proposed semi-custom design flow

they indicate the benefit of applying the proposed semi-custom design approach. There are many reasons why significantly higher gains can be expected for further scaled technology nodes beyond 20 nm (in the future, when applying the overall proposed approach):

- The considered designs in the experiments are relatively small, therefore it has not been very challenging for the automatic synthesis, placement and routing tool to find a good solution. However, for practical systems the circuits are much bigger (e.g. $> 10\times$ for one DSIP instance and $> 100\times$ for the digital baseband implementation) and therefore it is more likely that the conventional design flow leads to results that are less optimal. Since the proposed flow is explicitly based on design hierarchy and builds-up the layout in an explicit "divide-and-conquer" manner, it is more scalable and preserves the local and global efficiency.
- The experiments have been carried-out by using current 65/40 nm technology. However, for future DDSM technologies the relative importance of wires will be much higher [15].
- The applied semi-custom design flow is fully based on standard-cells, which are very generic. Nevertheless, by utilizing parameterizable custom cells for the most critical components, which are specifically designed for the targeted architecture template, a significant additional gain can be expected.
- By specifying routing constraints and by using a more flexible routing scheme, i.e. allow the routing of a single wire in horizontal and vertical direction on the same routing layer and not only uni-directional (river routing style), the length of the most important wires can potentially be further reduced.

- In the experiments the netlists have been designed manually and therefore mainly simple regular circuit structures have been chosen. However, for certain parts of the architecture a semi-regular netlist/layout may be more optimal. This is indicated by the fact that the netlist after automatic synthesis has typically always been smaller than the manually designed one.

The experience obtained during the application of the design flow also showed that the design time can be kept at an acceptable niveau. Clearly, future work is needed to unveil the full potential of the proposed design approach and to prove the assumptions. Importantly, the proposed layout-aware DSIP architecture template of this book is fully compatible to the proposed holistic design concept.

References

1. Anjum, O., Ahonen, T., Garzia, F., Nurmi, J., Brunelli, C., Berg, H.: State of the art baseband DSP platforms for software defined radio: a survey. EURASIP J. Wirel. Commun. Netw. **2011**(1), 5 (2011). doi:10.1186/1687-1499-2011-5
2. Ballhaus, W., Pagella, A., Vogel, C.: A change of pace for semiconductor industry? PricewaterhouseCoopers. Technology, Media and Telecommunications (2009)
3. Barale, F., Iyer, G.B., Perumana, B.G., Sen, P., Sarkar, S., Rachamadugu, A., Dudebout, N., Pinel, S., Laskar, J.: Pulse shaping and clock data recovery for multi-gigabit standard compliant 60 GHz digital radio. In: IEEE MTT-S International Microwave Symposium (MWSYM), pp. 908–911 (2010). doi:10.1109/MWSYM.2010.5517345
4. Baykas, T., Sum, C.S., Lan, Z., Wang, J., Rahman, M., Harada, H., Kato, S.: IEEE 802.15.3c: the first IEEE wireless standard for data rates over 1 Gb/s. IEEE Commun. Mag. **49**(7), 114–121 (2011). doi:10.1109/MCOM.2011.5936164
5. Bougard, B., Novo, D., Van der Perre, L., Catthoor, F.: Bridging the energy gap in size, weight and power constrained software defined radio: agile baseband processing as a key enabler. In: IEEE International Conference on Acoustics, Speech and Signal Processing (ICASSP), pp. 5384–5387 (2008). doi:10.1109/ICASSP.2008.4518877
6. Catthoor, F., Raghavan, P., Lambrechts, A., Jayapala, M., Kritikakou, A., Absar, J.: Ultra-Low Energy Domain-Specific Instruction-Set Processors, 1st edn. Springer (2010)
7. Chen, Z., Peng, X., Zhao, X., Xie, Q., Okamura, L., Zhou, D., Goto, S.: A 6.72-Gb/s 8pJ/bit/iteration IEEE 802.15.3c LDPC decoder chip. In: International Symposium on Integrated Circuits (ISIC), pp. 7–12 (2011). doi:10.1109/ISICir.2011.6131868
8. Dally, W.J., Balfour, J., Black-Shaffer, J.C., Harting, R.C., Parikh, V., Park, J., Sheffield, D.: Efficient embedded computing. Computer **41**(7), 27–32 (2008)
9. Diamantopoulos, D., Galiatsatos, P., Karachalios, A., Lentaris, G., Reisis, D., Soudris, D.: Configurable baseband digital transceiver for Gbps wireless 60 GHz communications. In: International Conference on Electronics, Circuits, and Systems (ICECS), pp. 192–195 (2011). doi:10.1109/ICECS.2011.6122246.
10. Encounter Digital Implementation System, Cadence. http://www.cadence.com/
11. Fasthuber, R., Li, M., Novo, D., Raghavan, P., Van der Perre, L., Catthoor, F.: Energy-efficient run-time scalable soft-output SSFE MIMO detector architectures. Transaction on high-performance embedded architectures and compilers (HiPEAC), Special Issue SAMOS 2009, vol. 5(3), pp. 1–20 (2011)
12. Fasthuber, R., Agrawal, P., Raghavan, P., Catthoor, F., Van der Perre, L.: A novel energy efficient wireless domain specific processor template. IEEE Trans. Circuits Syst. (TCAS) (2013, in preparation)

13. Guo, R., DeBrunner, L.S.: A novel fast canonical-signed-digit conversion technique for multiplication. In: IEEE International Conference on Acoustics, Speech and Signal Processing (ICASSP), pp. 1637–1640 (2011). doi:10.1109/ICASSP.2011.5946812
14. Hansen, C.: WiGiG: multi-gigabit wireless communications in the 60 GHz band. IEEE Wirel. Commun. **18**(6), 6–7 (2011). doi:10.1109/MWC.2011.6108325
15. IRC: ITRS Roadmap on Interconnect (2009)
16. Janhunen, J., Pitkanen, T., Silven, O., Juntti, M.: Fixed- and floating-point processor comparison for MIMO-OFDM detector. IEEE J. Sel. Top. Signal Process. **5**(8), 1588–1598 (2011). doi:10.1109/JSTSP.2011.2165830
17. Jayapala, M., Barat, F., Catthoor, F., Corporaal, H., Deconinck, G.: Clustered loop buffer organization for low energy VLIW embedded processors. IEEE Trans. Comput. **54**(6), 672–683 (2005)
18. Kanistras, N., Tsatsaragkos, I., Mahdi, A., Karagianni, K., Paliouras, V., Gioulekas, F., Lalos, E., Adaos, K., Birbas, M., Karaivazoglou, P., Koziotis, M., Perakis, M.: Digital baseband challenges for a 60GHz gigabit link. In: IEEE International Conference on Electronics, Circuits, and Systems (ICECS), pp. 346–349 (2011). doi:10.1109/ICECS.2011.6122284
19. Karlstrom, P., Zhou, W., Wang, C.H., Liu, D.: Design of PIONEER: a case study using NoGap. In: Asia Pacific Conference on Postgraduate Research in Microelectronics and Electronics (PrimeAsia), pp. 53–56 (2010). doi:10.1109/PRIMEASIA.2010.5604962
20. Li, M., Fasthuber, R., Novo, D., Van Der Perre, L., Catthoor, F.: Algorithm-architecture co-design of soft-output ML MIMO detector for parallel application specific instruction set processors. In: Design, Automation and Test in Europe (DATE), pp. 1608–1613, IMEC (2009)
21. Limberg, T., Winter, M., Bimberg, M., Klemm, R., Matus, E., Tavares, M.B., Fettweis, G., Ahlendorf, H., Robelly, P.: A fully programmable 40 GOPS SDR single chip baseband for LTE/WiMAX terminals. In: European Solid-State Circuits Conference (ESSCIRC), pp. 466–469. Technische Universtaet Dresden and Vodafone Chair Mobile Communications Systems (Fettweis) (2008)
22. Liu, H.M., Li, G.J., Yan, B., Li, Q.: A 100MHz digital down converter with modified FIR filter for wideband software-defined radios. In: International Conference on Electronics and Information Engineering (ICEIE), pp. V2-540–V2-544 (2010). doi:10.1109/ICEIE.2010.5559748
23. Marinkovic, M., Piz, M., Panic, G., Ehrig, M., Grass, E.: Performance evaluation of channel coding for Gbps 60-GHz OFDM-based wireless communications. In: IEEE International Symposium on Personal, Indoor and Mobile Radio Communication (PIMRC), pp. 994–998 (2010) doi:10.1109/PIMRC.2010.5671892
24. Medra, A.: Semi-custom design to reduce the interconnect energy in advanced technologies. Master thesis, Nile University, Egypt (2012)
25. Muller, J., Cathelin, A., Niknejad, A., Kaiser, A.: A FIR baseband filter for high data rate 60-GHz wireless communications. In: IEEE International Symposium on Circuit and Systems (ISCAS), pp. 1771–1774 (2010). doi:10.1109/ISCAS.2010.5537620
26. Niknejad, A.: Siliconization of 60 GHz. IEEE Microwave Mag. **11**(1), 78–85 (2010). doi:10.1109/MMM.2009.935209
27. Park, J., Jeong, W., Mahmoodi-Meimand, H., Wang, Y., Choo, H., Roy, K.: Computation sharing programmable FIR filter for low-power and high-performance applications. IEEE J. Solid-State Circuits **39**(2), 348–357 (2004). doi:10.1109/JSSC.2003.821785
28. Park, J.H., Richards, B., Nikolic, B.: A 2 Gb/s 5.6 mW digital LOS/NLOS equalizer for the 60 GHz band. IEEE J. Solid-State Circuits **46**(11), 2524–2534 (2011). doi:10.1109/JSSC.2011.2164137
29. Patil, D., Azizi, O., Horowitz, M., Ho, R., Ananthraman, R.: Robust energy-efficient adder topologies. In: IEEE Symposium on Computer Arithmetic (ARITH), pp. 16–28 (2007). doi:10.1109/ARITH.2007.31

30. Perahia, E., Cordeiro, C., Park, M., Yang, L.L.: IEEE 802.11ad: defining the next generation multi-Gbps Wi-Fi. In: IEEE Consumer Communications and Networking Conference (CCNC), pp. 1–5 (2010). doi:10.1109/CCNC.2010.5421713
31. Piz, M., Krstic, M., Ehrig, M., Grass, E.: An OFDM baseband receiver for short-range communication at 60 GHz. In: IEEE International Symposium on Circuit and Systems (ISCAS), pp. 409–412 (2009). doi:10.1109/ISCAS.2009.5117772
32. Portero, A., Talavera, G., Moreno, M., Carrabina, J., Catthoor, F.: Methodology for energy-flexibility space exploration and mapping of multimedia applications to single-processor platform styles. IEEE Trans. Circuits Syst. Video Technol. **21**(8), 1027–1039 (2011). doi:10.1109/TCSVT.2011.2129750
33. Psychou, G., Fasthuber, R., Hulzink, J., Husiken, J., Catthoor, F.: Subword handling in data-parallel mapping. In: Parallel Programming and Run-Time Management Techniques for Many-core Architectures (PARMA) (2012)
34. Raghavan, P., Lambrechts, A., Jayapala, M., Catthoor, F., Verkest, D., Corporaal, H.: Very wide register: an asymmetric register file organization for low power embedded processors. In: Design, Automation and Test in Europe (DATE), IMEC (2007)
35. Raghavan, P., Lambrechts, A., Jayapala, M., Catthoor, F., Verkest, D.: Distributed loop controller for multi-threading in uni-threaded ILP architectures. IEEE Trans. Comput. **58**(3), 311–321 (2009)
36. Rounioja, K., Puusaari, K.: Implementation of an HSDPA receiver with a customized vector processor. In: International Symposium on System-on-Chip (ISSOC), pp. 1–4 (2006). doi:10.1109/ISSOC.2006.322004
37. Rowen, C., Nuth, P., Fiske, S.: A DSP architecture optimized for wireless baseband. In: International Symposium on System-on-Chip (SOCC), pp. 151–156 (2009). doi:10.1109/SOCC.2009.5335658
38. Sheikh, F., Mill, M., Richards, B., Markovic, D., Nikolic, B.: A 1-190 MSample/s 8-64 tap energy-efficient reconfigurable FIR filter for multi-mode wireless communication. In: Symposium on VLSI Circuits, pp. 207–208 (2010). doi:10.1109/VLSIC.2010.5560297
39. Siligaris, A., Richard, O., Martineau, B., Mounet, C., Chaix, F., Ferragut, R., Dehos, C., Lanteri, J., Dussopt, L., Yamamoto, S.D., Pilard, R., Busson, P., Cathelin, A., Belot, D., Vincent, P.: A 65nm CMOS fully integrated transceiver module for 60GHz wireless HD applications. In: IEEE International Solid-State Circuits Conference (ISSCC), pp. 162–164 (2011). doi:10.1109/ISSCC.2011.5746264
40. Standard ECMA-387: high rate 60 GHz PHY, MAC and PALs. http://www.ecma-international.org/publications/standards/Ecma-387.htm
41. Standard WirelessHD. http://www.wirelesshd.org/
42. Suzuki, T., Yamada, H., Yamagishi, T., Takeda, D., Horisaki, K., Fujisawa, T., Van der Perre, L., Unekawa, Y.: High throughput and low power software defined radio using dynamically reconfigurable baseband processor. IEEE Micro **31**(6), 19–28 (2011). doi:10.1109/MM.2011.95
43. Thakkar, C., Lingkai, K., Kwangmo, J., Frappe, A., Alon, E.: A 10Gb/s 45mW adaptive 60GHz baseband in 65nm CMOS. In: Symposium on VLSI Circuits (VLSIC), pp. 24–25 (2011)
44. Tokumaru, M., Ikoma, H., Yamada, Y., Okamoto, K., Yamamoto, A., Shirakawa, Y.: A 14.6th-order 3.456GHz transmit baseband filter in 110nm CMOS for millimeter-wave communication systems. In: IEEE Custom Integrated. Circuit Conference (CICC), pp. 175–178 (2009). doi:10.1109/CICC.2009.5280872
45. van Berkel, C.: Multi-core for mobile phones. In: Design, Autom. and Test in Europe (DATE), pp. 1260–1265 (2009)
46. Vejanovski, R., Singh, J., Faulkner, M.: ASIC and DSP implementation of channel filter for 3G wireless TDD system. In: International ASIC/SOC Conference, pp. 47–51 (2001). doi:10.1109/ASIC.2001.954671

47. Verbruggen, B., Craninckx, J., Kuijk, M., Wambacq, P., Van der Plas, G.: A 2.6mW 6b 2.2GS/s 4-times interleaved fully dynamic pipelined ADC in 40nm digital CMOS. In: IEEE International Solid-State Circuits Conference (ISSCC), pp. 296–297 (2010). doi:10.1109/ISSCC.2010.5433925
48. Voronenko, Y., Püschel, M.: Multiplierless multiple constant multiplication. ACM Trans. Algorithms **3**(2), 11-es (2007). doi:10.1145/1240233.1240234

Chapter 6
Case Study 3: DSIP Architecture Instance for FFT Computation

6.1 Motivation, Related Work and Contributions

This section motivates this case study, summarizes related work on FFT baseband implementations and mentions the main contributions.

6.1.1 Context and Motivation

Pushed by Orthogonal Frequency-Division Multiplexing (OFDM) modulation-based standards on the one hand, and by frequency domain equalizer, i.e. in case of Single Carrier (SC) modulation, on the other hand, the computation of the Fast Fourier Transform (FFT) and the Inverse Fast Fourier Transform (IFFT) has a crucial role in modern baseband receivers. With the introduction of MIMO technology, the requirements and therefore the complexity for the (I)FFT computation has further increased. As discussed in Chap. 5, the 60 GHz band has a great potential in enabling very high data rate short-range wireless communication. However, the baseband implementation, and especially also the (I)FFT of the receiver, for 60 GHz standards is very challenging. In this chapter we focus on the (I)FFT for the IEEE 802.15.3c (WPAN) [3] and the IEEE 802.11ad (WLAN) [18] standards.

The 60 GHz communication standards define data rates up to 7 Gbps and strict latency requirements in the order of 3–6 μs. From this total latency budget, only a minor part can be spent on (I)FFT processing. Thus, (very) high performance and low latency computing is necessary. Especially to cope with the latter, the use of latest process technology is essential. To enable short-range high data-rate communication in mobile devices such as smartphones, the support for 60 GHz communication standards is also required there. Given the strict power budgets of these mobile devices, there is clear push for highly energy-efficient implementations of the complete 60 GHz wireless transceiver. Thus, this also effecting the implementation of the (I)FFT. The 60 GHz communication standards specify a

R. Fasthuber et al., *Energy-Efficient Communication Processors*,
DOI: 10.1007/978-1-4614-4992-8_6,

variety of (I)FFT blocksizes, which motivates the use of a flexible implementation. Since the (I)FFT is part of all current and emerging wireless communication standards and the NRE cost contributes significantly to the overall cost, the design of reusable components is essential. In summary, it is clear that there is a big demand for 60 GHz (I)FFT implementations that fulfill the 5 joint requirements of high performance, high flexibility, high energy efficiency, technology scaling-friendliness and high reusability.

6.1.2 Summary of Related Work

Several related ASICs [7, 12, 13], ASIPs [6, 10, 11, 21] and SDR processors [4, 8, 20] for the (I)FFT computation have been proposed. The ASICs are generally energy efficient and can cope with the strict latency requirements, but their lack of flexibility and reusability is the main issue. The latter is essentially for the cost-effective amortization of the high NRE costs. Some of the existing ASIP implementations are based on sub-threshold operation [23] or have a low-radix number split [17]. As a result, they cannot meet the throughput and latency requirements. Other processor-based solutions have been explored in [4, 8, 20], but they require a considerably large cycle count to compute a single (I)FFT. As a result, also they cannot meet the stringent timing requirements. Additionally, as it will be shown in Sect. 6.6, their energy consumption is clearly too high.

6.1.3 Main Contributions

In this case study we propose a novel (I)FFT implementation that is based on the proposed DSIP architecture template of Chap. 3. The proposal can fulfill the demanding specifications of the IEEE 802.15.3c (WPAN) and IEEE 802.11ad (WLAN) standards and achieves close to ASIC energy efficiency while offering the required degree of flexibility, high reusability and high technology scaling-friendliness. The comparison results with state-of-the-art ASIC, ASIP and SDR processor implementations motivate the proposed DSIP design approach. In contrast to case study 1 and 2, we leverage on the Target processor design tool [22] to implement the hardware and software.

6.2 Background on Driver

This section provides background information on the (I)FFT and motivates a flexible implementation.

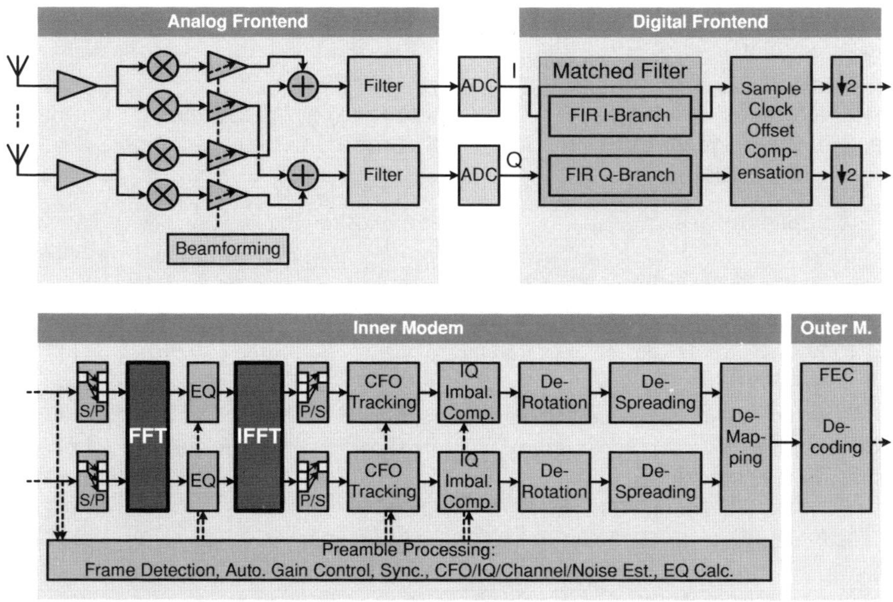

Fig. 6.1 Block diagram of the considered 60 GHz receiver. The front-end, the inner-modem and the outer-modem functionality is shown. The (I)FFT functional blocks is part of the inner modem and is situated before/after the equalizer

6.2.1 (I)FFT for High-Data Rate Standards

As shown in Fig. 6.1, the FFT/IFFT is typically part of the inner modem functionality of a receiver. In this case study we focus on the implementation of the (I)FFT for the IEEE 802.11ad and the IEEE 802.15.3c standards, which have many commonalities. The IEEE 802.15.3c standard specifies a FFT size of 512 points, which is also used in the IEEE 802.11ad standard as default size. The latter specifies additionally a low power SC PHY mode that uses cyclic blocks of length 64. Therefore, the IEEE 802.11ad standard leaves the choice whether to equalize with a 512 or 64-point FFT. In case of a significant high coherence bandwidth, the following alternative approach can be considered. Instead of computing the 512-point FFT, e.g. only a 128-point FFT is computed. Then, to compute the missing points in the frequency domain, linear interpolation (or even repetition) in the frequency domain is used. This approach is applicable for the channel estimation and the equalizer and results in a clear reduction of computational complexity. Importantly, since e.g. only 128 (instead of 512) time domain samples are sufficient to start with the channel estimation and equalizer computation, the latency specification can be easier met.

The throughput requirements differ between the SC and the OFDM mode:

- *SC mode:* The sampling frequency is specified with 1.76 GS/s, which results in a maximum time of 291 ns to compute the 512-point FFT. For the IFFT the same is valid.
- *OFDM mode:* Based on the sampling frequency of 2.64 GS/s, and taking the cyclic prefix length of 64 into account, the maximum time to compute the 512-point FFT is 218 ns.

The computational requirements are already challenging when considering only the throughput. However, the most constraining requirement is the Short Inter-Frame Space (SIFS) timing. This timing is defined in most IEEE 802.11 standards and relates to the time interval between the data frame and its acknowledgment. In the context of the IEEE 802.11ad standard the SIFS timing is defined with 3 μs. Since this timing includes also processing at the MAC layer, only a limited amount of it can be spend for the (I)FFT computation. Considering the relative computational complexity of the different functional blocks and the latencies for block buffering, we assume that 10 % of that time, i.e. 300 ns, for the (I)FFT computation seems reasonable. The worst case latency requirement is present in the SC mode, where both the FFT and the IFFT are in the critical timing path. Thus, here only 5 % of the SIFS timing, i.e. 150 ns, can maximally be spend for computing each, the 512-point FFT and the 512-point IFFT.

With a functional simulation of the IEEE 802.11ad standard in Matlab, which is also representative for the IEEE 802.15.3c standard, the quantization impact of the (I)FFT has been evaluated. The simulations have been performed by using the SC mode, 16-QAM, AWGN channel condition and LDPC coding (code rate 1/2) Fig. 6.2 shows the Bit-Error-Rate (BER) and Packet-Error-Rate (PER) performances for different number of internal (I)FFT quantization bits. As it can be seen, 8-bit internal quantization causes a major performance degradation. The 9-bit configuration causes only a minor degradation and with 10 bits close to floating-point performance can be observed. In the simulations we have seen that close to floating-point performance can also be achieved with the following configuration, which requires less data storage area: 8 bits for data input, internal word length growth up to 11 bits, scaling of the data output with saturation back to 8 bits. As it will be explained in Sect. 6.3.1, we use this configuration in the proposed implementation.

6.2.2 Motivation for a Flexible Implementation

The compliance with the targeted standards can be fulfilled by only supporting the (I)FFT block sizes 64, 128 and 512 points for the worst case throughput, latency and BER/PER specification, i.e. 3 modes. In a straightforward fashion, these 3 modes can be supported with 3 ASIC implementations. Nevertheless, a flexible (I)FFT implementation, which achieves almost equal energy efficient, would clearly be more cost effective. In order to improve the average energy efficiency, a

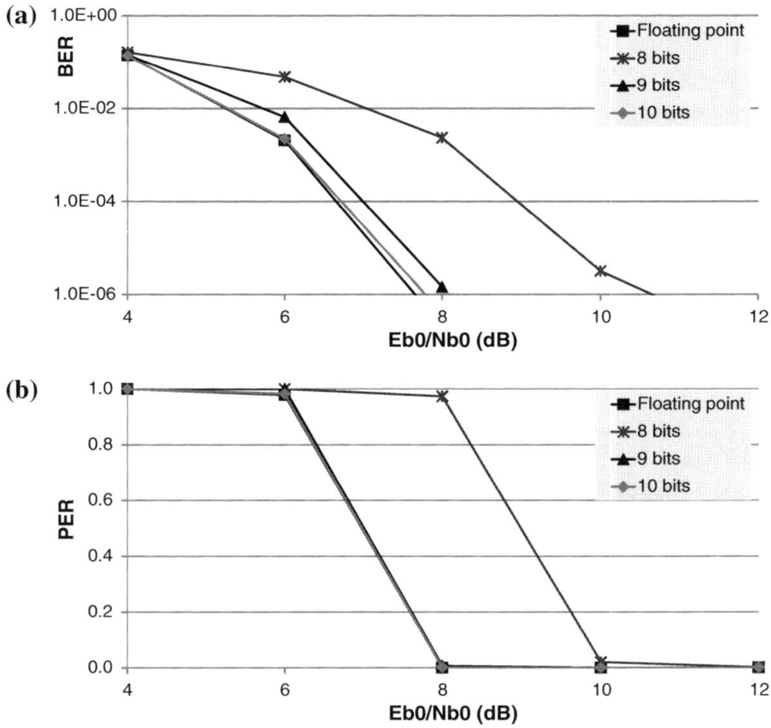

Fig. 6.2 **a** BER and **b** PER performance for different number of internal (I)FFT quantization bits. IEEE 802.11ad simulated mode: single carrier, 16-QAM, AWGN channel with LDPC coding (code rate 1/2). 9–10 bits are sufficient

flexible implementation will adapt to the dynamically varying actual requirements and not to the worst-case. For instance, the SC and the OFDM mode have different throughput requirements and that could be exploited. As mentioned above, in case of high coherence bandwidth, alternative (I)FFT computation schemes could be used. Also in case of a good channel condition, the quantization scheme could be changed. Thus, there are several reasons that motivate a flexible (I)FFT implementation. By adapting dynamically to the actual requirements, the average energy efficiency could even be better than the one of ASICs.

6.3 Algorithm Optimizations and Characteristics

In this section we first explain the algorithm that has been chosen for the proposed (I)FFT implementation. Then, we summarize its main characteristics that have been considered during design of the template architecture instance.

6.3.1 Algorithm Choice and Applied Optimizations

6.3.1.1 (a) Selection of (I)FFT Algorithm

In its generic form a FFT of length N is defined by

$$X(k) = \sum_{n=0}^{N-1} x(n) \times W_N^{nk} \tag{6.1}$$

where $0 \leq N$ and where W_N is the so called twiddle factor, denoting the Nth-primitive root of unity. Most algorithm optimization techniques share the same general strategy, namely first mapping the one-dimensional transform into a two or multi-dimensional representation, and then exploiting the congruence property of its coefficients to simplify the computation. Most commonly the FFT operation is broken down into radix-2 or radix-4 steps. The algorithm can be based on Decimation In Frequency (DIF) or Decimation In Time (DIT), which is an arbitrary choice of decomposing the larger FFT operations into smaller radix steps.

Based on the demanding throughput and latency requirements, we have opted for a 8-point sub-FFT operation (using radix-2 decomposition) based on DIT, which will be repetitively used to compute a 512-point FFT. The decomposition of the 512-point FFT into these sequence of 8-point FFTs is done by leveraging on the so called "cached" approach [2]. The Cached-FFT algorithm can be implemented for FFTs of any block size or radix [2]. It has been shown that implementations, which are based on the Cached-FFT algorithm, are significantly more efficient than implementations, which are based on the wide spread Cooley-Tukey algorithm [16]. The gains are present in both hardware and software implementations and are also proven for ASIP implementations [1]. The major reason for the higher efficiency of the Cached-FFT algorithm is the fact, that higher data locality is enabled.

Figure 6.3 shows the DFG of the Cached-FFT algorithm on an example. One can clearly see the decomposition into 8-point FFTs and permutations. The latter can be seen as row-column permutations of matrices with a size of 8×8 complex samples. Each parallel group of 8-point FFT operations is called epoch. Within an epoch there is a regular decomposition into operations which are based on local data, so called groups.

The Cached-FFT algorithm can also be used for the computation of the IFFT by the swapping real and imaginary part after each epoch. We have chosen to implement this (optional) swapping as part of the permutation after each epoch.

6.3.1.2 (b) Transformation of Multiplications

As shown in Fig. 6.4, the 8-point FFT, which is the basic processing element of the algorithm, is decomposed into 3 radix-2 steps. These steps are also called passes.

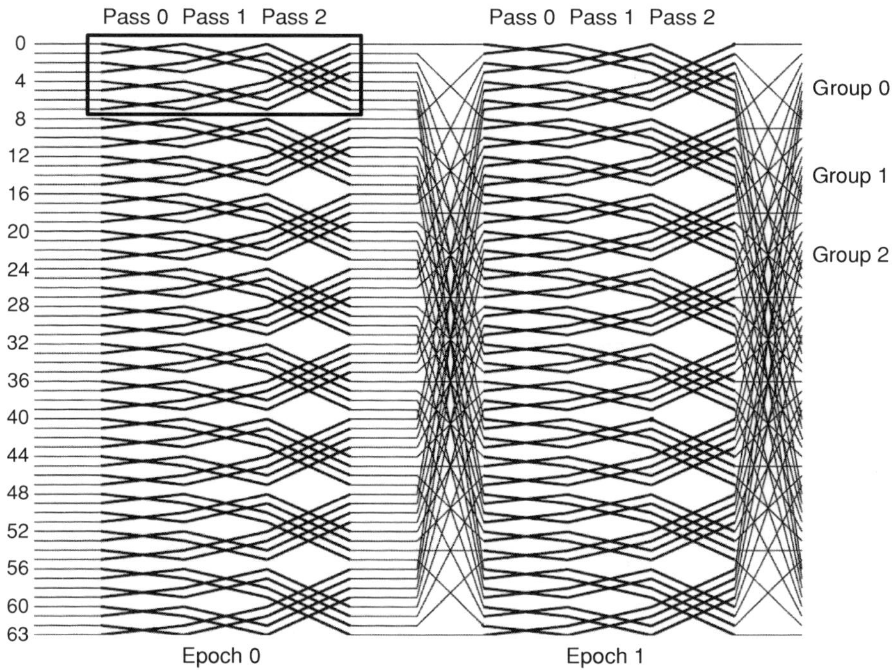

Fig. 6.3 Example of a data-flow graph that shows the 64-point Cached-FFT algorithm, which is composed of 8-point FFTs and permutations [2]

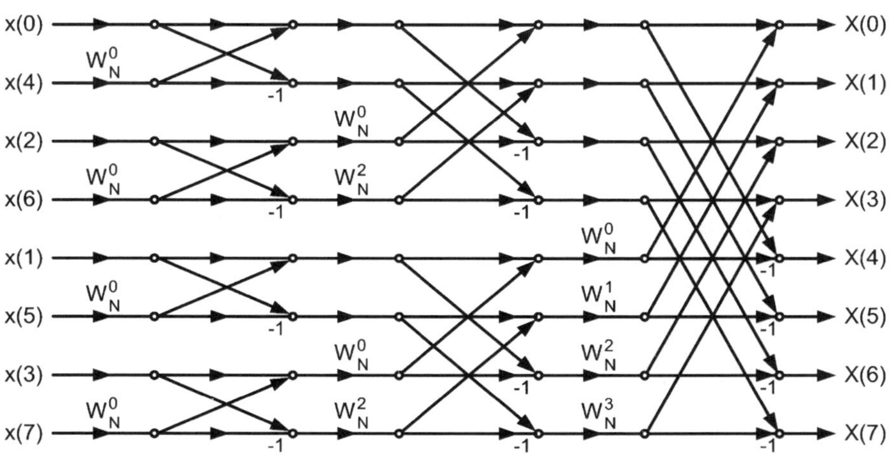

Fig. 6.4 Data-flow graph of the 8-point FFT, which is the basic processing element of the considered Cached-FFT algorithm. It consists of 3 radix-2 steps, so called passes

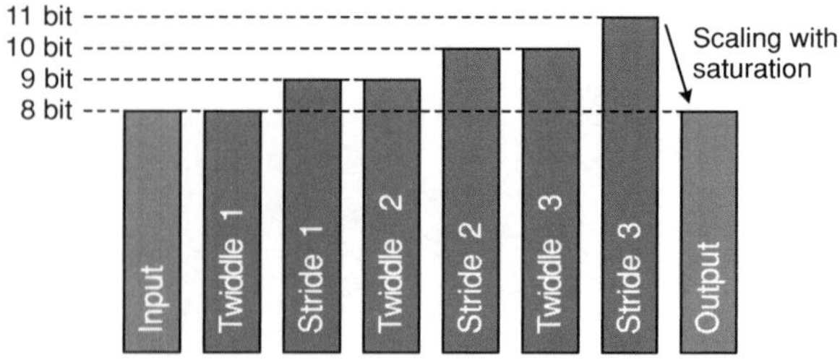

Fig. 6.5 Incremental word-length growth inside the 8-point FFT processing element

Each pass can be seen as a combination of multiplication (twiddle) stage and addition/summation (stride) stage. The multiplications of pass 1 and 2 are constant multiplications and can therefore be transformed to shift-add/sub operations. Nevertheless, for pass 0, variable multipliers are needed. The variable multiplications can be moved from the end to the beginning of the 8-point FFT, which enables to perform these operations with decreased word length. Hence, reduces complexity and decreases the latency.

To compute the 8-point FFT, the following sequence of stages can be identified: twiddle 1 (variable mult.), stride 1, twiddle 2 (shift-add/sub mult.), stride 2, twiddle 3 (shift-add/sub mult.) and stride 3. Only for the first twiddle stage the twiddle coefficients need be stored and loaded from memory. Since there is no feedback, these 6 stages of the 8-point FFT can be processed in pipelined fashion.

6.3.1.3 (c) Quantization Refinement

As mentioned in Sect. 6.2.1, we have opted for an 8-bit input/output data quantization and use internally more bits to avoid a significant BER/PER performance degradation. Figure 6.5 shows the chosen quantization scheme for the 8-point FFT computation. At every stride stage, we add an additional bit, resulting in maximal 11-bit internal word length. After the third stride stage, in order to go back to 8-bit word length, a scaling with saturation is performed. The Matlab simulation has shown that this configuration offers close to floating-point BER/PER performance. The fact that only 8 bits for the data representation outside the 8-point FFT are needed reduces the data storage requirements and reduces the complexity of the shuffler units.

6.3.2 Algorithm Characteristics

In the following we provide a summary of the main characteristics of the selected Cached-FFT algorithm.

6.3.2.1 Required Arithmetic and Logic Operations

In the considered algorithm arithmetic operations are only required within the 8-point FFT. All stride stages are based on basic arithmetic operations (BAOs), i.e. additions and subtractions. Additions and subtractions are also required to perform the constant multiplications (MCO)s of the last two twiddle stages. Variable multiplications (MVOW)s are only present in the first twiddle stage. The word length scaling in the last stage requires a constant multiplication (MCO). All other operation types, i.e. MVON, TSO and BLCO, are not present. The fact that additions and subtractions, which are simple operations, are dominating indicates that the algorithm can efficiently be implemented.

6.3.2.2 Required Word-Length Support

The variable multiplication in the first twiddle stage is performed with 8 bits. All additions and subtractions are performed with 8–11 bits. Except for the computations within the 8-point FFT, data transfers and data storages with 8-bit (sub-)word length is needed.

6.3.2.3 Data Dependencies and Parallelization

The following main parallelization options exist in the considered Cached-FFT algorithm: (1) parallel computation of real and imaginary part, (2) parallel computation within the 8-point FFT, (3) parallel computation of different groups and (4) parallel computation of the different epoch stages. We exploit the parallelization options (1)–(3) to meet the stringent throughput and latency requirements.

6.3.2.4 Data and Control Flow

The data-flow structure of the algorithm is completely deterministic and regular. No conditions are present. The algorithm is heavily loop-dominated.

6.3.2.5 Data Storage Requirements

A distinction between (1) input data, (2) intermediate data between epochs, (3) epoch internal data, (4) twiddle factors and (5) output data can be made. For the (1) input data as well as for the (5) output data a storage of the FFT block size is needed. Considering the 512-point FFT with 8-bit data quantization and complex data, this results in 8 kbits of memory for the input data and of 8 kbits of memory for the output data. To save on memory space, the output data could be stored on the same location as the input data (in-place). Nevertheless, to cope with the tight latency requirement, we apply a streaming-based processing where the input and output data are stored in different memories. Thus, as soon as a line from the input data memory has been read and provided to the arithmetic datapath, it can be overwritten with the input data of the subsequent FFT. Because of the permutations that need to be done across the whole block size, for the (2) intermediate data between epochs again at least a storage of 8 kbits is required. We consider to distribute this storage across the parallel processing datapaths of the processor. The amount of required (3) epoch internal data storage depends on how the the arithmetic datapath for the computation of the 8-point FFT will be implemented and on how many epochs are computed in parallel. Considering that an epoch is computed fully parallel and a pipeline register is present after each stage, at least 6 (stages) $\times$ 8 (FFT sub-size) $\times$ 9.5 (quantization, average number of bits) $\times$ 2 (complex data) $=$ 912-bit storage is needed. For the computation of the 512-point FFT, at least $1 + 8 + 64 = 73$ different complex (4) twiddle factors are required. Thus, a minimum storage of 1168 bits needs to be present.

6.4 Proposed DSIP Architecture Instance

In this section, we propose a DSIP architecture template instance which is well suited for the targeted (I)FFT application domain, i.e. it can compute 64, 128 and 512-point FFTs. In the following the hierarchical architecture is explained in a top-down manner.

6.4.1 Top Level

Figure 6.6 shows the top level view of the proposed architecture instance. It consists of one or multiple homogeneous clusters and the Shared Top Level Control (STLC).

6.4.1.1 Clusters

Across clusters, Data Level Parallelism (DLP) and Instruction Level Parallelism (ILP) is exploited. With the number of instantiated clusters the maximal achievable throughput can be scaled.

6.4.1.2 Shared Top Level Control

The STLC consists of the L1 master control with the corresponding L1 IMEM Direct Memory Access (DMA), the L1 IMEM and the DMEM LUTU. The L1

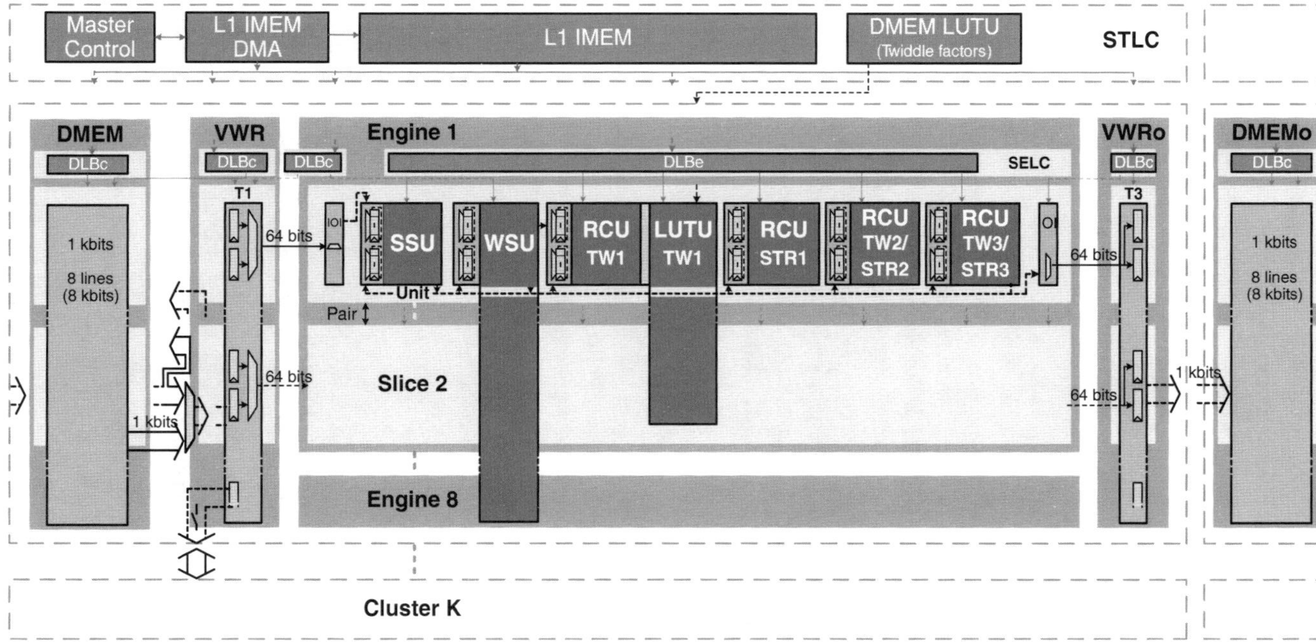

Fig. 6.6 **Top** level view of the proposed DSIP architecture instance for perform. FFT computations. With the number of clusters K the throughput can be scaled

master control has the full control of the processor. All instructions that are provided to the clusters are stored in the L1 IMEM. The DMEM LUTU contains all twiddle factors. The required twiddle factors will be loaded to the LUTU TW1s, i.e. present in the slices, perform the FFT computation takes place. Since all clusters are homogeneous and execute the same set of instructions, the L1 IMEMs can be shared amongst them. The memories of the STLC are currently implemented with D-FFs and D-latches.

6.4.2 Cluster Level

A cluster consists of a DMEM, a VWR, 8 engines, a VWR for the output (VWRo) and the L0 Cluster Level Control (CLC). The latter consists of multiple cluster level DLBs (DLBc)s. In the Fig. 6.6 also the DMEM output (DMEMo) is depicted. The output of the (I)FFT is written to this memory, which would typically be the DMEM of the subsequent architecture instance. Given that 8 engines are present, each engine includes 2 slices, each slice includes 8 slots, and each slot is 8-bit wide, the datapath between the DMEM and the VWR (and the VWRo and the DMEMo) is 1024 bits wide.

6.4.2.1 Data Memory

The DMEM, the DMEMo, the VWR and the VWRo have a wide interface of 1024 bits. A wide interface enables to reduce the number of accesses and also the width of the read and write address. This decreases the control energy. The DMEM has a size of 8 kbits (8 words), which enables to store the data of a complete 512-point FFT. It is currently implemented with D-FFs.

6.4.2.2 Very Wide Register

The VWR, which is characterized by an asymmetrical interface [19] (1024 bits vs. 128 bits), replace the conventional register file. A VWR of type T1 is employed for the data transfers from the DMEM to the engines. At every read access data is transferred to both slices of an engine, therefore the width of the read port is $2 \times 64 = 128$ bits. Since for the considered (I)FFT mapping data needs to be moved from the DMEM to engines, but not back, only 1 VWR is present. Nevertheless, to enable more flexibility, further VWRs of different types could be added (e.g. like done in the MIMO and FFT filter instances). The VWR is implemented with standard-cell D-FFs and has a size of 1024 bits.

6.4.2.3 Engines

To meet the targeted throughput and latency requirements, 8 engines are employed within a cluster, which enable to compute a 64-point FFT in parallel. Each engine includes 2 paired slices, i.e. one for real and one for imaginary data. Amongst the 8 homogeneous engines and utilized slices, DLP and ILP can be exploited. The engine level will be explained further below.

6.4.2.4 Very Wide Register Output

Once the output data is computed, it is written to the VWRo. The VWRo is implemented as type T3. It has 8 narrow write interfaces with 128 bits each and 1 wide read interface with 1024 bits. Like the VWR, the VWRo is implemented with standard-cell D-FFs and has a size of 1024 bits.

6.4.2.5 Data Memory Output

Like the DMEM, the DMEMo has a size of 8 kbits (8 words) and is currently implemented with D-FFs.

6.4.2.6 Cluster Level Control

As it can be seen in Fig. 6.6, the CLC consists of several cluster level DLBs (DLBc)s to control the the DMEM, the VWR, the WSU and the VWRo (potentially also the DMEMo if not considered in the subsequent architecture instance). This separation prevents a merging of all loops to a common loop nest, as it would be required for a centralized loop buffer [15]. Thus, the number of No OPerations (NOP)s, the storage and the energy is reduced. The DLBcs are currently implemented with D-FFs, however, they can potentially be replaced with D-Latches.

6.4.3 Engine Level

An engine consists of 2 paired slices, which share the same control. One slice is used for the computation of the real part and one slice for the computation of the imaginary part of the complex data. The slices of an engine are tightly interconnected to avoid long latencies. The slice level will be explained further below.

6.4.3.1 Slices

In each slice the heterogeneous arithmetic and shuffler units are present. The internal slice structure will be explained further below.

6.4.3.2 Engine Level Control

In the current implementation the Shared Engine Level Control (SELC) consists of a single distributed loop buffer (DLBe), which provides instructions to all units. Nevertheless, a split into multiple DLBes can potentially reduce the instruction storage size and energy consumption, therefore this optimization should be considered in the future. The DLBe is currently implemented with D-FFs, however, an implementation with D-Latches may be possible as well.

6.4.4 Slice Level

The currently implemented slice is customized for the targeted Cached-FFT algorithm, i.e. the arithmetic operations are computed within Reusable Custom Unit (RCU)s. This approach was chosen to ensure that the very demanding latency requirements can be met, but the drawback of this solution is the reduced flexibility, i.e. support for different algorithms. As mentioned in Sect. 6.3.2.1, in the considered algorithm, most operations are additions and subtractions, and only few variable multiplications are present. This is also the case for most other (I)FFT algorithms. Therefore the arithmetic operations could also be computed with a combination of Basic Arithmetic Unit (BAU) and Multiplier Unit (MU). An implementation of such a slice would enable a higher flexibility and could be considered as an alternative option. Nevertheless, whether the targeted latency constraint can be met with such a solution, has still to be verified. Potentially the latest available process technology in general-purpose flavor would have to be used. We want to emphasize that this important investigation should be part of future work. The current slice implementation is based on hardware SIMD. Future work should also consider the option of employing software SIMD. A more flexible/programmable implementation would enable a higher reusability and a better adaptability to the actual requirements (see Sect. 6.2.2).

Figure 6.7 shows the slice that has been implemented for this work. As it can be seen, it consists of a SSU, a WSU, a RCU TW1 (computations of the first twiddle stage), a LUTU TW1 (twiddle factor storage), a RCU STR1 (first stride stage), a RCU TW2/STR2 (second twiddle/stride stage) and a RCU TW3/STR3 (third twiddle/stride stage). The datapath contains 8 slots and is 64-bit wide. Each unit (except the LUTU TW1) has individual input registers (also referred as pipeline registers). Data from the VWR can be written to the SSU. The (lower) interconnection network enables to exchange data amongst units and to write the output

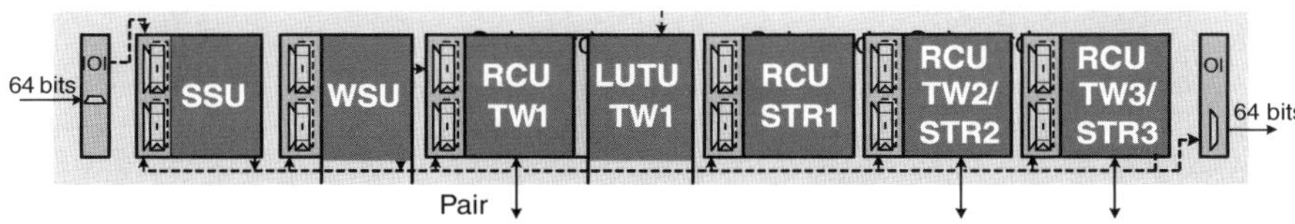

Fig. 6.7 Overview of a slice. A slice is either used to compute the real or the imaginary part of the complex data. The pairing of two slices enables to compute complex data efficiently. The fully pipelined datapath contains 8 slots and is 64-bit wide

data to the VWRo. Certain units have a direction connection to the paired slice, hence, that data can be exchanged without utilizing the WSU. With the proposed slice, an 8-point FFT can be computed fully parallel in a pipelined fashion.

6.4.4.1 Subword Shuffler Unit (SSU)

For the considered Cached-FFT algorithm, two essential data shuffle operations are inherently required:

1. *Bit reversal:* It refers to the input (or output) re-ordering which is necessary to enable in-place operations while processing the recursive stages of the (I)FFT algorithm. Note, the name bit reversal means literally changing the bits in an index vector, which points to the actual data vector. The actual operation is the re-ordering of the elements in the data vector. The number of elements in the data vector corresponds to the FFT block size. This operation is supported by utilizing the SSU (small distance shuffling) together with the WSU/VWR (long distance shuffling). Certain local re-ordering operations, which are necessary within a 8-point FFT, are also done within the arithmetic units. Note, the flexibility offered by the WSU and VWR enables to support different FFT block sizes.
2. *Permutation:* It is a specific data shuffle operation that is performed between the different epoch stages, which can be seen in Fig. 6.3. This permutation, which can be seen as a row-column permutation of 8×8 words, is entirely performed in the SSU. A graphical representation of this operation is shown in Fig. 6.8. Since this operation is deterministic and used in the same form for all considered FFT block sizes, the permutation pattern is hard-coded within the SSU, thus only 1 instruction is needed.

The VWR and units of the slice can write to the local input registers of the SSU.

6.4.4.2 Word Shuffler Unit

As mentioned above, the Word Shuffler Unit (WSU) is used together with the SSU to support the bit reversal operations. The WSU can be seen as two 8×8 cross-

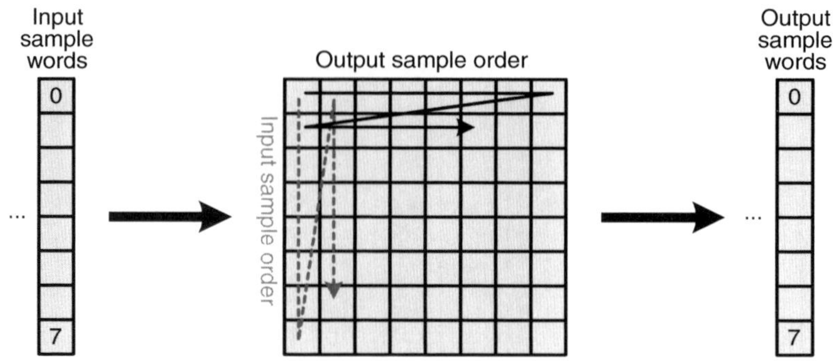

Fig. 6.8 Row-column permutation that is supported with the SSU. This operation is performed between the different epoch stages of the Cached-FFT algorithm

bars that connect the slices of the same type (real/imaginary) across the 8 engines. The output of the WSU can directly be provided to the RCU TW1 (i.e. without using the lower interconnect network).

6.4.4.3 Twiddle Unit 1 (RCU TW1)

This unit is used to perform the variable complex twiddle factor multiplications of the first stage (see Sect. 6.3.1), i.e. $(a + bi) \times (c + di)$, whereby a/b represents the input data and c/d represents the twiddle factor. The implementation of the RCU TW1 is depicted in Fig. 6.9a. To exchange the data between the real and imaginary slice of an engine, a direction connected is present.

6.4.4.4 Twiddle Storage 1 (LUTU TW1)

In this look-up table unit the twiddle factors, which are needed by the RCU TW1, are stored. As shown in Fig. 6.6, the required twiddle factors are loaded from the shared DMEM LUTU to the RCUs TW1.

6.4.4.5 Stride Unit 1 (RCU STR1)

The RCU STR1 is a simple unit that performs local re-ordering as well as additions and subtractions. As depicted in Fig. 6.9b, the units across slices do not need to be connected. In contrast to the twiddle and stride stage 2 and 3 (see Sect. 6.3.1), the functionalities of the twiddle and stride stage 1 have not been combined into a single unit. The reason for that design decision is related to the delay balancing of the pipeline.

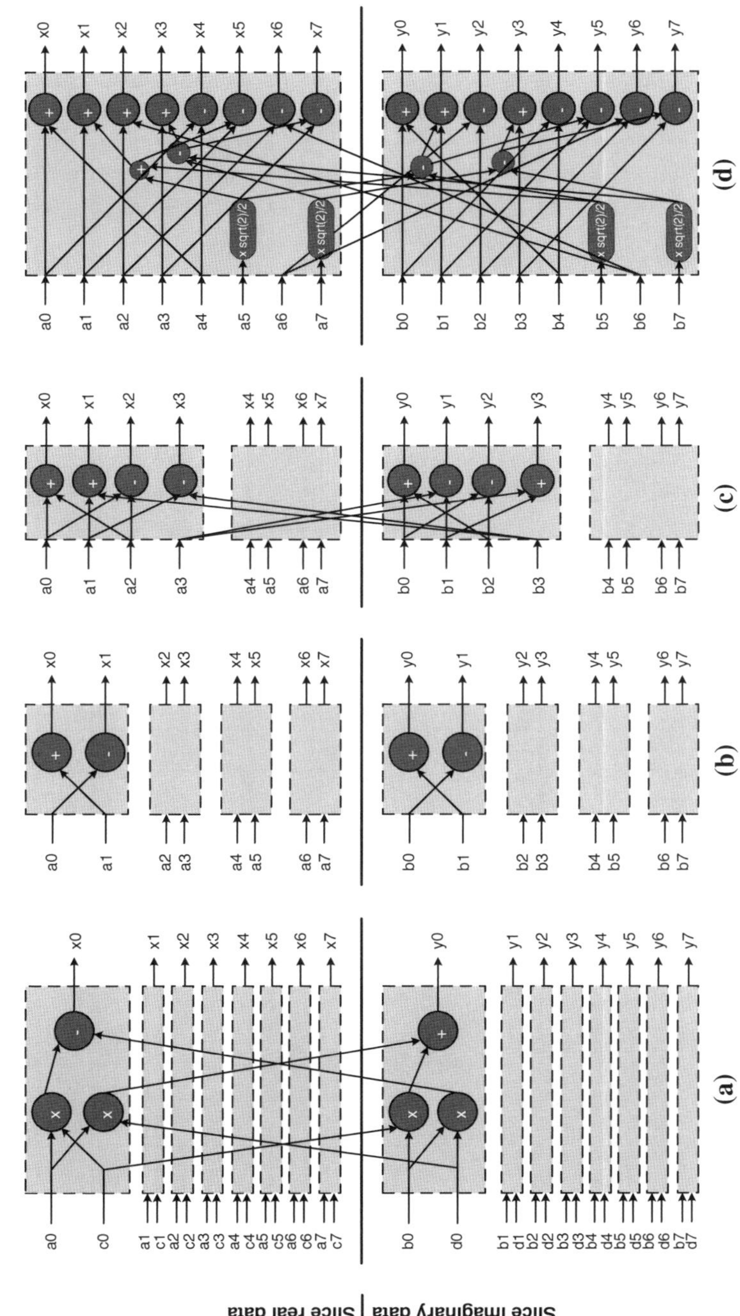

Fig. 6.9 Arithmetic units of the proposed DSIP FFT architecture instance. Most units are connected across the two slices of an engine in order to reduce latency. **a** RCU TW1. **b** RCU STR1. **c** RCU TW2/STR2. **d** RCU TW3/STR3

6.4.4.6 Twiddle/Stride Unit 2 (RCU TW2/STR2)

The RCU TW2/STR2 arithmetic unit performs the computations of the second twiddle and stride stage that is part of the 8-point FFT. As mentioned in Sect. 6.3.1, the constant multiplications have been transformed to shift-add/sub operations, therefore no multipliers are needed. Figure 6.9c shows the implementation of this unit (note, the hard-wired shifts are not shown), which requires a direction connection between the two slices of an engine.

6.4.4.7 Twiddle/Stride Unit 3 (RCU TW3/STR3)

The computations of the last twiddle and stride stage of the 8-point FFT are performed with the RCU TW3/STR3. Like for the RCU TW2/STR2, no multipliers are present and the real and imaginary slice of an engine are directly connected within this unit. The implementation of this unit is shown in Fig. 6.9d. The scaling back to 8-bit word length is also performed here.

6.5 Software Mapping and Hardware Implementation

In this section the utilized processor design tool, the mapping of the considered Cached-FFT algorithm onto the proposed architecture instance and the hardware implementation results are shown and discussed.

6.5.1 Utilized Processor Design Tool Suite

The architecture instances of case study 1 and 2 have been implemented by writing VHDL RTL code by hand, and the software has been manually written in a low-level hardware language. Because of the high design freedom, this design approach enables to obtain very efficient implementations, but at the cost of high design time. The design time can significantly be reduced by leveraging on a processor design tool suite which enables to model the architecture in a high-level description language, and includes a C-compiler, an instruction set simulator as well as a RTL generator. The Target tool suite [22] offers this functionality and was therefore used during this case study.

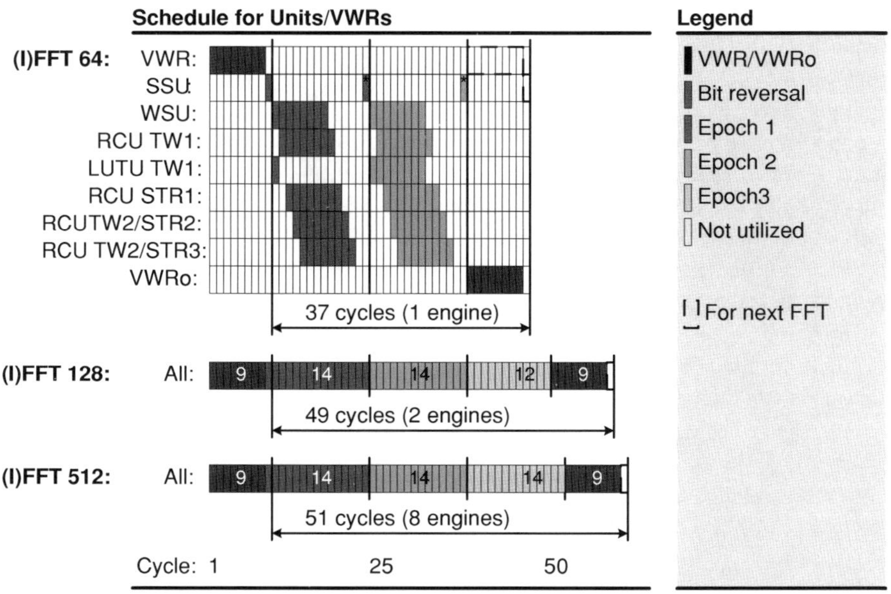

Fig. 6.10 Overview of the Cached-FFT scheduling. For the 64-point FFT also the activity of the VWR(o) and the units is shown. For subsequent FFT computations, the writing from the VWR to the units and the writing of the units to the VWR out can be done in parallel. Depending on the FFT block size, a different number of engines is utilized to compute the output of a single FFT

6.5.2 Software Mapping and Scheduling

The considered Cached-FFT algorithm has been mapped for 3 different FFT block sizes onto the proposed DSIP architecture instance. For the computation of the FFT output, the following steps are performed: (1) write input data to the DMEM, (2) transfer input data from the DMEM to the VWR, (3) transfer data from the VWR to the SSU, (4) initial bit reversal for epoch 1, (5) compute epoch 1, (6) initial bit reversal for epoch 2, (7) compute epoch 2, (8) initial bit reversal for epoch 3 (only for 128 and 512-point FFT), (9) compute epoch 3 (only for 128 and 512-point FFT), (10) transfer output data to the VWRo and (11) transfer data from the VWRo to the DMEMo. From this sequence it can be noticed that intermediate data does not need be written back to the VWR/DMEM. Instead, it is kept in the storage that is present in the SSU, which is the case for all FFT block sizes. Since for FFTs with bigger block sizes also more engines (hence also more SSUs) are used, the larger intermediate data can be distributed across all utilized SSUs.

Figure 6.10 depicts the scheduling for the steps (3)–(10). The activity of the VWR(o) and the units is shown for the mapping of the 64-point FFT, which utilizes only one engine. From the activity of the units it can clearly be seen that the processing is done in a pipelined fashion. The reading from the VWR (step 3) and the writing to the VWRo (step 10) can be done in parallel in case multiple

FFTs are computed in a sequence, i.e. streaming, which is normally the case. Because of the fixed twiddle coefficients in the epoch 1 stage, only a single load from the LUTU TW1 to the RCU TW1 is needed. Since the output of the preceding epoch is required before the processing of the next epoch can take place (feedback), the WSU and each arithmetic unit have to wait for 6 cycles (NOPs). Thus, there is still optimization potential and future work should investigate in mapping alternatives to enable a higher hardware utilization. The presence of NOP cycles also motivate the use of distributed engine-level loop buffers.

As it can be seen from Fig. 6.10, depending on the FFT block size, 1, 2 or 8 engines are utilized to compute the output of a single FFT. Inactive engines can be power-gated. To maximize the throughput, 8 64-point FFTs or 4 128-point FFTs can be computed in parallel. Since in these configurations the total required data storage is not higher than in the case of the 512-point FFT, the employed data memories are large enough. To compute FFTs of block size 64, 128 and 512 in a streaming fashion, 37, 49 and 51 respectively clock cycles are needed. The maximum latency in clock cycles, which includes DMEM(o) and VWR(o) accesses, is 51, 63 and 65 respectively.

6.5.3 Hardware Implementation and Results

The proposed DSIP architecture instance has first been described in the high-level hardware description language nML, and then the corresponding VHDL RTL code has been automatically generated by the Target tool suite [22]. Afterwards, the VHDL RTL code has been synthesized using Cadence RTL Compiler Physical, automatically placed and routed with Cadence Encounter (EDI). For the energy estimation, ModelSim and Cadence Encounter Power System (EPS) have been used. The reported power/energy numbers are based on post-layout gate-level simulations (full mapping) with annotated parasitics. TSMC 40 nm Low-power (LP) low-threshold-voltage standard-cell technology has been employed. For the current implementation, mostly D-FF-based memories have been used.

The resulting implementation (layout) can operate at a clock frequency of 440 MHz (WC corner). An instance with 1 cluster (=8 engines), including all data and L0/L1 instruction memories as well as the DMEM LUTU and even the DMEMo, consumes a total area of 0.751 mm^2. With 62.5 % the area utilization (standard-cell vs. chip area) is rather low. The main reason for the rather low area utilization is the combination of standard-cell based memories and the application of the conventional standard-cell design flow. As motivated in the appendix of Chap. 5, the proposed semi-custom design flow can solve this issue. In future designs also the employment of memory macros should be considered, since they consume typically less area than standard-cell based implementations. Figure 6.11a shows the area breakdown for the fore-mentioned instance. As it can be seen, 58 % of the area is consumed by the slices, the rest by the data storage and control. Roughly half of the

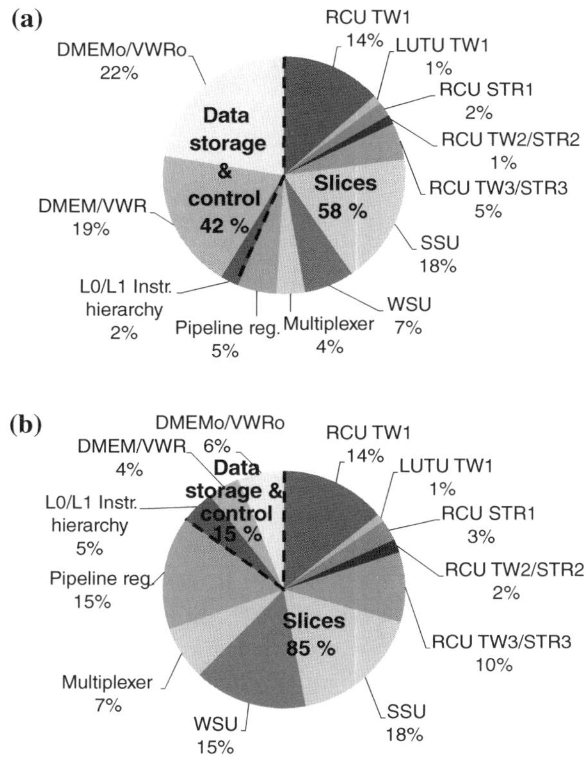

Fig. 6.11 **a** Area and **b** power breakdown of the proposed DSIP FFT architecture instance with one cluster. For the power breakdown the computation of the 512-point FFT is considered. L0/L1 instruction hierarchy with DMEM LUTU (content for LUTU TW1) is included. In addition, also the DMEMo is included, which would typically be the DMEM of the subsequent DSIP architecture instance and therefore considered there (see Fig. 6.6)

area is occupied by the memories/registers. The data memory requirements are rather high since the FFT is a block-based functional block.

For the current instance we assume that the whole input and output data of the FFT is buffered in the DMEM and DMEMo respectively, which consume the largest part of memory footprint. Considering the fact that each FFT input word is only read once from the DMEM (and can then immediately be overwritten), such a high amount of (buffer) storage is not required in case the input data is provided “just” in time and the output data is read as soon as it is available (streaming processing with minimal latency). Thus, depending on the rest of the platform, a significant reduction of the total area may be possible.

Table 6.1 shows the power consumption of the proposed architecture during the computation of different FFT sizes operating at 440 MHz. Including all data and L0/L1 instruction memories, the total power consumption for the 64, 128 and 512-point FFT are 13.21, 24.56 and 85.61 mW respectively. As desired, the total power consumption increases almost linearly with the FFT size, i.e. with the number of utilized engines. Figure 6.11b shows the power breakdown for the 512-point operation mode. As it can be seen, 85 % of the total power is spend in the slices, thus, the “overhead” of the DMEM(o)/VWR(o) data storage and control is rather low. Nevertheless, since only $14 + 3 + 2 + 10 = 29$ % of the total power is

Table 6.1 Power consumption for different FFT sizes (all data and L0/L1 instruction memories are included)

FFT size	64	128	512
Number of active engines	1	2	8
Leakage power (mW)	0.34	0.35	0.36
Dynamic power (mW)	12.87	24.21	85.25
Total power (mW)	13.21	24.56	85.61
Total power/number of active engines (mW)	13.21	12.28	10.70

The power consumption increases almost linearly with the FFT size

Table 6.2 Throughput and latency for different FFT sizes

FFT size	64	128	512
Number of required engines/FFT	1	2	8
Max. number of parallel FFTs/cluster possible	8	4	1
Number of clock cycles/FFT (streaming)[a]	37	49	51
Throughput for compute only 1 FFT/cluster (GS/s)	0.761	1.149	4.417
Throughput for compute max. nr. FFTs/cluster (GS/s)	6.089	4.598	4.417
Nr. clk. cycles to compute FFT (VWR rd.→VWRo wr.)[a]	46	58	60
-//- (DMEM wr.→VWR wr.+VWRo rd.→DMEMo wr.)	<5	<5	<5
Nr. clk. cycles to compute FFT (total)	51	63	65
Latency (total) (ns)	116	144	148

The architecture template instance includes 1 cluster/8 engines and operates at 440 MHz clock frequency

[a] See schedule of Fig. 6.10

consumed by the arithmetic units, the power consumption of the shuffler units (and generally of the multiplexers) as well of the pipeline (local) registers is considerably high. It is expected that the power consumption of the shufflers (very wire dominated) can significantly be reduced by applying the proposed semi-custom design flow. A better balancing of the pipeline stages (and potentially also merging of stages) can help to reduce the power consumption in the pipeline registers. We considers these optimizations as future work.

6.5.4 Throughput and Scalability

Table 6.2 summarizes the achievable throughputs and latencies for different FFT block sizes. The numbers are based on the proposed scheduling of Sect. 6.5.2, 440 MHz clock frequency and an architecture instance with 1 cluster/8 engines. For computing only one FFT at a time, the achievable throughput's for the 64, 128 and 512-point FFT are 0.761, 1.149 and 4.417 GS/s respectively. The throughput can be scaled linearly with the number of employed engines and clusters. The maximal throughput of 6.089 GS/s is achieved by computing eight 64-point FFTs in parallel within one cluster. Considering also the clock cycles that are required to move the data through the DMEM(o)s and VWR(s), a maximal latency of 148 ns for the computation of the 512-point FFT can be calculated. For the 64-point and 128-point FFT the latency is shorter.

As mentioned in Sect. 6.2.1, for the compliance with the two targeted 60 GHz communication standards IEEE 802.11ad and IEEE 802.15.3c, a maximal throughput of 2.64 GS/s (OFDM mode) is required and the latency budget of 150 ns (SIFS timing) must not be exceeded. Given the implementation results of the proposed DSIP architecture instance design, these two specifications can clearly be fulfilled. To further reduce the latency (in case this would be beneficial for the implementation of other functional blocks of the critical processing chain), TSMC 40 nm General-Purpose (GP) instead of LP technology could be used (enables higher clock frequencies). Beside the throughput and latency results, also the results on area and power consumption are very acceptable. Based on the real arithmetic operations shown in Fig. 6.9, the proposed architecture offers under fully utilization a high energy efficiency of 4.11 GOPS/mW and a high area efficiency of 486 GOPS/mm^2. Taking the amount of required real operations to compute a 512-point FFT with the Cooley-Tukey algorithm, and the 512-point FFT throughput that was achieved on the proposed DSIP architecture instance, into account, a high area and energy efficiency of 2.32 GOPS/mW and 264 GOPS/mm^2 respectively can be calculated.

Depending on the maximal required throughput, an architecture instance with one or several clusters can be employed. At run-time, with the number of activated clusters and engines, the throughput can dynamically be scaled. The architecture supports the FFT and IFFT block sizes 64, 128 and 512. It is not only possible to compute multiple FFTs of the same block size in parallel, but also to perform the parallel computation of different FFT sizes on the same architecture instance. For instance, four 64-point FFTs and two 128-point FFTs could be computed in parallel on the same cluster. This enables a certain degree of flexibility which can be exploited to increase the average energy efficiency. As mentioned in Sect. 6.2.1, depending on the coherence bandwidth, the FFT size could dynamically be adapted. For run-time situations/modes during which the latency and throughput requirements are relaxed, (global) power gating can be applied to reduce the power consumption (turn the processor only on for a short time). Because of the current slice implementation, which is mostly based on RCUs and on hardware SIMD, the offered flexibility for run-time scalability on the algorithm-level is still rather limited. For instance, by employing more generic BAUs that are based on software SIMD, not only the algorithm or the computations of the algorithm, but also the quantization scheme could dynamically be adapted. Nevertheless, future work should investigate in the achievable gain of such a design approach.

6.6 Comparison and Discussion

In this section we compare the proposed DSIP architecture instance to ASIC and programmable implementations from literature.

Table 6.3 Overview of the FFT ASIC references that are used for comparisons

Reference name	[12]	[13]	[7]
FFT architecture	memory based	memory based	MDF[a]
FFT algorithm	DIF[b], radix-16	DIF[b], mod. radix-16	DIF[b], mod. radix-32
FFT size	512	512	512
Word length (bits)	12	12 (blk.-floating point)	11
SQNR (dB)	40	57	35
Technology (nm)	90	90	90 nm/1.2 V
Area (mm^2)	2.46	0.93	0.78
Power (mW)	103.5	42	92.8
Clock (MHz)	324	324	310
Throughput rate(GS/s)	2.59	2.59	2.5

[a] Multi-path Delay Feedback (MDF)

[b] Decimation in Frequency (DIF)

6.6.1 ASIC References

We have selected 3 FFT ASIC references, which target the same application domain and standards, for comparison [7, 12, 13]. Their characteristics are summarized in Table 6.3.

6.6.2 Flexible References from Literature

We have selected 4 programmable FFT implementations for comparison. Depending on their architecture style, they are denoted as ASIP1, SDR1, SDR2 and SDR3. ASIP1 is a FFT processor (customized with butterfly instruction) which is based on the Tensilica Xtensa platform [10]. In contrast to ASIP1, SDR1-SDR3 are more generic baseband processors that are also used for the computation of other functional blocks. SDR1 refers to the Block Processing Engine (BPE), which is a mixed-grain vector processor, and which has also been designed based on an architecture template [8]. It also includes a butterfly instruction. We denote the Sandblaster SB3500 communication processor as SDR2 [4]. The Tensilica ConnX BaseBand Engine (BBE), which is a 3-way VLIW architecture, is denoted as SDR3. This processor also includes a butterfly instruction. A summary of implementation-related information of these processors can be found in Table 6.4. To provide an idea on their efficiency, we show their scaled area and power consumption for a throughput target of 2.5 GS/s.

6.6.3 Comparison (Normalized, Pessimistic)

Since 512 points is the default block size defined in the targeted 60 GHz standards, and all ASIC references support only this size, we use the 512-point FFT

Table 6.4 Overview of related flexible/programmable implementations from literature

Reference name	ASIP1 [10]	SDR1 [8]	SDR2 [4]	SDR3 [20]
Platform/Architecture	Tensilica Xtensa	BPE[a]	SandBlaster	Tensilica BBE
Supported FFT sizes	128–1024	16–1024	64–2048	512–8192
Technology (nm)	130	65	65	65
Area (Kgates)	147	n.a.	n.a.	430
Area (mm^2)	n.a.	n.a.	n.a.	1.1 (cell)[b]
Nr. clk. cycles to comp.1024-p. FFT	4526	5180	2198	1812
Power (uW/MHz)	190	272	n.a.	n.a.
Estimation for 1024-point FFT:				
Consid. throughput requirem. (GS/s)	2.5	2.5	2.5	2.5
Required clock frequency (GHz)	11.1	12.7	5.4	4.4
Area consumption (mm^2)	n.a.	n.a.	n.a.	12.1 (cell)[b]
Power consumption (mW)	2109	3454	n.a.	n.a.
Estimation for 512-point FFT:				
Latency to comp. FFT at 2 GHz clkf. (ns)	1020	1167	495	410

The power consumption for a throughput target of 2.5 GS/s and a 1024-point FFT is estimated. Thereby it is assumed that the power scales linearly with the throughput and that the technology is not changed (same as indicated in the table). For the latency estimation, the clock cycle count was reduced from the 1024-point to the 512-point FFT by considering the relative computational complexity

[a] Block Processing Engine (BPE)

[b] Only cell area of core (without memories)

implementation/mapping for comparison. In case for processor implementations the power or area consumption has not been specified for that block size but for other ones, we scale their throughput proportionally to their given FFT sizes and use the given power and area numbers for comparison. For instance, from 1024 to 512 points, we assume that the throughput would increase by factor 2 (1024/512)$\times$(10/9 stages)=2.22.

To enable a fair comparison, we do not only consider the same FFT size, but also scale all references to the same technology, i.e. 40 nm. Thereby we use the following factors to scale between two technology nodes [9, 14]: 1.2 for clock freq./throughput, 2 for area (if not already specified in GE), 1.5 for energy/power. After the common FFT size and the technology has been considered, we normalize the area and power numbers to throughput. The resulting numbers are more in favor of the references. For example, by experience it has been observed that the gain of scaling a conventional design (not optimized for technology) to a newer technology is typically less than the one assumed. Also, for the processor references, to achieve the targeted throughput, the architecture would have to be significantly changed, e.g. many more pipeline registers would have to be inserted, and therefore the area and power would not scale linearly.

Figure 6.12a shows the relative area comparison under these for us pessimistic assumptions. It can be seen that ASIC3 is the most area efficient design and

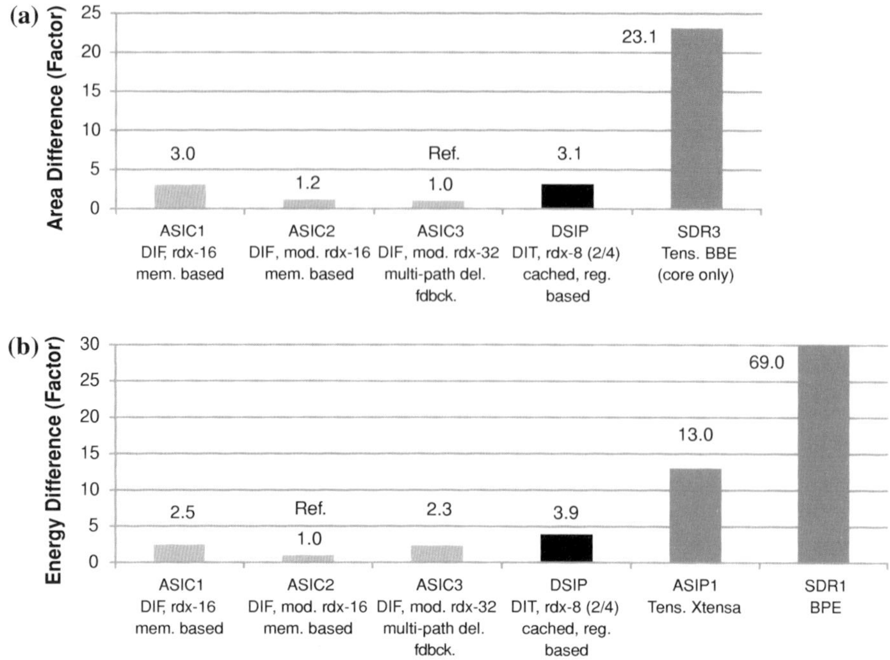

Fig. 6.12 Area and energy comparison of different FFT implementations. FFT size, technology and throughput are normalized. ASIC1-3: [7, 12, 13], ASIP1: [10], SDR1-3: [4, 8, 20]. Estimations for the ASICs/ASIPs/SDRs are based on optimistic assumptions (i.e. in favor of the references). **a** Area comparison. **b** Energy comparison

amongst the 3 ASICs there is a maximum difference of factor 3. Compared to the best ASIC, the proposed DSIP architecture instance has an area penalty of factor 3.1, however, it consumes basically the same area as ASIC1. The comparison with the SDR processor SDR3 (Tensilica BBE) shows that the proposed DSIP architecture instance is at least a factor 7 more area efficient (in the reference certain memories are even not included). Thus, the area consumption of the proposed design is positioned between ASICs and SDR processors.

In Fig. 6.12b the relative energy comparison is shown. ASIC2 offers the highest energy efficiency. It is interesting to observe that the most area efficient ASIC is not the most energy efficient one. ASIC1 and ASIC3 are more than a factor 2 worse in energy efficiency compared to ASIC2, so ASIC2 is clearly an exception. Compared to the best ASIC, the proposed DSIP architecture instance has an energy penalty of factor 3.9. Compared to ASIC1 and ASIC3, the difference is less than a factor 2. This relative comparison shows that the processor references from literature have a clear drawback in terms of energy efficiency. Our proposed design consumes more than a factor 3 less energy than the ASIP reference and even more than a factor 17 less energy than the SDR processor reference. These results clearly demonstrate the high energy-performance efficiency of the proposed solution, largely closing the gap between instruction-set processors and custom ASICs.

6.6.4 Discussion

6.6.4.1 Proposal Versus ASIC

As shown in Fig. 6.12 and mentioned above, the maximum area and energy penalty of the proposed DSIP architecture instance compared to the dedicated ASIC solutions is within a factor 3.1 and 3.9 respectively. The overhead of the proposed design in respect to the ASICs has several causes. One major reason is clearly the architectural overhead for enabling flexibility. For instance, as it can be seen in Fig. 6.11, 29 % of the total area and 40 % of the total energy is consumed by the WSU, SSU and the multiplexers, which enable flexibility. In the ASICs these connections are hardwired and therefore cheaper in terms of implementation cost. Nevertheless, by applying the proposed semi-custom design flow, which ensures a more regular placement and routing, it can be assumed that this overhead for shuffling and multiplexing can dramatically be reduced. In addition, it will enable a higher area utilization factor. A major difference comes also from the fact that the ASICs are using a higher radix value, which enables more efficient FFT implementations. Furthermore, our current implementation uses D-FF-based standard-cell memories. Clearly, compared to the hard-macro SRAMs that are used in the ASIC designs, this makes a big difference in terms of area and energy consumption. Also, since we have applied the conventional design flow for obtaining the layout, the placement and routing of the memories has not been done very efficiently (see appendix of Chap. 5). Beside the application of the proposed semi-custom design flow, many other optimizations can be carried out to bridge the gap between the ASICs and the proposed DSIP architecture instance. A list of optimization opportunities is provided in Sect. 6.6.4.3.

The ASIC references are fixed to a block size of 512 points, while the proposed DSIP architecture instance supports additionally 64 and 128 points. As indicated in Sect. 6.2, these smaller FFT sizes can be particular useful in case of the channel estimation, which is in the latency critical processing path. By dynamically (depending on the channel condition) switching from the 512 to the 128 or 64-point FFT, the average energy consumption can be further reduced. Note, when using 60 GHz short range communication, the communication devices are typically kept quite close to each other so it can be assumed that the channel is most of the time in a good condition. Thus, there is a good reason why this gap in terms of energy efficiency can be further minimized. Especially the benefit of higher flexibility and reusability, which reduce the NRE cost and the risk of failure, make the proposed design as the preferred option.

6.6.4.2 Proposal Versus Other Processors

As indicated in Table 6.4, the processor references from literature consume more than 2 W when computing the 1024-point FFT for 2.5 GS/s throughput. For a 512-point FFT at 2.5 GS/s throughput, and scaled implementations to 40 nm, the ASIP1 reference would consume about 150 mW and the SDR1 reference about 820 mW. Considering the typical power budget for a complete wireless transceiver (analog and digital) to be in the order of 1 W [5], and the fact that for 60 GHz transceivers the ADC/DAC as well as the LDPC decoder take a major part of this budget, 150 mW for the FFT is unacceptably high. Beside the unacceptably high power consumption, the processor references from literature are clearly not able to fulfill the low-latency requirement of the targeted 60 GHz standards (150 ns). As it can be seen from Table 6.4, even if these processors would be clocked with 2 GHz (which seems rather unrealistic even for 16 nm technology), the latency for computing the 512-point FFT output would be higher than 400 ns. Thus, the considered ASIP and SDR processor references are definitely not suited for the targeted 60 GHz application domain. To a large extend their high cost for enabling flexibility (instruction hierarchy, highly multiplexed datapath) and for accessing data (memory with narrow ports, low data locality) is responsible for that.

In contrast, with a reasonable power consumption of 48.5 mW at 2.5 GS/s for the 512-point FFT, and because of fulfilling the tight SIFS latency requirement, the proposed DSIP architecture solution of this book seems feasible and very promising.

Reasons for the higher efficiency compared to the processor references include (1) implementation of constant multiplications with shift-add/sub operations, (2) higher hardware utilization achieved with the employed units and the schedule, (3) optimized interconnect with reduced number of multiplexers, (4) efficient data hierarchy with wide read/write ports and (5) efficient control hierarchy. Besides, the utilized algorithm and quantization scheme contribute positively in achieving high efficiency.

6.6.4.3 Scope for Optimizations to Fully Close the Gap

As mentioned throughout this chapter, still many options are available to further optimize the energy-performance efficiency of the proposed design. Some of them will have a relatively small impact on the FFT case study but would be more effective for other DSIP architecture designs. We believe that, when combined applied, they will allow to fully close the gap with the custom ASICs in terms of energy-performance efficiency. In the following the proposed optimizations are summarized:

- Adjust the memory size of the DMEM and DMEMo to the actual requirements (currently there are dimensioned for worst case). For instance, if the FFT input data is provided "just in time", the buffering of the whole FFT input vector is not needed. In the ideal case this optimization can reduce the total area by 35 %.

- As shown in the schedule of Fig. 6.10, the arithmetic units of a slice are inactive for 6 cycles between the computations of two subsequent epochs, which is caused by the data dependency of the pipeline. By overlapping the computations of different parallel epochs, the datapath can potentially be better utilized. This optimization would ideally lead to a gain of 18 clock cycles for the 512-point FFT. Besides, the reading from the VWR can already be done while the last epoch is computed, and the writing to the VWRo can be overlapped with the computation of the first epoch. Thus, this would lead to a gain of 8 clock cycles. Considering both scheduling optimizations, 26 out of the 51 clock cycles could be saved, which increases the area and energy efficiency by 50 %!
- Replace the D-FF-based standard-cell memories with D-latch-based standard-cell memories and/or with hard-macro memories (e.g. based on SRAM or ideally based on novel technologies such as RRAM). Realistically about 15 % of total area can be saved.
- Apply the semi-custom design flow that is proposed in the appendix of Chap. 3. Because the architecture has a very regular parallel structure (e.g. "slots" and slices are replicated), it can be assumed that a very efficient semi-custom layout can be made. This would be beneficial in multiple ways. Currently 25 % of the total area and 33 % of the total power is consumed by the two shufflers, i.e. WSU and SSU. By reducing the length of the related wires, the area and power consumption of these units can significantly go down. Furthermore, because currently standard-cell-based memories are used, a more regular placement of the cells would also lead to a significant reduction of power consumption (see experimental results in the appendix of Chap. 5). In addition to the fore-mentioned benefits, also the area utilization factor would most-likely increase from currently 62.5 % to about 85–95 %, thus, reducing the total area by more than 22 %.
- For the current implementation TSMC 40 nm Low-Power (LP) technology is used. Because of the low leakage characteristic, low-power technology is beneficial for designs in which large parts of the area is mostly inactive, e.g. memory-dominated designs. However, compared to General-Purpose (GP) technology, the gate delays are longer and therefore the achievable clock frequency is lower. Since in the proposed FFT DSIP architecture instance high clock frequency is important (reduces latency), and most architecture parts are rather active, the implementation with GP technology is probably a better choice. Because of the higher clock frequency, less parallel engines are needed for the same throughput. Thus, it increases the area efficiency. In return, the drawback of GP technology is the increased leakage power. From Table 6.1 it can be seen that the leakage power is currently very low (e.g. 2.6 % of the total power in case only 1/8 engines are used). Taking also the results of case study 1 and 2 into account, which are based on GP technology, it can be assumed that the contribution of leakage power to the total power would not exceed 5 % even if GP technology is used for this design. Hence, the use of GP technology should be strongly considered.
- As shown in Fig. 6.11b, the local registers of the units, which are here referred as pipeline registers (since the data in the slices flows in a pipelined fashion)

contribute with 15 % to the total power consumption. This percentage is rather high. As it can be seen in Fig. 6.9, the complexity of the different arithmetic units, and especially the number of operators in the critical paths, is rather varying amongst them. Therefore, by merging simple pipeline stages (e.g. RCU STR1 and RCU TW2/STR2) and/or by performing a better delay-balancing, the power consumption in the pipeline registers can be reduced. Note, a reduced number of pipeline stages would also be beneficial in regard to the feedback-related scheduling problem mentioned further above.

- The contribution of the control to the total power consumption is with 5 % rather low. Nevertheless, especially given the current schedule of the units in which wait cycles are present between the computation of subsequent epochs, the employment of several distributed loop buffers at the engine-level can make sense. Currently only one loop buffer is used for controlling all units of an engine.
- Mainly because of the demanding latency requirement, the current slice architecture is based on RCUs and hardware SIMD, i.e. strongly customized for the considered FFT algorithm. Nevertheless, as motivated in Sect. 6.4.4, it would make sense to replace the RCUs with BAUs and MUs, since this would enhance the flexibility and reusability. The possible increase of latency can probably be compensated by the use of GP technology. By enabling the exploitation of multiple subword sizes, a software SIMD datapath would enable to trade-off BER/PER performance versus energy consumption at run-time. In addition, a software SIMD datapath would further increase the reusability of the proposed FFT architecture instance, since future communication standards may require different quantization schemes.
- Exploit the flexibility of the proposed design to potentially reduce the average energy consumption by dynamically adapting to the actual requirements (instead of operating permanently in worst case mode). This concept, which is based on system scenarios, has been explained in Sect. 1.3.6 and some potential run-time optimizations for the FFT case study have been indicated in Sect. 6.2. However, a more fundamental and global application analysis should be performed to identify further optimization options.

As it can be seen from this list, the current implementation leaves still a big room for optimizations and significant gains. Thus, it can be assumed that, by applying the above mentioned optimizations, the remaining area and energy efficiency gaps to the best ASIC references would not exceed a factor 2 (for worse case mode).

6.7 Conclusions

This chapter introduces a processor which is well-suited for the computation of the 64, 128 and 512-point (I)FFT in 60 GHz communication systems. The scalable architecture is based on the proposed DSIP architecture template of Chap. 3 and fully supports the demanding latency and throughput requirements specified by the IEEE 802.11ad and IEEE 802.15.3c standard.

The implementation results in TSMC 40 nm technology, with the mapping of a 512-point FFT, show a high energy and area efficiency of 2320 MOPS/mW and 264 GOPS/mm^2 respectively. For the computation of a 512-point FFT at 4.417 GS/s, with a total latency of 148 ns, 85.6 mW total power are consumed. An instance with 1 cluster, which consists of 8 engines and occupies an area of 0.751 mm^2 (including all memories), offers a maximal throughput of 6.089 GS/s when 8 parallel 64-point FFTs are executed.

FFT ASIC references from literature are a factor 1.1–3.1 more area efficient and a factor 1.6–3.9 more energy efficient than the proposed design. Nevertheless, given the many options for further optimizations, it can be assumed that these gaps can be reduced to a factor less than 2 for both area and energy. Compared to state-of-the-art processor solutions, the proposed implementation is at least a factor 7 more area and a factor 3 more energy efficient. Beside these huge penalties in terms of area and energy, the considered processor references from literature are by far not able to meet the demanding latency requirements imposed by 60 GHz systems.

For the highest specified throughput of 2.64 GS/s (OFDM mode), the proposed design consumes 52 mW power during the computation of the 512-point FFT. Referring to the conclusion of the FIR filter case study (see Sect. 5.7), we can also confirm with the results of this FFT case study that a programmable processor solution for the complete mobile 60 GHz baseband engine appears to be feasible, when based on the proposed DSIP architecture style.

With this case study we have also demonstrated that currently available commercial processor design tool suites can be used to model the proposed DSIP architecture template of this book (at least to a large extend). Included features, such as C-compilation, cycle-accurate instruction set simulation and automatic generation of RTL code, significantly contribute in reducing design time.

References

1. Atak, O., Atalar, A., Arikan, E., Ishebabi, H., Kammler, D., Ascheid, G., Meyr, H., Nicola, M., Masera, G.: Design of application specific processors for the cached FFT algorithm. In: IEEE International Conference on Acoustics, Speech and Signal Processing (ICASSP), vol. 3, pp. III-III (2006). doi:10.1109/ICASSP.2006.1660832
2. Baas, B.: A generalized cached-FFT algorithm. In: IEEE International Conference on Acoustics, Speech and Signal Processing (ICASSP), vol. 5, pp. 89–92. IEEE (2005). doi:10.1109/ICASSP.2005.1416247
3. Baykas, T., Sum, C.S., Lan, Z., Wang, J., Rahman, M., Harada, H., Kato, S.: IEEE 802.15.3c: the first IEEE wireless standard for data rates over 1 Gb/s. IEEE Commun. Mag. **49**(7), 114–121 (2011). doi:10.1109/MCOM.2011.5936164
4. Beheshti, B.: On performance of LTE UE DFT and FFT implementations in flexible software based baseband processors. In: IEEE Long Island Systems, Applications and Technology conference (LISAT), pp. 1–4 (2009). doi:10.1109/LISAT.2009.5031556
5. van Berkel, C.: Multi-core for mobile phones. In: Design, Automation and Test in Europe (DATE), pp. 1260–1265 (2009)

6. Bo, Y., Han, J., Zou, Y., Zeng, X.: A low power ASIP for precision configurable FFT processing. In: Asia-Pacific Signal Information Processing Association Annual Summit and Conference (APSIPA ASC), pp. 1–4 (2012)
7. Cho, T., Lee, H.: A High-speed low-complexity modified FFT processor for high rate WPAN applications. IEEE Trans Very Large Scale Integr. Syst. **21**(1), 187–191 (2013). doi:10.1109/TVLSI.2011.2182068
8. Cupaiuolo, T., Lo Iacono, D.: A flexible and fast software implementation of the FFT on the BPE platform. In: Design, Automation and Test in Europe (DATE), pp. 1467–1470 (2012). doi:10.1109/DATE.2012.6176598
9. Fasthuber, R., Li, M., Novo, D., Raghavan, P., Van der Perre, L., Catthoor, F.: Energy-efficient run-time scalable soft-output SSFE MIMO detector architectures. In: Transaction on High-Performance Embedded Architectures and Compilers (HiPEAC), Special Issue SAMOS 2009, vol. 5, no. 3, pp. 1–20 (2011)
10. Guan, X., Fei, Y., Lin, H.: Hierarchical design of an application-specific instruction set processor for high-throughput and scalable FFT processing. IEEE Trans on Very Large Scale Integr. Syst. **20**(3), 551–563 (2012). doi:10.1109/TVLSI.2011.2105512
11. Heo, K., Cho, S., Lee, J., Sunwoo, M.: Application-specific DSP architecture for fast Fourier transform. In: IEEE International Conference on Application-Specific Systems, Architectures and Processors (ASAP), pp. 369–377 (2003). doi:10.1109/ASAP.2003.1212860
12. Huang, S.J., Chen, S.G.: A green FFT processor with 2.5-GS/s for IEEE 802.15.3c (WPANs). In: International Conference on Green Circuits and Systems (ICGCS), pp. 9–13 (2010). doi:10.1109/ICGCS.2010.5543105
13. Huang, S.J., Chen, S.G.: A high-throughput Radix-16 FFT processor with parallel and normal input/output ordering for IEEE 802.15.3c Systems. IEEE Trans. Circuits Syst. **59**(8), 1752–1765 (2012). doi:10.1109/TCSI.2011.2180430
14. Janhunen, J., Pitkanen, T., Silven, O., Juntti, M.: Fixed- and floating-point processor comparison for MIMO-OFDM detector. IEEE J. Sel. Top. Signal Process. **5**(8), 1588–1598 (2011). doi:10.1109/JSTSP.2011.2165830
15. Jayapala, M., Barat, F., Catthoor, F., Corporaal, H., Deconinck, G.: Clustered loop buffer organization for low energy VLIW embedded processors. IEEE Trans. Comput. **54**(6), 672–683 (2005)
16. Jayasumana, G., Loeffler, C.: Searching for the best Cooley-Tukey FFT algorithms. In: IEEE International Conference on Acoustics, Speech and Signal Processing (ICASSP), vol. 12, pp. 2408–2411 (1987). doi:10.1109/ICASSP.1987.1169923
17. Liu, G., Feng, Q.: ASIC design of low-power reconfigurable FFT processor. In: International Conference on ASIC (ASICON), pp. 44–47 (2007). doi:10.1109/ICASIC.2007.4415563
18. Perahia, E., Cordeiro, C., Park, M., Yang, L.L.: IEEE 802.11ad: defining the next generation multi-Gbps Wi-Fi. In: IEEE Consumer Communications and Networking Conference (CCNC), pp. 1–5 (2010). doi:10.1109/CCNC.2010.5421713
19. Raghavan, P., Lambrechts, A., Jayapala, M., Catthoor, F., Verkest, D., Corporaal, H.: Very wide register: an asymmetric register file organization for low power embedded processors. In: Design, Automation and Test in Europe (DATE). IMEC (2007)
20. Rowen, C., Nuth, P., Fiske, S.: A DSP architecture optimized for wireless baseband. In: International Symposium on System-on-Chip (SOC), pp. 151–156 (2009). doi:10.1109/SOCC.2009.5335658
21. Tang, S.N., Liao, C.H., Chang, T.Y.: An Area- and energy-efficient multimode FFT processor for WPAN/WLAN/WMAN systems. IEEE J. Solid-State Circuits **47**(6), 1419–1435 (2012). doi:10.1109/JSSC.2012.2187406
22. Target ASIP development toolsuite (inc. IP Designer, IP Programmer), Target Compiler Technologies. http://www.retarget.com/
23. Zainal, M., Yoshizawa, S., Miyanaga, Y.: Low power FFT design for wireless communication systems. In: International Symposium on Intelligent Signal Processing and Communications Systems (ISPACS), pp. 1–4 (2008). doi:10.1109/ISPACS.2009.4806724

Chapter 7
Front-End Design Flow: Bridging the Algorithm-Architecture Gap

7.1 Motivation and Issues, Overview of Proposal and Contributions

7.1.1 Context and Motivation

The design of a mobile SDR system requires researchers and engineers from different design communities. As mentioned in Sect. 1.2.4, in general, the following communities can be identified: (1) system, functionality and algorithm designers, (2) architecture and platform designers and (3) back-end designers. Because each community is focused on its on field, which is already very complex on its own, communities do typically not understand the issues and constraints of other communities very well. As a consequence of this cultural gap, the design productivity and the resulting implementation efficiency is suffering. Therefore it very important to link the design communities stronger together. Because of the proposed template design approach of this book, and the high reusability that this approach is offering, a good basis for establishing strong links is created.

In this chapter we focus on the link between the algorithm and architecture designers, i.e. between the fore-mentioned design communities (1) and (2). As motivated previously, to achieve high energy efficiency, it is very essential that the algorithms and architectures are well matched. Thus, the application of an algorithm-architecture co-design flow is essential. However, since the algorithm transformation space and the architecture design space can be very large, this can be a non-trivial task. Therefore, depending on the targeted architecture style and the inherently enabled design freedom, the establishment and the application of such an algorithm-architecture co-design approach can be very challenging [6, 12].

R. Fasthuber et al., *Energy-Efficient Communication Processors*,
DOI: 10.1007/978-1-4614-4992-8_7,

7.1.2 Algorithm-Architecture Co-Design for Traditional Architecture Styles

Whether an algorithm-architecture co-design flow can be established/applied or not depends essentially on the considered architecture style. In the following we review the traditionally applied design flows and the main issues regarding co-design.

7.1.2.1 Traditional Standard-Cell ASIC

For an ASIC, the algorithm and the architecture is very tightly coupled, i.e. they represent each other. Thus, changes in the algorithm will directly translate to changes in the ASIC architecture. Because of this strong link, basic hardware cost models, e.g. area/energy of a multiplier, are typically sufficiently accurate for the initial algorithm design and optimization phase. For a specific algorithm and for given specifications on throughput and latency, a rather limited set of feasible ASIC architecture options exist. Automated HLS tools, as referred in Sect. 2.5.2, can generate different ASIC architectures and thus, support the designers with the evaluation of different architecture trade-offs. To generate the set of possible architectures, e.g. parallel versus serial architecture, reuse of operators, etc., code transformation algorithms, such as the one of Spiral for multiplications [11], are used within HLS tools. Because HLS tools can also be coupled with logic synthesis and back-end design tools, the automatically obtained results at architectural level can be very accurate, i.e. basically no architecture-technology gap exists. For this reason, in general, the establishment and application of an algorithm-architecture co-design flow for ASICs is a limited challenge.

7.1.2.2 Traditional ASIP

The application of an algorithm-architecture co-design flow for ASIPs is essential to achieve high implementation efficiency. However, the establishment/application of such a co-design flow for ASIPs causes a big challenge [6, 12]. The main cause for this big challenge is the combination of the following: (1) the big architecture design space (ASI/HA design freedom), (2) the inaccurate architecture cost models (because of the ASI/HA design freedom and the employed "unpredictable" standard-cell back-end design flow), (3) the big algorithm transformation space and (4) the fact, that multiple strongly heterogeneous algorithms, in case of ASIP (BP), which may not even be fully defined at architecture design time, have to be considered.

In the general system design flow, the algorithm design phase is preceding the architecture design phase. Therefore, during the algorithm design phase, the architecture is not yet defined. For this reason, accurate architecture/implementation cost models are not available. Basic hardware cost models, like in the ASIC approach, can be used as a guideline. However, because they do not capture the strong influence of scheduling and because the energy and area of ASIPs is often anyway dominated by the memories and the control, such basic models are typically not very meaningful. As proposed in [1, 2], a High-Level Estimation (HLE) flow could be applied to evaluate different algorithm options. However, especially for ASIPs, where the datapath is customized with ASIs, the accuracy of such a flow is also low. For this reason, to obtain efficient solutions, iterations between the algorithm and architecture design phases are needed. The feedback of the algorithm mapping provides input for algorithm transformations.

ASIP architectures are typically designed based on architecture templates and/or processor design tools. To enable tool support, the architecture template/processor design tool pre-defines/constraints the general architecture structure, i.e. Instruction Control/Storage Hierarchy (ICH/ISH), Data Storage Hierarchy (DSH) and interconnect of cores/FUs. This approach limits the architecture design space, which has the benefit of reducing the co-design complexity. However, if the pre-defined general architecture structure is not well-suited for the targeted algorithms, the achievable energy efficiency is also constraint. Despite the pre-defined general architecture structure, typically the following architecture design decisions, which inherently involve a trade-off between flexibility and efficiency, can still be influenced: (1) Which algorithms should directly be implemented in hardware, i.e. with HAs? (2) Which algorithm segments should be implemented in HW, i.e. with ASIs? (3) How many operators should an ASI include (complexity)? (4) How many ASIs should be designed? (5) How many cores/FUs should the processor include? (6) Which FU should include which ASI? (7) How many SIMD slots (within each FU)? (8) Size of instruction and data memories? Because of the big design space, these design decisions are clearly not obvious. Therefore, typically the following type of tools are used to guide these decisions: (1) code profiling tools, e.g. to identify often executed code segments which are candidates for hardware implementation, (2) code mapping/compilation tools, e.g. to obtain hardware utilization feedback for a certain architecture configuration and to identify bottlenecks, and (3) hardware complexity/cost estimation tools, e.g. to estimate the achievable clock frequency of the processor. The architecture design is clearly simplified by the use of such tools. Nevertheless, before the fore-mentioned tools can be used, the code of the targeted algorithms must be available.

After an initial architecture has be designed and the algorithms have been mapped, the resulting information, e.g. scheduling bottlenecks, can be used as an input for further algorithm optimizations. Thus, the link to the algorithm design phase is basically enabled, but in order to find an optimal solution, many iterations between both design phases can be required. For a big algorithm and architecture design space, a convergence within a limited design time may even be impossible.

A further issue related to the algorithm-architecture co-design of ASIPs can be caused by the link between architecture and back-end/technology. The fore-mentioned hardware complexity/cost estimation tools, which are responsible for this link, are often either accurate and slow or inaccurate and fast. The long run-time of accurate tools makes the exploration of a big architecture design space infeasible. Inaccurate tools, which may e.g. neglect the influence of wires, may lead to sub-optimal design decisions, which may again result in multiple design iterations.

Considering the above mentioned issues, the algorithm-architecture co-design of ASIPs is clearly a remaining challenge. The essential cause for this problem is the large design freedom on the algorithm and on the architecture side. Recently, we have proposed an algorithm-architecture co-design flow/guideline that tackles this challenge [7]. Nevertheless, the practical implementation remains still an issue. Especially the algorithm-side of the co-design flow of [7] can largely be reused for the proposed DSIP design approach of this book.

7.1.2.3 Traditional DSIP/DSP

Traditional DSIP/DSP architectures are typically either not customized at all (only a predefined architecture is chosen) or only a coarse customization for the set of targeted algorithms is performed (e.g. memory sizes, number of cores). Because also ASIs are not included, the architecture design space is very limited. As a consequence of this limited architecture design space, which enables high reusability, the hardware for the possible architectures is basically available in form of an IP. Therefore accurate cost models can be generated and can already be used during the algorithm design phase. Because of the very limited architecture customization, the applied design flow does not qualify as algorithm-architecture co-design flow. Basically only the algorithm is optimized so that it fits the underlying architecture.

7.1.3 Proposed Measures to Enable an Effective Algorithm-Architecture Co-Design Flow

In this book, we propose to combine several measures to simplify the algorithm-architecture co-design challenge. Importantly, despite the reduction of the design space, we take care that this measures are not harming the achievable implementation efficiency of designs. In the following, the considered measures, which can be seen as part of the overall proposed DSIP design approach, are summarized:

1. The most important measure is the use of an architecture template with a pre-defined medium grain customization space. The architecture template enables high reusability, which is an essential criterion. Because of the pre-defined medium grain customization space, the architecture design space is limited. This is important, because it enables the characterization with meaningful cost models and the employment of automated design flows. Since the proposed DSIP architecture template and the associated pre-defined medium-grain parameter space are based on the requirement analysis of the targeted

application domain, we assume, that the restricted architecture design space does basically not decrease the optimality of the obtainable architecture instances.
2. The proposed DSIP platform is implemented with multiple architecture instances. Therefore, for each architecture instance, only a limited set of algorithms, which have typically also similar characteristics, have to be considered.
3. As explained in Sect. 3.1.2 and Sect. 3.6, we envision to characterize the architecture template design space with cost models. To obtain these cost models, we apply the proposed semi-custom back-end design approach introduced in the appendix of Chap. 3. Because the resulting cost models are based on actual hardware implementations, i.e. they capture the joint constraints of architecture and back-end/technology, they are reasonably accurate. To reduce the template architecture design space, i.e. to reduce the amount of cost models and points in the cost models, solutions, which are not pareto optimal, are discarded. Thus, the cost models are accurate and capture only relevant design points.
4. We leverage on the fore-mentioned cost models during the algorithm design and optimization phases. Therefore, the resulting algorithms are already well-suited for the targeted architecture style. To guide decisions of algorithm designers, the architecture cost models could be combined with HLE flows [1, 2]. HLE flows, which estimates the area/energy implementation cost for different algorithms, can be applied on Matlab code.
5. To ensure that even the system and functionality designers take the hardware constraints into account, i.e. to ensure consistency and to avoid inefficiencies, we propagate the cost models further up in the general design flow. However, for this community, the architecture cost models will have to be translated to models with higher abstraction. For instance, cost models for system and functionality designer could capture the area/energy costs of different DSIP architecture instances, e.g. instances that support different FFT sizes and instances that support different filtering operations. The proposed model to quantify flexibility, see Sect. 3.6.3, will also help in that.
6. Establish consistent design flows around the template which guide design decisions. By leveraging on cost models, not only meaningful design decisions can be made, but the design flows can also be automated. This is an essential criteria for high productivity.

In this book, we consider system, functionality and algorithm design out of scope. However, we strongly encourage to further investigate in design flows, which leverage on the fore-mentioned cost models, to establish strong links to this design community.

Nevertheless, to contribute in bridging the algorithm-architecture co-design gap, we propose a well-suited DSIP architecture template instantiation design flow in this chapter. We assume that preceding steps, such as algorithm design/optimization and partitioning, have already been performed. During the proposed design flow, minor algorithm transformations can be applied. The design flow starts with one or multiple targeted algorithms and the system/hardware specifications, e.g. throughput and flexibility requirements. After applying the proposed design flow, beside the architecture instance, also a rough well-suited algorithm mapping will be available. Referring to Fig. 2.5, this means, with the proposed design flow, the algorithm, architecture and software design steps are linked together. Thus, the proposal contributes significantly in bridging the cultural gaps.

7.1.4 Main Contributions

The main contribution of this chapter is the proposed DSIP architecture template instantiation design flow, which includes algorithm-architecture co-design optimizations. Because it establishes a strong link, it can significantly contribute in bridging the cultural gap between algorithm and architecture designers. The underlying DSIP architecture template and the medium-grain pre-defined design space simplify the design challenge. We demonstrate the proposed flow on the MIMO detector and on the FIR filter case studies of Chaps. 4 and 5 respectively.

7.2 Proposed Architecture Template Instantiation Design Flow

In the following we propose a design flow, which leads, for a given set of targeted algorithms and system specifications, to an efficient DSIP architecture template instance. The design flow, which includes the algorithm-architecture co-design concept, was specifically made for the proposed architecture template. However, certain concepts may be reusable for other architectures.

Referring to the general system design flow of Sect. 2.3, we assume that the functionality design phase, the main algorithm design phase (including quantization) and the platform partitioning design phase have been completed. As a result, for each functional block of the platform, the targeted algorithms as well as the system/hardware specifications are available. As mentioned in Sect. 7.1.3, we envision that the above mentioned phases are driven by cost models. The purpose of this cost models is to ensure that the designed functionality, including the designed algorithms, can efficiently be implemented with the proposed DSIP platform concept.

The proposed architecture template instantiation design flow focuses on the implementation of a single functional block. Since the platform consists of multiple blocks, it has to be executed multiple times. As shown in Fig. 7.1, the design flow starts with the instance specific and the template specific inputs. The instance specific inputs, which correspond to the functional block specific inputs, are the

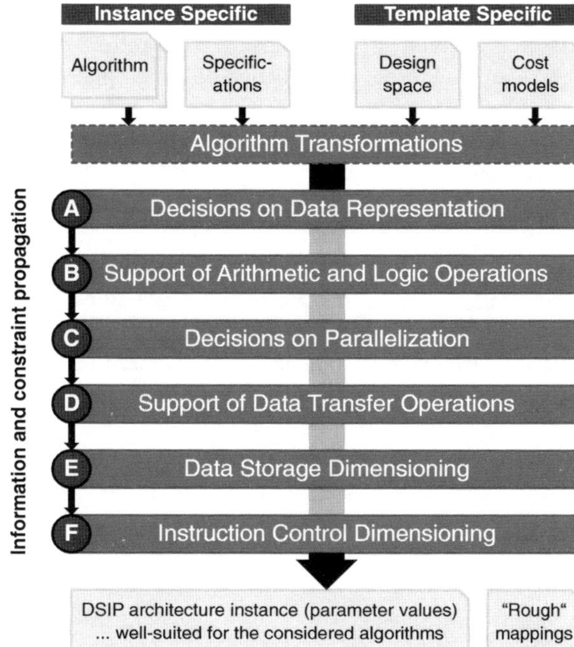

Fig. 7.1 Proposed flow to design a DSIP template architecture instance for a certain functional block. At each step, as many as possible design decisions should be made. However, if not sufficient information for making an optimal design decision is available, the decision should be postponed. The order of the steps A–F corresponds to the order of the template parameters in Fig. 3.16

targeted algorithms and the system/hardware specifications. The targeted algorithms should be represented in form of C/Matlab code (to enable simulations), in form of Control Data Flow Graphs (CDFG)s and in form of loop structures. The main system/hardware specifications are (1) the required flexibility, i.e. the different modes that need to be supported, (2) the budget on the maximal acceptable SNR degradation, which is required for I/O-approximate algorithm transformations (quantization), (3) the throughput and latency requirements and (4) the implementation constraints, e.g. maximal hardware area. The template-specific inputs are the template architecture design space, i.e. the parameters and the value ranges, and the cost models. Figure 3.16 summarizes the envisioned architecture parameters of the proposed DSIP template. Examples for parameters are the number of slices per engine and the SIMD type of an unit. An example for a cost model is a graph that shows the relationship between the subword size and the delay of a HardSIMD BAU. In general, cost models provide a guideline for design decisions and enable to estimate implementation results in advance. Preferably, the value ranges of the template parameters are fully defined, characterized and associated with cost models. As mentioned in Sect. 3.6, we consider this task as future work.

Figure 7.1 depicts the 7 steps of the proposed design flow. It can be noticed that the order of these design flow steps has been used before, e.g. for the description of the domain-specific requirements in Sect. 3.3 and for the description of the architecture template parameters in Fig. 3.16. The chosen order considers essentially the dependencies of the major steps. Thus, at every step, certain design decisions can be made and therefore the overall architecture design space is systematically reduced. To keep the exploration space small and therefore manageable, the general goal is to decide on as many as possible template parameter values at every step. Nevertheless, it can occur that at a certain step certain information, which would be required for making an optimal design decision, is not available. In this case, the design decision should be postponed to a later step and the already available information should be propagated. We refer to this technique as unidirectional information and constraint propagation. It avoids the occurrence of "chicken-and-egg" problems (e.g. phase coupling in compiler) and therefore very cost-effective results can be obtained without causing an explosion in design time. At the end of the design flow, all the parameter values of the template are decided and therefore the architecture instance is fully described. In addition, because the architecture customization is tightly coupled with considerations regarding mapping, rough algorithm mappings of the considered algorithms are available. By combining this mapping related information with the cost models, basically a rough energy estimation is possible. Although the proposed design flow targets mostly homogeneous instances, e.g. all clusters are the same, it is not limited to that. In the following the 7 main steps of the proposed design flow are briefly explained.

7.2.1 Algorithm Transformations

In this initial step, the targeted algorithms are optimized/prepared for an efficient mapping onto the proposed DSIP architecture template. For I/O-approximate transformations, the SNR degradation budget is considered. The following main types of optimization/transformation techniques are applied: (1) transformations to make the data-flow structure more deterministic, to enable an efficient parallelization, to reduce data storage requirements and to reduce the number of operations, (2) input value scaling to ensure deterministic signs (for SoftSIMD), (3) conversion of complex operators to primitive operators that are directly supported, (4) refinement of the quantization, e.g. to 4 bit granularity, and (5) optimization of multiplications. We assume that most of these techniques have already applied during the algorithm design phase and that here only a refinement is performed.

The first type of optimization/transformation technique is traditionally also applied for other processor architectures. It includes the concept of Data Transformation and Storage Exploration (DTSE) [3]. The DSIP architecture is mainly designed for deterministic data flows, therefore it is essential to apply techniques that ensure determinism, such as predication [5, 10]. The data storage requirements and the number of operations may be reduced by considering different algorithm variants and by merging of code parts.

To enable a mapping onto the proposed SoftSIMD datapath, the sign of signal values must be known at design/compile time (determinism). By performing a dynamic range analysis and by scaling the input values accordingly, this can normally be ensured.

In the conventional ASIP design approach ASIs can be added to the datapath. Therefore, the algorithms can generally contain any type of operation. In contrast, in the proposed DSIP design approach of this book, by default, only primitive operations are supported. For this reason, complex operations have to be decomposed to primitive operations. As depicted in Fig. 7.2, thereby several trade-offs may appear. At this stage, it may not be possible to decide, which of the decompositions are the optimal ones. If this is the case, it is better to propagate the option space to main step B and to postpone the decision. In general, after this transformation technique, typically the Basic Arithmetic Operations (BAO)s are computationally dominant.

The architecture template supports only fixed-point operations, which are generally more efficient than floating-point operations. Since algorithms are typically designed with floating-point representation, a conversion to fixed-point representation is necessary. As previously mentioned, we assume that the considered algorithms have already been quantized and that a fine-grain quantization, i.e. any random number of bits can be chosen, has been performed [8, 9]. However, as motivated in Sect. 5.4, in a processor hardware, it is cost-ineffective to support many different word lengths with a granularity of one bit. Therefore, the word lengths of all signals have to be refined. A reasonable granularity could be 4 bits (e.g. 8 bits, 12 bits, 16 bits etc.).

Finally, also the multiplications are optimized. Constant multiplications are identified and for the variable multiplications, the possible value ranges of the multipliers are identified. The binary constants are typically converted to CSD format and the number of required shift-add/sub iterations is reduced.

7.2.2 Decisions on Data Representation

After the previous initial step, the algorithms consist of primitive operations and for each operation the word lengths of the signals are known. The first task of this main step A is to determine, how many operations, per operation type, are performed with a certain word length. This information is used to assign a relative importance amongst word lengths of a certain operation type. For example, if only 1 out of 100 BAOs has to be performed on 24-bit data, then a low weight for 24-bit word length is assigned. The relative importance is used to decide, which of the occurring word lengths should directly be supported in hardware and for which word lengths a rounding to the next higher supported word lengths is acceptable. Depending on the number of targeted word lengths, a decision on the type of datapath, i.e. HardSIMD or SoftSIMD, can be made. If an operation on a higher word length has to be supported but it occurs very seldom, then this operation should be computed in a multi-cycle fashion, i.e. the support is not affecting the efficiency of the most occurring operations.

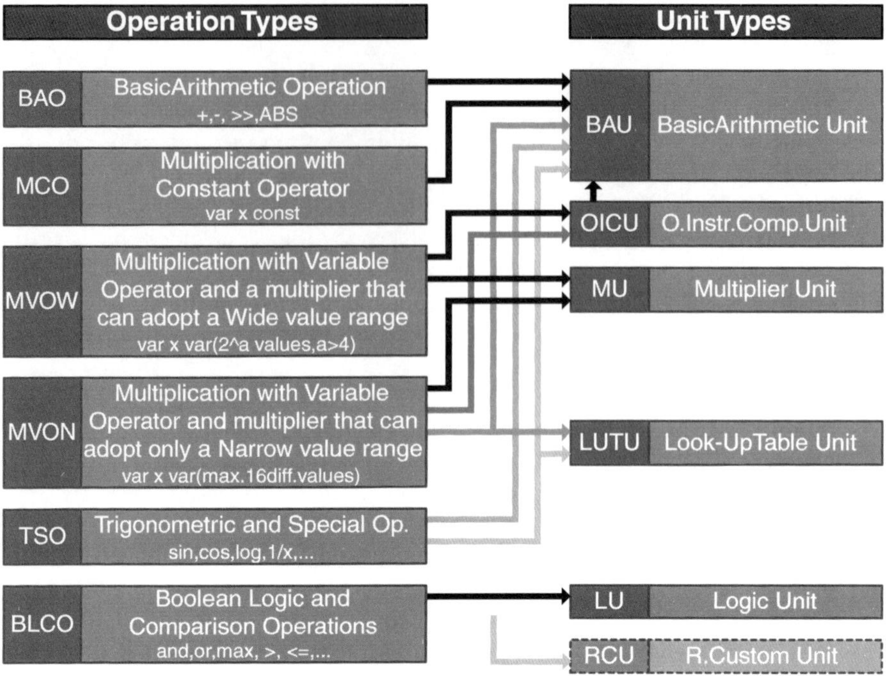

Fig. 7.2 Mapping of arithmetic and logic operation types onto arithmetic and logic unit types

7.2.3 Support of Arithmetic and Logic Operations

In this main step B, for all arithmetic and logic operations, the mapping onto supported unit types is decided. Figure 7.2 provides an overview of the possible mappings. The mapping of the BAOs, the MCOs and the BLCOs is simple, since only one option is considered. However, the mapping of other operation types involves trade-offs, which should be evaluated at this step. For the evaluation of such trade-offs, the area cost for a particular unit and the relative importance of an operation type could be considered. As depicted in Fig. 7.2, the support of MVOW/MVONs can also be enabled by a combination of BAU and Online Instruction Computation Unit (OICU). For more information on this option we refer to [4].

In case a certain complex operation has not yet been decomposed into primitive operations, see initial algorithm transformation step, then this has to be done here. As mentioned in Sect. 3.5.4, in exceptional cases it can occur, that the design and employment of a Reusable Custom Unit (RCU) is crucial for achieving high efficiency. For such a case, a more complex reusable operation is directly implemented in hardware and therefore a decomposition is not needed. We want to emphasize that such cases will occur very seldom and typically transformations can be applied to avoid them.

In addition to the type of employed units, a decision on the datapath sub-operators of the units should be made. For instance, the BAU could implement a shift-add/sub datapath or a shift-add-add/sub datapath. The LU could support only logic operations, e.g. AND, OR, or also comparison operations, e.g. $\geq$, $=$. An analysis of the DFGs of the considered algorithms will provide the necessary information. To enable high reusability, the template should support only a few pre-defined unit configurations. After this step, the required type of units and their datapath sub-operators are known.

7.2.4 Decisions on Parallelization

During this main step C, the enabled parallelization, i.e. the number of slots, units, slices, engines and clusters, is determined. Thereby the throughput and latency specifications as well as the achievable clock frequency are considered. A general separation between (1) analysis, characterization and assignment of parallelization options and (2) dimensioning of the architecture instance can be made.

7.2.4.1 Analysis, Characterization and Assignment

First, all possible data parallelization options are determined. Thereby mainly loops, but also loop bodies are considered. Especially the parallelization of the dominant operation type, which is typically the BAO, is important. Second, the identified data parallelization options are characterized. Metrics for the characterization are (1) data dependency, (2) frequency of data exchange, (3) limitation on execution order, (4) maximal parallelization factor, (5) possible limitations on exploit-ability when considering different modes and (6) requirements on control (only DLP or DLP/ILP). As third step, the granularities of the parallelization options are evaluated and parallelization options are assigned to certain hierarchy levels in the DSIP architecture template. As mentioned in Sect. 3.3.3, in general, a differentiation between fine-grain and coarse-grain parallelization is possible. A fine-grain parallelization is indicated by a strong data dependency and a high data exchange frequency. For instance, a parallel computation of the real and the imaginary part of a variable complex multiplication can be considered as fine-grain. An example for a coarse-grain parallelization is the computation of the baseband functionality for multiple antennas in parallel. In this case, the data dependency and the data exchange frequency is low. As shown in Fig. 3.9, the DSIP architecture template offers a scalability across multiple hierarchy levels. The inner levels, i.e. slots, units and slices are more suited for fine-grain parallelization whereas the outer levels, i.e. engines and clusters, are more suited for coarse-grain parallelization. The information on the granularities and on the requirements of the control together with the considerations regarding throughput and latency are then used to assign the characterized data parallelization options to

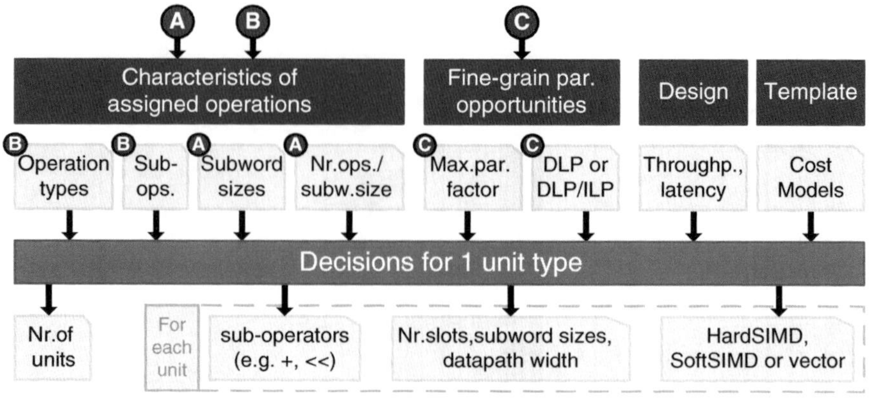

Fig. 7.3 Considered information for deciding on the parallelization of one unit type. If the assigned operation types differ strongly (e.g. required subword sizes, DLP vs. ILP/DLP), it can make sense to employ multiple units of the same type

the best-suited hierarchy levels in the DSIP architecture template. Typically only the most promising data parallelization options, e.g. with high regularity, are exploited.

7.2.4.2 Dimensioning

In this step, the actual decision on the number of slots, units, slices, engines and clusters, is made. Thereby we leverage on the following two orders of consideration: (1) the degree of parallelization of inner hierarchy levels are decided first (and then the outer hierarchy levels) and (2) the most important unit types are considered first (and then the less important ones). Because of the domain-specific characteristics and the employed architecture concepts, the considered order of (1) is a good choice: Since the data flow is rather deterministic and since the data dependency at higher hierarchy levels, e.g. clusters, is lower, typically all data transfer and storage bottlenecks can be resolved. However, the parallelization at lower hierarchy levels, where the data flow is more irregular, is more challenging. In addition, since the loop bodies are typically rather small, it is very important to enable an efficient schedule in which NOPs are avoided.

Inner hierarchy levels: In the following, we assume that the operations, which are mapped onto the BAUs, e.g. BAOs and MCOs, are computationally dominant. Thus, we first dimension the BAUs before we dimension all other units. Figure 7.3 provides a general overview of the considered information and the output of the unit dimensioning. The first category provides information on the characteristics of the operations that have to mapped onto the BAUs. This includes the subword sizes, which need to be supported, and the relative importance of a certain subword size. In addition, also the information on the preferred sub-operators, see main step B, is provided. The second category of input information includes the set of fine-grain parallelization options that have to be assigned to units. Thus, the information on maximal parallelization factors and the requirements regarding control. Besides the fore-mentioned information, the throughput and latency requirements as well as cost models are considered.

Based on this input first a decision on the number of units per type is made. For instance, in the following example it makes sense to employ two units of the same unit type, but with different configurations: The first type of operation has to be performed on 8-bit subwords, DLP can be exploited, a maximal parallelization factor of 16 is possible and the operations are dominant. The second type of operation requires 32-bit subwords, a combination of DLP/ILP is needed, the maximal parallelization factor is 2 and operations of this type occur rather seldom. A mapping of both operation types onto the same unit instance would be sub-optimal, because the hardware requirements for the second operation type would degrade the achievable performance of the first operation type. Once the number of units is decided, each unit instance is configured. For the example of above, the first unit instance could employ 16-way 8-bit HardSIMD and the second unit instance 2-way 32-bit vector operations.

After the previous step, the units of the most important unit type have been dimensioned. As a next step, the datapath width of the slice is decided. For that, typically the datapath width of the most important unit instance is used. For the example of above, this would be 16×8 bits=128 bits. All other units have to be matched to this slice datapath width. This means, if the internal datapath width of a unit is smaller, multiplexers to select subwords are required. E.g. for the second unit instance, which has an internal datapath of 2×32 bits=64 bits, a 2-to-1 multiplexer for each input operand is needed.

In the next step, the parallelization of all other required arithmetic and logic unit types, e.g. MU, LU, LUTU, is determined. As shown in Table 3.1, the breakdown for the different operation types is typically rather unbalanced. Hence, employing the same degree of parallelization for each unit type would lead to a low hardware utilization and therefore to a low area efficiency. Thus, the operation breakdown should be considered to decide on the parallelization degree of each unit type/instance individually. At this step, it can also be decided whether a sharing of certain resources, e.g. a big LUT which is seldom used, across slices/engines would make sense.

Outer hierarchy levels: After the previous steps, the units of a slice have been dimensioned. The design flow continues by deciding on the parallelization of higher hierarchy levels. Thereby again the characterized and assigned parallelization options as well as the throughput requirements are considered. In general, the information and constraints from the inner hierarchy levels are propagated to the outer hierarchy levels.

Decisions that are taken during this main step C have a strong influence on the enabled flexibility of the architecture instance. Therefore it is important to consider the targeted set of algorithms and to dimension the architecture so that all of these algorithms can efficiently be mapped. It is very likely that in the future new algorithms, which are not yet available, should be mapped onto the same architecture instance. Since their exact characteristics cannot be considered during the architecture instance design, it could be possible that this new algorithms cannot efficiently be mapped. However, we have experienced that the algorithms of a certain functional block have all similar characteristics. For this reason, we assume that the fore-mentioned case will rarely occur. In general, for the dimensioning, the trade-off between enabled flexibility and efficiency has to be considered.

Since all subsequent steps rely on the design decisions that have been made so far, it is recommended to perform a rough scheduling of the targeted algorithms at this stage. Potential issues, e.g. because of data dependencies, should be resolved before continuing. Because the mapping has implicitly been considered throughout the flow, a rough schedule may be obtained with little effort.

7.2.5 Support of Data Transfers

Since all the arithmetic and logic datapaths have been defined and since all the parallelization options, including the characteristics on data exchange, have been assigned to hierarchy levels, the basic data flow is know. Thus, decisions regarding the enabled data transfers can be made. Also here, first the inner and then the outer hierarchy levels are considered. A group of units could share the same local registers so that the requirements on the slice-internal interconnect network are reduced. Typically in can be assumed that the input data is correctly packet when it is written to the DMEM. In general, by analyzing the data flow, which is typically streaming-based, all the parameters of Fig. 3.16d can be decided. Again, here, as well as in the sub-sequent steps, trade-offs between flexibility and efficiency are involved.

7.2.6 Data Storage Dimensioning

At this main step, as depicted in Fig. 3.16e, the sizes of the LUTs within the units and the shared L1 LUT memory, the number of local registers, the number of VWRs, the number of VWR entries per slice and the number of DMEM lines are determined. Some of the decisions, e.g. the number of VWRs (which is linked to data transfers), may have already been made before. Also here, a rough schedule enables to make good decisions. By combining the information on the number of VWR entries per slice and on the number of slices within a cluster, the word length of the wide datapath, which is between the DMEM and the VWR, can be calculated.

Since for certain functional blocks (typically block-based) the DMEM area/ energy can be very significant, it makes sense to roughly evaluate especially the background memory requirements already at the beginning of the design flow and to consider them during the different instance design steps.

7.2.7 Instruction Control Dimensioning

We perform the control dimensioning as last step because the considered algorithms are typically deterministic and data dominated. In addition, during the main step C, by deciding e.g. on the number of units and engines, essential decisions regarding control have already been made. In this last step, we refine the control and dimension all related memories. For instance, condition handling and synchronization may still need to be considered. The number of employed DLBs could still be refined and the control bits have to be assigned accordingly. For the DLBs, the sizes of the control tables and the instruction memories have to be decided. Thereby, the lengths of the loop bodies and the number of different instructions within a loop should be considered. Further, a decision, whether multiple clusters will share the same part of the L1 IMEM, has to be made. Finally, the sizes of the L1 IMEM and the L1 DMA IMEM are dimensioned.

7.3 Application on Case Study 1: MIMO Detector

In this section we apply the proposed flow to design an architecture instance for MIMO detection. The result is the architecture instance that is proposed in Chap. 4 . During the application of the flow, only one algorithm and one quantization scheme is considered. Nevertheless, the application of the flow, when multiple algorithms and/or quantization schemes are targeted, is similar. The main difference is that the joint requirements need to be determined and more flexibility/efficiency trade-offs will be involved. The summary in Sect. 4.3.2 is based on the following analysis.

7.3.1 Algorithm Transformations

In our previous work, a coarse-grain floating-point to fixed-point conversion has been performed. The signals which represent the constellation points are represented with 3 bits, all other signals with 16 bits. The 3-bit representation is only effecting the MVON operations. A strength of the architecture template is the support of SoftSIMD. However, because of the applied coarse-grain quantization, the benefit of SoftSIMD cannot be exploited. We assume that the implementation efficiency of this design can be increased by performing a more fine-grain and a more aggressive quantization. However, we consider this as future work. As mentioned in Sect. 4.3.1, the algorithm has been heavily optimized. The optimized algorithm is composed of primitive operations, is dominated by BAOs and has a completely deterministic data-flow structure. Therefore it is well-suited for the mapping onto the proposed DSIP architecture template.

7.3.2 Decisions on Data Representation

Because of the coarse-grain quantization, only a limited space for design decisions exists. Since all operations are performed on 16-bit data, a 16-bit datapath will be employed. The quantization of the multiplier for MVON operations, which is represented with 3 bits, is considered later. Due to the limited subword sizes, the choice for SoftSIMD is discarded.

7.3.3 Support of Arithmetic and Logic Operations

The considered MIMO detection algorithm has the following operation breakdown, which is especially representative for the most complex modes: 66 % BAO, 0 % MCO, 1 % MVOW, 24 % MVON, 0 % TSO, 9 % BCLO. Constant multiplications (MCO) and special operations (TSO) are not present. Most operations are BAOs, which have a rather low implementation complexity. BAOs are mapped onto the BAU. Beside the BAOs, a significant amount of MVONs are present (24 %). Especially because multiplications are more expensive than basic operations, it is important to support them efficiently. As shown in Fig. 7.2, MVON operations can either be mapped onto the MU, onto a combination of BAU and OICU or onto a combination of BAU and LUTU. As motivated in Sect. 4.4, mainly because of the small 3-bit multiplier quantization, the combination of BAU and LUTU is the best option. All possible 3-bit multiplier values are encoded in CSD , decomposed to shift-add/sub instructions and stored in the LUTU. Thereby the required 3-bit quantization is considered. At run-time, depending on the actual multiplier value, the corresponding sequence is looked-up and loaded to an instruction buffer. Then, the actual multiplication is performed with the BAU. By mapping BAOs and MVONs onto the BAU, the BAU becomes clearly the most important unit. For the algorithm, mostly shift-add/sub-add/sub operations need to be executed. Therefore we enable this computation on the BAU in one clock cycle. Beside BAOs and MVONs, MVOWs and BCLOs need to be supported. For both operation types, the mapping is already pre-determined. MVOW operations are mapped onto the MU and BCLO operations are mapped onto the LU. The LU needs to support comparison operations.

7.3.4 Decisions on Parallelization

7.3.4.1 Analysis, Characterization and Assignment

Table 7.1 shows the analysis, characterization and assignment of different parallelization options. The parallelization across MIMO symbols is well-suited for

Table 7.1 Analysis, characterization and assignment of parallelization options (case study 1)

Parallelization option	Across MIMO symbols	Re || Im Data	Across constellations	List Gen. || LLR Gen.	Within list/ LLR Gen.
Data dependency	No	Low	Low	Avoidable	High
Data exchange frequency	No exchange	Low	Low	Low to high (buffering)	High
Limitation on exec. order	No	No	Basically no	1. List gen., 2. LLR gen.	Strict execution order
Max. parallel. factor	Basically unlimited	2	m_{Nt}= 16/64-QAM	List gen.: 1, LLR gen.: 6 to $N_t \times 6$	Strongly varying (1 to $N_t \times 6$)
Scaling issue	No	No	Basically no	Lin. versus quad. complexity (N_t)	Lin. versus quad. complexity (N_t)
SIMD exploitable	Yes (buffering)	Basically yes	Yes	Not well	Not well
Chosen level	Cluster	Engine	Slot	-	-

the cluster level, since basically no dependencies are involved. We can leverage on multiple parallel clusters to increase the throughput. The parallel computation of the real and the imaginary part with two engines is well-suited because (1) the data exchange frequency is rather low, (2) the engines can operate completely synchronous and (3) the data storage requirements can be reduced (an engine requires mostly either real or imaginary data). A parallelization across slices would be more difficult, since the instructions for the real and the imaginary part are slightly different. For the slot level, i.e. inside a unit, the parallelization across m_{Nt} constellations is selected. In this hierarchy level, for all BAOs, SIMD can be exploited. Due to the high dependencies and/or scaling issues, the other options are not well-suited for parallelization.

7.3.4.2 Dimensioning

Inner hierarchy levels: The operations on the BAU are clearly dominant, therefore the BAU is first considered. To decide, whether one or multiple BAUs should be employed, the characteristics of the assigned operations are analyzed. Both, BAOs and MVONs, operate on 16-bit data, so no difference in subword size exists. The sub-operation, i.e. shift-add/sub-add/sub, is suited for both. The parallel computation of BAOs can be done on a SIMD datapath. However, for MVONs, because each slot has to execute an individual sequence of shift-add/sub-operations, a vector datapath is required. So in general, it can make sense to employ two BAUs, one for SIMD and one for vector operations. Nevertheless, because only the support for one subword size is required, a BAU, which supports SIMD and vector operations, is not much more complex than a BAU that supports only SIMD operations (basically only the control for the different slots need to be modified, the datapath and the achievable clock frequency remains the same). Because of data dependencies and scheduling issues, the instantiation of two BAUs would potentially lead to a lower area utilization. For this reason, we decide to employ only one BAU within a slice. As mentioned above, we can perform a parallel computation across m_{Nt} constellations. For the targeted modulation schemes 16/64-QAM, m_{Nt} corresponds to 16/64. The choice for the parallelization factor involves a trade-off between achievable throughput/latency and hardware utilization. A high parallelization factor leads to a high throughput and a low latency, but if the factor is above 16, the hardware cannot be fully utilized in 16-QAM mode. Because the throughput can be increased by employing multiple clusters, the specification on latency (per MIMO symbol), is most important. Based on (1) the number of operations that are required to compute one MIMO symbol, (2) the type of operations and the mapping onto unit types, (3) the knowledge that the real and imaginary part will be computed in parallel (with two engines), (4) the general architecture and the typical latencies for data transfers and (5) the typically achievable clock frequency (considering that the BAU with the chosen datapath determines the maximal delay), we can estimate the expected latencies for different parallelization factors. The maximal latency, for the computation of a demanding MIMO 4×4 64-QAM near-ML mode MIMO detection, for which we pessimistically assume 200×64 clock cycles (with a parallelization factor of (1) and a clock frequency of 700 MHz, results in 18 μs. With a parallelization factor of 8, the latency is reduced to 2.3 μs, which seems reasonable for standards such as IEEE 802.11ac and LTE. Thus, on the basis of this analysis and considerations, we employ a BAU with a 8-way 16-bit vector datapath. This results in a slice-datapath width of 128 bits.

After the parallelization of the dominant unit type is known, i.e. the datapath width of the slice is fixed, the parallelization of all other unit types, i.e. MU, LUTU and LU, is decided. The operation breakdown, considering the most important mode and the defined mappings onto unit types, is 79 % BAU, 1 % MU, 13 % LUTU and 7 % LU. As it can be seen, it is rather unbalanced. Choosing the same parallelization factor for all unit types is not a good choice, because it would lead to a low area utilization. Inside a unit, instead of 8 parallel 16-bit datapaths, only one or a few could be implemented and shared across the 8 I/O-slots. In this case, the computation for the 8 subwords is done sequentially and requires several cycles. This approach enables to achieve a higher area utilization and thus, a higher area efficiency. Nevertheless, the sharing of sub-datapaths increases the latency of the vector operation, which can lead to a sub-optimal schedule and/or to increased storage requirements. In addition, the customization for the specific operation breakdown may limit the reusability of the architecture for different algorithms. For this reason, the trade-off between customization, i.e. achieved area efficiency, and flexibility has to be carefully evaluated. Once a more accurate schedule is available, the decisions may be refined.

In this work, we choose the following parallelization factors: The utilization of the MU is very low and the area of a multiplier is comparatively high. For this reason, we employ only one MU, which includes 4 parallel 8-bit multi-precision multipliers which are shared amongst the 8 slots. A 16-bit multiplication requires 8 clock cycles and a 8-bit multiplication requires 2 clock cycles for the complete 8-entry vector. The MVONs in the list generator are different from the ones in the LLR generator. In order to avoid a frequent reloading of the memory of a shared LUTU, two LUTUs are employed. They are denoted as LUTU1 and LUTU2. The content of LUTU2 is different for computations on real and on imaginary data. Because each engine operates either on real or on imaginary data, only the corresponding content needs to be stored. For each LUTU, only one LUT is employed and shared amongst all 8 slots. Since the 3-bit multiplier operand is available before the actual multiplication has to take place, the latency for the table look-up can largely be hided. Because of the rather low utilization and the limited functionality requirements, only one LU is instantiated. In contrary to the MU and the LUTUs, each of the 8 slots has an individual sub-datapath. The main reasons for not sharing resources across slots are (1) the required hardware resources for performing logic and comparison operations are anyway low, (2) the latency of the max-operation, which is frequently needed, is reduced and (3) in case the LU is used for condition handling, no additional latency is introduced.

Outer hierarchy levels: Based on the analysis of the parallelization options, a parallel computation of real and imaginary part across two engines is well suited. A cluster with two engines, each with one slice, has a total datapath width of 256 bits. This seems a reasonable granularity. A higher throughput, per cluster, can be achieved by employing engines with multiple slices. Nevertheless, this would have the consequence that (1) the granularity of a cluster is decreased and that (2) the delay for VWR types, which share data amongst slices, is increased. The latter could cause the issue that the critical path is moved from the BAU to the VWR. Instead of scaling the throughput with multiple slices, we scale it with multiple clusters. The parallelization across MIMO symbols fits clearly in the category of coarse-grain. Since engines include individual L0 controls, a linear scaling can be assumed and power gating could be applied. Compared to the option of instantiating multiple slices, more control hardware is needed. However, since the area for the L0 control is low compared to the datapath and the L1 IMEM, this overhead is acceptable.

7.3.5 Support of Data Transfers

To enable the broadcasting of one subword to all the 8 slots, which is required for comparison operations, a simple SSU with one stage is employed. The LU requires one clock cycle, whereas the MU and the LUTUs require several clock cycles to compute the output of a vector. Therefore, we denote the LU as single-cycle unit and we group the MU and the LUTUs together and denote them as multi-cycle units. Since the multi-cycle units (1) need multiple cycles before the require a new input vector and produce a new output vector and (2) are typically not fully active, they can share the same local registers. Because of this sharing, less local register files are present within a slice and thus, the complexity of the lower interconnect network and the area is reduced. The BAU and the SSU, which are the most important units, can write their output data directly to all other units. All other units can write their output data only via the VWRs to other units. This choice, which considers the typical data-flow patterns, reduces the requirements on multiplexers.

The slice of engine 1 operates on real data whereas the slice of engine 2 operates on imaginary data. For a complex multiplication, a data exchange between the two slices is required. Since this operation occurs rather frequently and the data exchange via the VWRs or a WSU would introduce latencies, we enable a direct connection between both BAUs. This combination is denoted as pair. We employ 3 VWRs of different types. T1 enables slice-specific data transfers from the DMEM to the VWR and T3 enables data transfers in the opposite direction. T2 serves for sharing data, e.g. the input data or the channel matrix, amongst slices. Because two VWR read ports towards the slices are present (T1 and T2), the slice-internal upper interconnection network needs to include multiplexers between the VWRs and the local registers. Since clusters operate independently from each other, no connection between them is required.

7.3.6 Data Storage Dimensioning

To store the instruction sequences, the LUTs of LUTU1/2 require a size of $8\times6=48$ bits and $8\times3\times6=144$ bits respectively (8 possibilities for a 3-bit multiplier value, 6 bits for encoding the shift-add/sub instruction). We assume the same amount for the shared L1 LUT memories. For the BAU 4, for the SSU 1, for the LU 2 and for the MU/LUTUs 3 local registers are instantiated. Two additional local registers are used for output data. This decision is based on a rough schedule. As mentioned above, 3 VWRs are employed. For each slice, 2 register entries per VWR are assigned. Thus, a VWR has 4 register entries and therefore a total word length of $4\times128=512$ bits. The VWRo consists of 1 VWR with 512 bits. The required DMEM size depends on the maximal number of antennas that should be supported. For MIMO $2\times2/4\times4/8\times8$, 8/10/14 DMEM lines, which correspond to 4/5/7 kbits, are needed (per cluster).

7.3.7 Instruction Control Dimensioning

Because the data-flow structure of the algorithm is completely deterministic and does not contain conditions, no hardware for condition handling and synchronization is required. The algorithm is loop-dominated, therefore the loop buffer concept is well exploitable. We employ for each unit an individual DLB and for a cluster in total 4 DLBs. The individual loop bodies, for the different DLBs, are rather small. For engine level DLBs, between 4 and 16, for cluster level DLBs, up to 20 different instructions are required. All DLBs combine the instructions and the loop control information in one memory (DLB type direct). Within an engine, the total control word is 62 bits wide and the total L0 loop buffer storage is 744 bits. For the L0 cluster level control, 360 bits for instruction storage are considered. Since all clusters execute the same instructions, the L1 IMEM can be shared amongst them. Therefore the requirements on the L1 storage are rather low.

7.4 Application on Case Study 2: FIR Filter

In this section we demonstrate the application of the proposed design flow on the FIR filter, which results in the architecture template instances HW-SIMDi and SW-SIMDi, which are proposed in Chap. 5.

7.4.1 Algorithm Transformations

The motivation for the FIR direct form structure and the applied algorithm transformations can be found in Sect. 5.3. As it can be seen in Fig. 5.4f, the result is a DFG which includes 5 shift-add/sub operations. For a comparison, we show the optimized FIR transposed form structure, which is based on the same constant coefficients, in Fig. 7.4. With the multiplier-block generator of Spiral [11], the multiplications have been optimized across filter taps. Afterwards, the word length has been limited to 12 bits, left shifts have been converted to right shifts and the number of operations have been reduced. The result is the DFG of Fig. 7.4, which includes 6 instead of 5 shift-add/sub operations. The better opportunities for

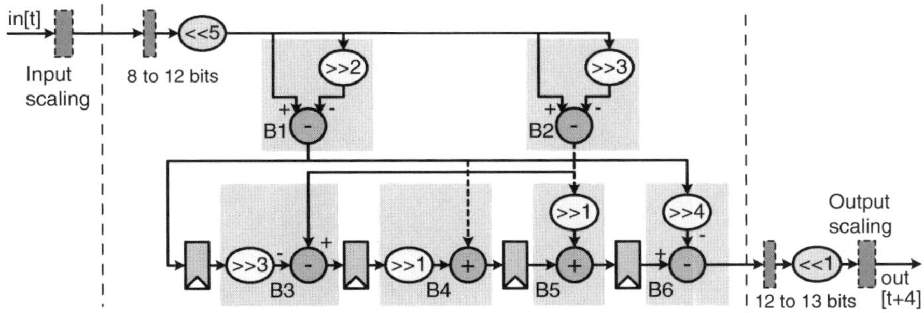

Fig. 7.4 5-tap FIR filter in transposed form structure. Multiplications across taps have been optimized. Optimized for 12-bit word length and right shift-add/sub operations. Compared to the direct form structure, one more shift-add/sub operation is required

parallelization and the reduced number of operations motivate our choice for the direct form structure. Regarding the number of operations, the direct form structure is a good choice for symmetrical filters, because the number of multiplications can be reduced by close to 50 %. However, if non-symmetrical filters are considered, the optimization space for the direct form structure is limited. In this case, the transposed form structure is better, because it enables the optimization across filter taps. Especially if the number of non-zero bits in the coefficients is high, this broad optimization will significantly reduce the operation count. Importantly, on the proposed DSIP architecture, both, direct form and transposed form filters can be mapped. In this work we focus on the symmetrical 5-tap filter and the optimized direct form structure of Fig. 5.4f.

7.4.2 Decisions on Data Representation

For the 5-tap FIR filter, 8 bits for the input data and 12 bits for the internal data are required. After the conversion of the MCOs to shift-add/sub operations, all required operations are of type BAO. We implement the arithmetic datapath of the HW-SIMDi with 12 bits. For the arithmetic datapath of the SW-SIMDi, the subword sizes 8, 12, 16 and 24 bits are chosen. The support of the subword sizes 16 and 24 bits ensures, that also filters with higher orders (which will typically require a higher precision) can be mapped. All the selected subword sizes are multiple of 4 bits. This property can be exploited to reduce the implementation cost. With the chosen subword sizes, a 48-bit datapath can be fully utilized.

7.4.3 Support of Arithmetic and Logic Operations

BAOs are mapped onto the BAU. The BAU implements the basic shift-add/sub operation $\pm(A \gg 0..7) \pm B$, which is suited for BAOs and MCOs. Because the negation of input A and B can be programmed, the mapping of all CSD encoded numbers is possible. Beside the right shift, also left shift is required. We assign this operation to the SSU.

7.4.4 Decisions on Parallelization

7.4.4.1 Analysis, Characterization and Assignment

Table 7.2 provides an overview of the possible and assigned parallelization options. Because the computations can be performed independently from each other, a parallel computation of the real and the imaginary part with two clusters is

Table 7.2 Analysis, characterization and assignment of parallelization options (case study 2)

Parallelization option	Re || Im Data	Across Complete Filter DFG	Across Taps	Mult. || Add. of Delay Chain
Data dependency	No	Only input samples	Yes	Yes
Data exchange frequency	No exchange	Only broadcast of input samples	High	High
Limitation on exec. order	No	Yes, if output needs to be in order	Yes, prev. tap must be comp. first	1. Mult., 2. Add.
Max. parallel. factor	2	Only limited by available input data (latency)	Nr. taps	2..2×Nr. taps
Scaling issue	No	For high filter orders: w/o expl. symm.: no, w. e. sym.: local reg.	Yes, especially when par. across slots	Yes, e.g. nr. non-zero bits in coefficients
SIMD exploitable	Yes	Yes	No (diff. coefficients)	Basically yes
Chosen level	Cluster	$\geq$ Slot	–	–

a good choice. In addition, across slots, the outputs, i.e. for different time samples, of the filter structure are computed in parallel. A shuffling/distribution needs to ensure that each slots receives the required input samples. The number of parallel slots can be adjusted within a unit and then further scaled with multiple slices, engines and clusters. For the clusters, we consider a multiple of 2.

7.4.4.2 Dimensioning

Inner hierarchy levels: Since the left shifting will be supported with the SSU and all other arithmetic operations are performed with the BAU, only a BAU and a SSU are required. For a SoftSIMD datapath width of 48 bits, the subword organizations 6×8, 4×12, 3×16 and 2×24 bits enable a full utilization. Clearly, also with a multiple of 48 bits, e.g. 96 bits, 144 bits, the same is valid. However, considering the support for the whole shuffling range, the complexity of the SSU increases significantly with higher datapath width. For this reason, we chose a datapath width of 48 bits for the slice of the SW-SIMDi. Because it enables a fair direct comparison, we chose the same datapath width also for the slice of the HW-SIMDi. Considering the required subword size of 12 bits for the filter, a slice offers a parallelization factor of 4.

Outer hierarchy levels: Assuming that the output of the DFG can be computed in 5 clock cycles and a clock frequency of 900 MHz, a slice offers a throughput of 0.72 Gsps. Thus, for the targeted 3.52 Gsps, 5 parallel slices are required. Multiple slices can be instantiated (1) within an engine, (2) with multiple engines, (3) with multiple clusters or (4) as a combination of (1)–(3). The choice is essentially a trade-off between flexibility, efficiency, granularity for scalability and latency (for a wider DMEM, the serial to parallel conversion requires more clock cycles). Considering also the required bandwidth from the VWRs to the slices, we select two slices per engine and two engines per cluster, which seems to be a good compromise. For 48×4×2=384-bit wide DMEMs/VWRs, the serial to parallel conversion requires 32 clock cycles. Thus, the achieved latency is clearly below the required 167 clock cycles, which is specified in Sect. 5.2.1. The throughput can be scaled with multiple clusters, which have to be multiples of 2 (complex number).

7.4.5 Support of Data Transfers

As shown in Fig. 7.5, the 4 slots of a slice operate on slightly delayed input samples. Slot 1 on *in*[*t*], slot 2 on *in*[*t*+1], slot 3 on *in*[*t*+2] and slot 4 on *in*[*t*+3]. This input order ensures that also the output order is correct. For the HW-SIMDi, the input samples are packed as 12-bit subwords in 48-bit words. For the SW-SIMDi, additionally the packing of 8-bit subwords is possible. Assuming that each input subword is only present in one input datapath word, the SSU needs to be capable of combining/merging subwords from two datapath words. The SSU for the SW-SIMDi must also support the resizing of subword sizes. The functionality of the SSU is implemented with multiple stages, which enable different data-flow patterns.

Inside a slice, a bidirectional connection between the BAU and the SSU is employed. In addition, both units can read to/write from the VWRs/VWRo. Considering the 12-bit input subwords and the fore-mentioned order, 2 input datapath words lead to 5 different SSU output datapath words. Since the BAU requires a new SSU-output datapath word at every clock cycle, a slice has to access the VWR every 2.5 clock cycles. Thus, for the 4 slices of a cluster, a minimum of 2 VWR read ports are required. Because at least two slices require the same input samples, we employ a VWR type that enables a broadcasting to slices. Each of the two read multiplexers is either assigned to slice 1 (A) or slice 2 (B) of an engine. If one engine is not used, the other engine can leverage on both read ports. We instantiate a second VWR that enables the storing of intermediate data and the data transfer to the DMEM. This ensures, that also filters with higher

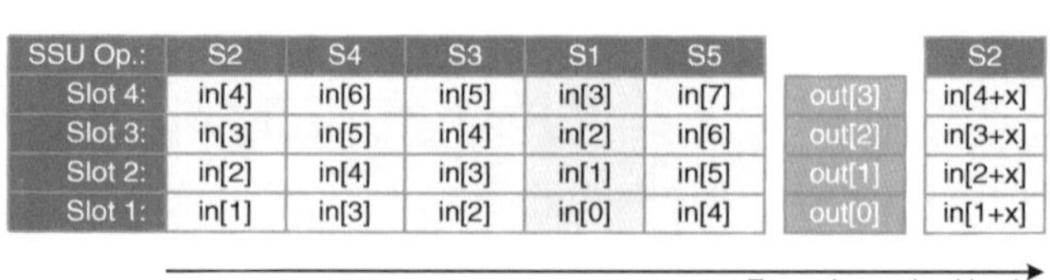

Fig. 7.5 Execution and input sample order within a slice. The labels S1–S5 correspond to the SSU operations in Fig. 5.4f

orders, which may require more intermediate storage, can be mapped. A connection between clusters is not needed. The data routing to/from the clusters is done outside of the considered DSIP architecture instances.

7.4.6 Data Storage Dimensioning

For the SSU, at least two local registers, which can be written from the VWRs, are required. The BAU requires at least two local registers for the two inputs. In total, we instantiate 4 local registers for the SSU and 3 for the BAU. As mentioned above, two VWRs are employed. We choose two register entries per slice, so 8 48-bit datapath words per VWR can be stored. This choice results in a DMEM/VWRs/VWRo datapath word length of $8\times48=384$ bits. If the data can be provided at the required time at the input of the processor, it can directly be written to the VWRs. Thus, in this case, the DMEM is not needed. Nevertheless, to ensure that the processors remain flexible, we instantiate 2 DMEM lines for the HW-SIMDi and 4 DMEM lines for the SW-SIMDi.

7.4.7 Instruction Control Dimensioning

Like for the considered MIMO detection algorithm, the data flow is of streaming nature and completely deterministic. We employ 2 DLBs per engine and 4 DLBs per cluster. For the DLBs of the engine, we store the instructions for loop control and the actual datapath instructions in different IMEMs (indirect DLB type). Since all clusters execute the same set of instructions, they can share the same L1 IMEM. The chosen DLB and L1 IMEM sizes are mentioned in Sect. 5.4.

7.5 Conclusions

To achieve high implementation efficiency, it is very essential that the algorithms and the architectures are co-optimized for each other. For traditional ASICs, this is inherently the case. For traditional DSIPs/DSPs, basically only algorithm optimizations can be applied. For traditional ASIPs, co-optimization are essential. However, the establishment/application of an algorithm-architecture co-design flow for ASIPs is very challenging. The main reason for this is the big algorithm transformation space together with the big architecture design space. The latter is caused by the ASI/HA design freedom. A further reason is the rather weak link between different system design phases. For instance, because of the ASI/HA design freedom and the "unpredictable" standard-cell back-end design flow, no accurate/stable architecture cost models can be made.

In contrast, the establishment/application of an algorithm-architecture co-design flow for the proposed DSIP architecture template is implicitly enabled by the taken design approach. The most important element, which enables a co-design flow with clearly bounded complexity, is the employment of an architecture template with a pre-defined medium-grain architecture design space. Because the architecture design space can be pre-characterized with accurate cost models (accurate, because of the proposed "predictable" semi-custom back-end design flow), a strong link to the functionality/algorithm design phase can be created. As a consequence, the designed functionalities and the corresponding algorithms are well-suited for an efficient implementation with the proposed DSIP architecture style.

This chapter focuses on the architecture design phase of the overall algorithm-architecture co-design flow. The proposed design flow supports the design of efficient DSIP architecture template instances. The flow starts with the targeted algorithms (or their characteristics) and the system/hardware specifications. Throughout the unidirectional flow, which comprises the propagation of constraints, certain parameter values are decided at each step. At the end of the flow, a well-suited template architecture instance, together with rough mappings of the considered algorithms, are available. The architecture instances for all 3 case studies (in this chapter only in detail shown for the MIMO detector of Chap. 4 and for the FIR filter of Chap. 5) have successfully been designed by applying the proposed design flow.

The proposed design flow of this chapter, which complements the proposed DSIP architecture template of this book, provides a good starting point for bridging the cultural gap between algorithm and architecture designers. However, it has to be emphasized, that future work has to further mature the proposal. For instance, the application of the design flow for multiple algorithms has not yet been demonstrated. Investigations in design automation, which leverage on cost models for evaluating trade-offs, are also a necessity.

References

1. Berset, T., Catthoor, F.: Technical note TN-11-WATS-TIP2: high level estimation for independent component analysis. Technical report, Holst Centre, IMEC-NL (2012)
2. Catthoor, F.: Slide set that explains the high-level estimation flow, IMEC. Available on request
3. Catthoor, F., Danckaert, K., Wuytack, S., Dutt, N.: Code transformations for data transfer and storage exploration preprocessing in multimedia processors. IEEE Des. Test Comput. **18**(3), 70–82 (2001)
4. Fasthuber, R., Raghavan, P., Catthoor, F.: An Enhancement for enabling variable multiplications on a general shift-add/sub datapath (2013, in preparation)
5. Li, M., Bougard, B., Novo, D., Van Thillo, W., Van der Perre, L., Catthoor, F.: Adaptive SSFE Near-ML MIMO detector with dynamic search range and 80–103Mbps flexible implementation. In: IEEE Global Telecommunication Conference. (GLOBECOM), pp. 1–5. IMEC (2008). doi:10.1109/GLOCOM.2008.ECP.632

6. Li, M.: Algorithm and architecture co-design for software defined radio baseband. Ph.D. thesis, KU Leuven (2010)
7. Li, M., Appeltans, R., Amin, A., Torrea-Duran, R., Cappelle, H., Hartmann, M., Yomo, H., Kobayashi, K., Dejonghe, A., Van Der Perre, L.: Overview of a software defined downlink inner receiver for category-E LTE-advanced UE. In: IEEE International Conference on Communication (ICC), pp. 1–5 (2011). doi:10.1109/icc.2011.5963387
8. Novo, D.: Exploiting adaptive precision in software defined radios. Ph.D. thesis, KU Leuven (2010)
9. Novo, D., Li, M., Fasthuber, R., Raghavan, P., Catthoor, F.: Exploiting finite precision information to guide data-flow mapping. In: Design Automation Conference (DAC), pp. 248–253 (2010)
10. Smith, A., Nagarajan, R., Sankaralingam, K., McDonald, R., Burger, D., Keckler, S., McKinley, K.: Dataflow predication. In: IEEE/ACM International Symposium on Microarchitecture (MICRO), pp. 89–102 (2006). doi:10.1109/MICRO.2006.17
11. Voronenko, Y., Püschel, M.: Multiplierless multiple constant multiplication. ACM Trans. Algorithms **3**(2), 11-es (2007). doi:10.1145/1240233.1240234
12. Zhang, N.: Algorithm/Architecture co-design for wireless communication systems. Ph.D. thesis, University of California, Berkeley (2001)

Chapter 8
Conclusions and Future Work

8.1 Conclusions

8.1.1 A Clear Need for New Design Approaches for Wireless Application Platforms

To enable emerging next-generation smartphones, which have a high usability and which can cost effectively be realized, high computational performance, high energy efficiency, high flexibility, high reusability and technology scaling-friendliness need to be combined. However, because these metrics have counter-acting effects, e.g. the highest energy efficiency is only achievable without flexibility, realizing a system that can sufficiently fulfill all of these requirements is an enormous and an increasing challenge. State-of-the-art design approaches, such as the employment of an ASIC, an ASIP or a conventional DSIP platform, are suited in fulfilling a few of these metrics, but not all of them combined. For the most promising approaches, such as the ASIP and the DSIP design approach, the insufficient energy-efficiency (compared to the potential represented by the ASIC style) and the rather weak link to technology are the major issues. As a consequence, new design approaches, which can sufficiently fulfill the joint requirements, are urgently needed.

8.1.2 Proposed Design Approach for Wireless Application Platforms

In this book, we propose a new DSIP architecture template and design approach to tackle the fore-mentioned challenge in the wireless baseband application domain [8–10, 12]. The core of this work is the energy-efficient scaling-friendly wireless domain-specific architecture template, which is based on innovative architecture concepts, such as Software SIMD, VWR and DLB. By considering common

R. Fasthuber et al., *Energy-Efficient Communication Processors*,
DOI: 10.1007/978-1-4614-4992-8_8,

domain-specific algorithm properties, a good match between algorithms and architecture template is ensured. The technology-aware design of this architecture template, which focuses on wire scalability, enables an efficient implementation even in very advanced technologies. To increase the productivity, the proposed architecture template is complemented with an algorithm-architecture co-design flow (see Chap. 7) and with an architecture-layout design flow (see appendix of Chap. 3 and [17]).

8.1.3 Results are Very Promising

The proposed design approach has been evaluated with three case studies. For the first case study, a DSIP architecture template instance for advanced MIMO detection, i.e. required for emerging LTE/WLAN systems, was designed [5–7, 11]. The design of the second case study is well-suited for high-throughput FIR filtering in multi-Gbps 60 GHz systems [13]. The designed FFT instance, introduced as third case study, targets emerging WLAN and 60 GHz systems. Although all designs offer high computational performance, flexibility and reusability, they can achieve a high energy efficiency of 3300 MOPS/mW (MIMO detector), 970 MOPS/mW (FIR) and 2320 MOPS/mW (FFT) in TSMC 40 nm standard-cell technology. With these results, the proposed implementations offer close to standard-cell ASIC energy efficiency. By dynamically trading-off BER performance and energy consumption at run-time, the average energy consumption of the proposals could even beat the ones of ASICs. Further, by applying the wire-friendly semi-custom design approach proposed in the appendix of Chap. 3, the energy efficiency of the resulting implementation can be very close to the potential technology limits, which is clearly desired (see Sect. 1.2). The energy and area comparison with state-of-the-art programmable solutions shows that the proposed designs are at least a factor 2–3 more efficient. For traditionally employed SDR processors, the differences are even much higher. Therefore, the proposed DSIP architecture template design approach of this book is very promising.

8.1.4 Main Messages

The main general messages of this book can be summarized as follows:

- *The design of wireless communication systems, which fulfill the joint requirements, will become much more challenging in the future:* The main reasons for that are: (1) the trend towards very high data rates, many standards and functionalities, (2) the slow growth of available battery capacity, (3) the need for more advanced technology nodes and the big limitations that come with it, and (4) the fast evolving and competitive/cost-sensitive handset market. The following 4 subsequent messages play a key role in tackling this challenge.

(1) *New process technologies beyond 40 nm are crucial, but they have to be employed/considered in an effective/sufficient manner:* The continuation of technology scaling and the employment of new technology features, such as FinFET devices, RRAMs or STT-MRAMs, and 3D stacking, are essential requisites for future designs. However, to leverage on the benefit of these new technologies, they have to be employed/considered in an effective/sufficient manner.
(2) *Cross-optimizations, and therefore inter-disciplinary cooperations, are becoming more crucial:* For the successful design of future systems, which implies better optimizations across all levels of the design space, researchers/engineers from different communities and disciplines will have to work closer together. Thus, holistic and consistent design flows are a necessity.
(3) *Sufficient design automation support and reuse across a broad range of matching application sub-domains is becoming more important than ever:* This is caused by the increasing complexity of systems and the consequences on NRE costs.
(4) *It is crucial to choose the best-suited overall design trade-offs:* Design objectives, such as flexibility and high performance, are conflicting. This is the reason why an ASIC is more energy efficient than a DSP. Thus, it is very important to choose the best-suited trade-offs in the design space.

- *Traditional architecture styles and associated design approaches will no longer be suited:* This is the case because no existing solution can fulfill the joint requirements of high computational performance, high energy efficiency, high flexibility, high reusability and technology scaling-friendliness in a sufficient manner.
- *The proposed design approach of this book is very promising:* This is the case because the joint requirements can be fulfilled sufficiently. Nevertheless, it is clear that more research (and development) is required before this work can be fully used in a commercial context. Especially the design automation aspect is currently a weakness.

8.1.5 Main Contributions

The main contributions of this book are summarized in Sect. 1.5.2.

8.2 Future Work

8.2.1 Remaining Challenges and Tasks

The main remaining technical challenges and tasks can be summarized as follows:

- *Definition and characterization of template parameter space:* In this book the list of possible parameters has been shown Sect. 3.6.2, but the parameter ranges have not been defined. It is still a challenge to evaluate the different parameters, to characterize different architecture options with meaningful cost models and to clearly define the instance customization space. For this task, to enable scalability and reproducibility, the architecture should be fully implemented with the proposed semi-custom layout style introduced in the appendix of Chap. 3 and in [17].
- *Evaluation of architecture template for more functional blocks:* In this book we have evaluated the architecture template for a MIMO detector, a FIR filter and a FFT. Nevertheless, to further prove the domain-specificness of the proposed architecture template, more functional blocks should be implemented. Currently, the design of a FEC, i.e. LDPC and Turbo decoder, architecture instance is ongoing. The initial results look very promising and indicate, like the results of this book, close to standard-cell ASIC energy efficiency. Because this functional block is rather data memory dominated, we expect, that the area efficiencies will also be rather close to the ones of ASICs. Furthermore, the design of a more generic FFT architecture instance should be considered.
- *Evaluation on flexibility/efficiency trade-off of architecture instances:* The proposed architecture instances of this book were evaluated in detail while considering only one algorithm. Although the architectures would enable the mapping of different algorithms, it has not been fully evaluated yet, how well other algorithms would perform on them. Especially for architecture instances, which need to support multiple rather heterogeneous algorithms efficiently, it is important to consider the joint requirements during the design. This trade-off between customizing for a specific algorithm, i.e. increasing efficiency, and making an instance more flexible, i.e. enabling the mapping of many algorithms, should be further evaluated. Thereby also the benefit of Software SIMD should be further explored.

  The above mentioned challenges and tasks, which are mostly related to scalability, are largely considered in the ongoing research of the IMEC SSET-CSI group.
- *Partitioning and platform design for multiple standards and modes:* In Sect. 3.1 the envisioned DSIP architecture platform was explained. Although the employed functionality and functional blocks are rather similar, it is still unclear how efficiently the proposed platform can handle multiple standards and modes. Especially the partitioning, and deciding on the degree of customization for each instance, is a challenge.

  The currently ongoing PhD research of Prashant Agrawal (IMEC) is tackling this challenge [1–3].
- *Integration of new low-energy memory technologies:* As mentioned in Sect. 3.4 , emerging embedded NVMs, such as RRAMs or STT-MRAMs, can significantly reduce the energy of the data and instruction storage hierarchy. However, the integration of such new memory technologies into the DSIP architecture template has not been considered yet.

  The currently ongoing PhD research of Manu Komalan (UCMadrid) [15] and Matthias Hartmann (IMEC) is addressing this.

- *Measures to ensure high reliability:* As mentioned in Sect. 1.2.2, with the scaling towards DDSM technologies, the reliability of circuits is becoming a big concern. In Sect. 3.6.1 it was mentioned that the proposed DSIP architecture template has certain attributes that simplify the effective mitigation of the reliability issue. However, future work is required to characterize reliability and to integrate mitigation countermeasures into the template.
  The currently ongoing PhD research of Halil Kukner (IMEC) [16], Mohamed Sabry (EPFL) [20, 21] and Dimitrios Rodopoulos (NTUAthens) [19] is tackling this task.
- *Implementation of run-time scalability:* In the course of the case studies, we have strongly motivated the exploitation of a dynamic run-time manager. However, the implementation of a run-time manager for the wireless domain, including the required monitors, is still missing.
  Current research is focusing on run-time/system scenarios, which may lead to promising proposals for the implementation of the run-time manager. We refer to the PhD work of Elena Hammari [14], Iason Filipopoulos and Mladen Skelin from NTNU Trondheim.
- *Further refinement and automation of front-end design:* In this book, the proposed architecture template instantiation flow, introduced in Chap. 7, was manually applied. However, to increase the productivity and to evaluate different trade-offs, which is especially important when considering multiple algorithms per instance, tool support is essential. In addition, because of the employed disruptive architecture concepts, such as DLBs, VWRs and the combination of shared VWRs and local registers, currently no suitable compiler for the proposed architecture template exists. For the case studies, a manual software scheduling was performed. However, this approach is clearly not scalable for bigger designs. Thus, the need for a simulation/RTL generation framework and a suitable compiler is prevailing.
  Current effort is ongoing to enable a tool support via the Target Compiler framework [24]. We cooperate with other Universities, such as IIT Delhi with the PhD work of Namita Sharma [22, 23] and UCMadrid with the PhD work of Carlos Juega, to tackle the remaining research challenges.
- *Further refinement and automation of back-end design:* The experimental results of the proposed semi-custom design flow (see appendices of the Chaps. 3 and 5 as well as in [17]) are promising. However, further experiments, e.g. to find the best trade-off between manual and automatic placement, are needed. Besides, also the design of custom-cells should be considered. For the related experiments that have been carried-out so far, parts of the proposed semi-custom back-end design flow have been automated. However, for the support of the whole architecture template, especially when considering the medium-grain customization space, advanced tools are clearly needed. To solve the existing research challenges, more cooperations are envisaged.

The proposed front-end design approach (see Chap. 7) and the proposed back-end design approach (see appendix of the Chap. 3 and in [17]) enable strong links between different design communities. However, further work, e.g. to define the right abstraction layers and models, is a necessity for solving the cultural challenges.

8.2.2 Planned Continuation of Work

The work of this book is being exploited and further matured in the context of the IMEC SSET-CSI group. For the long-term, the implementation of the whole wireless platform, using the proposed approach, is planned. As mentioned above, currently, the design of a FEC architecture instance is ongoing. Besides, the Target Compiler framework is being extended for more efficiently enabling compiler and modeling support. Also, the further application of the proposed semi-custom design approach is planned. The proposed DSIP architecture template is being referred as BLOX-ADRES.

Beside the exploitation in the wireless domain, it makes sense to reuse the proposed approach, or at least parts of it, for other application domains. For instance, the mobile graphics and multimedia domain have similar requirements as the wireless domain, therefore a high reuse should be possible. Nevertheless, also the exploitation in ultra low power domains, e.g. in the biomedical sensor node domain, is interesting. For example, results from the Holst Center in IMEC-NL, which have implemented biomedical algorithms on the FEENECS architecture of [4], which is the basis of this work, are promising [4, 18].

References

1. Agrawal, P., Fasthuber, R., Raghavan, P., Vander Aa, T., Ahmad, U., Van der Perre, L., Catthoor, F.: High level analysis of trade-offs across different partitioning schemes for wireless applications. In: IEEE Workshop on Signal Processing Systems (SIPS) (2011)
2. Agrawal, P., Sugand, K., Palkovic, M., Raghavan, P., Van der Perre, L., Catthoor, F.: Partitioning and assignment exploration for multiple modes of IEEE 802.11n modem on heterogeneous MPSoC platform. In: Euromicro Conference on Digital System Design (DSD), pp. 608–615 (2012). doi:10.1109/DSD.2012.102
3. Agrawal, P., Raghavan, P., Hartmann, M., Sharma, N., Van der Perre, L., Catthoor, F.: Early exploration for platform architecture instantiation with multi-mode application partitioning. In: Design Automation Conference (DAC) (2013)
4. Catthoor, F., Raghavan, P., Lambrechts, A., Jayapala, M., Kritikakou, A., Absar, J.: Ultra-Low Energy Domain-Specific Instruction-Set Processors, 1st edn. Springer (2010)
5. Fasthuber, R., Li, M., Novo, D., Van Der Perre, L., Catthoor, F.: Novel energy-efficient scalable soft-output SSFE MIMO detector architectures. In: International Conference on Embedded Computer Systems (IC-SAMOS) IMEC (2009). doi:10.1109/ICSAMOS.2009.5289228
6. Fasthuber, R., Li, M., Novo, D., Van Der Perre, L., Catthoor, F.: Exploration of soft-output MIMO detector implementations on massive parallel processors. J. Sig. Proc. Syst. (JSPS) **64**(1), 75–92 (2010). doi:10.1007/s11265-010-0499-0

7. Fasthuber, R., Li, M., Novo, D., Raghavan, P., Van der Perre, L., Catthoor, F.: Energy-efficient run-time scalable soft-output SSFE MIMO detector architectures. Trans. High-Perform Embed Architect Compilers (HiPEAC) (Special Issue SAMOS 2009) **5**(3), 1–20 (2011)
8. Fasthuber, R., Raghavan, P., Catthoor, F.: A programmable low energy massive-parallel architecture for wireless communication systems. In: Workshop in DATE: Designing for Embedded Parallel Computing Platforms (Architecture Session) (2011)
9. Fasthuber, R., Raghavan, P., Catthoor, F.: An highly-efficient processor template for wireless communication systems. In: DATE PhD Forum (2012)
10. Fasthuber, R., Agrawal, P., Raghavan, P., Catthoor, F., Van der Perre, L.: A Novel Energy Efficient Wireless Domain Specific Processor Template. IEEE Trans. Circuits Syst. (TCAS) (2013, in preparation)
11. Fasthuber, R., Raghavan, P., Catthoor, F.: A Scalable MIMO detector processor with Near-ASIC energy-efficiency. IEEE Trans. Very Large Scale Integr. Syst. (TVLSI) (2013)
12. Fasthuber, R., Raghavan, P., Catthoor, F.: An Enhancement for enabling variable multiplications on a general shift-add/sub datapath (2013, in preparation)
13. Fasthuber, R., Raghavan, P., Van Thillo, W., Catthoor, F.: An energy-efficient software SIMD processor for very high throughput 60 GHz PHY FIR. J. Sig. Proc. Syst. (JSPS) (2013, in preparation)
14. Hammari, E., Catthoor, F., Huisken, J., Kjeldsberg, P.: Application Of medium-grain multiprocessor mapping methodology to epileptic seizure predictor. In: NORCHIP, pp. 1–6 (2010). doi:10.1109/NORCHIP.2010.5669489
15. Komalan, M., Hartmann, M., Gomez Perez, J.I., Tenllado, C., Artes Garcia, A., Catthoor, F.: System level exploration of Resistive-RAM (ReRAM) based hybrid instruction memory organization. In: Memory Architecture and Organization Workshop (MeAOW) (2012).
16. Kukner, H., Weckx, P., Raghavan, P., Kaczer, B., Catthoor, F., Van der Perre, L., Lauwereins, R., Groeseneken, G.: Impact of duty factor, stress stimuli, and gate drive strength on gate delay degradation with an atomistic trap-based BTI model. In: Euromicro Conference on Digital System Design (DSD), pp. 1–7 (2012). doi:10.1109/DSD.2012.77
17. Medra, A.: Semi-custom design to reduce the interconnect energy in advanced technologies. Master thesis, Nile University, Egypt (2012)
18. Novo, D., Kritikakou, A., Raghavan, P., Van der Perre, L., Huisken, J., Catthoor, F.: Ultra low energy Domain Specific Instruction-set processor for on-line surveillance. In: IEEE Symposium on Application Specific Processors (SASP), pp. 30–35 (2010). doi:10.1109/SASP.2010.5521151
19. Rodopoulos, D., Mahato, S.B., de Almeida Camargo, V., Kaczer, B., Catthoor, F., Cosemans, S., Groeseneken, G., Papanikolaou, A., Soudris, D.: Time and workload dependent device variability in circuit simulations. In: IEEE International Conference on IC Design Technology (ICICDT), pp. 1–4 (2011). doi:10.1109/ICICDT.2011.5783193
20. Sabry, M., Ayala, J., Atienza, D.: Thermal-aware compilation for register window-based embedded processors. IEEE Embed. Syst. Lett. (ESL) **2**(4), 103–106 (2010). doi:10.1109/LES.2010.2081343
21. Sabry, M., Coskun, A., Atienza, D., Rosing, T., Brunschwiler, T.: Energy-efficient multiobjective thermal control for liquid-cooled 3-D stacked architectures. IEEEE Trans. on Comput.-Aided Des (TCAD) **30**(12), 1883–1896 (2011). doi:10.1109/TCAD.2011.2164540
22. Sharma, N., Sahula, V., Ravikumar, C.: Energy aware task scheduling for soft real time systems using an analytical approach for energy estimation. Int. J. Adv. Stud Comput, Sci Eng (IJASCSE) (2013)
23. Sharma, N., Vander Aa, T., Agrawal, P., Raghavan, P., Ranjan Panda, P., Catthoor, F.: Data memory optimization in LTE downlink. In: IEEE International Conference on Acoustics, Speech and Signal Processing (ICASSP) (2013)
24. Target ASIP development toolsuite (inc. IP Designer, IP Programmer), Target Compiler Technologies. http://www.retarget.com/

Index

0-9

60 GHz, 1, 2, 18, 73, 74, 81, 171–173, 206, 219, 242, 246, 248, 280

A

Algorithm, 5, 7, 10, 14, 19, 28, 30, 31, 34–39, 41, 70, 71, 74, 75, 80, 83–85, 100, 116, 118, 120, 141, 144, 175, 189, 223, 224, 226, 237, 251, 252, 254–258, 265, 266, 268, 271, 272, 282
Algorithm-architecture, 14, 36, 72, 122, 167, 251–254, 256, 275, 276
Algorithm design, 33–35, 37, 38, 252, 253, 255
Algorithm optimization, 13, 14, 72, 137, 138, 141, 223
Application domain, 1, 32, 43, 112, 154, 228, 242, 246, 255, 279, 284
Architecture, 6–13, 15–17, 35, 40, 97, 107, 121, 145, 180, 208, 228, 256
Architecture design, 7, 9, 13, 33, 40, 42, 76, 78, 107, 126, 129, 141, 208, 209, 251–255, 257, 258, 276
ASIC, 6, 12, 14, 15, 28, 29, 31, 35, 40, 43, 47, 51, 52, 57, 93, 138, 159, 161, 164, 172, 202, 203, 205, 207, 219, 220, 242, 244, 245, 246, 248, 252, 275, 280, 282
ASIP, 6–8, 12, 14, 16, 28–32, 41, 43, 52–54, 56, 57, 70, 138, 220, 246, 252, 253, 275
Autonomy, 3, 4

B

Back-end, 10, 14, 18, 69, 124, 125, 128, 131, 207, 283
BAO, 80, 82, 86, 89, 90, 106, 113, 144, 179, 227, 259, 260, 266, 267, 273
Battery capacity, 4, 280
BAU, 90, 97, 102, 103, 105, 113, 115, 119, 127, 149, 151, 179, 185, 190, 192, 197, 210, 262, 267, 273
BER simulation, 34, 37, 74, 138, 140, 143, 157, 165, 280
BLCO, 81, 82, 92, 114, 119, 144, 154, 260
BLOX-ADRES, 284

C

Cached, 97, 99, 101, 146, 181, 183, 184, 224, 230, 232, 237
Cluster, 97, 99, 101, 146, 181, 183, 184, 230
Co-design, 14, 31, 72, 175, 194, 251–256, 280
Cognitive radio, 27
Comparison, 17, 18, 81, 92, 122, 127, 138, 141, 150, 159, 162, 165, 166, 199, 200, 202, 203, 241–244, 269, 273
Compilation, 31, 33, 36, 41, 46
CSD, 89, 90, 152, 177, 206, 266
Cultural gap, 10, 11, 128, 251

D

Datapath, 9, 90, 94, 31, 42, 46, 49, 70, 77, 78, 87, 101, 106, 114–116, 123, 129, 146, 149, 151, 159, 164, 176, 177, 183, 188, 198, 199, 232, 261, 264, 268, 274
DDSM, 5, 8, 9, 50, 56, 69, 71, 75, 124, 127, 138, 207
Design approach, 1, 9, 11–13, 16, 17, 29, 32, 36, 38, 40, 52, 53, 69, 70, 73, 74, 124, 125, 143, 207, 214, 236, 254, 259, 279–281, 283
Design flow, 13, 14, 16, 18, 19, 33, 36–38, 41, 47–51, 57, 76, 108–111, 122, 124–126, 128–130, 131, 202, 207–210, 212–214, 238, 240, 245, 247, 251, 252, 254, 257, 263, 264, 272, 276, 283
Digital frontend, 17, 27, 69

R. Fasthuber et al., *Energy-Efficient Communication Processors*,
DOI: 10.1007/978-1-4614-4992-8,

DLP, 42, 43, 55, 82–84, 92, 97, 100–102, 109, 114, 115, 146, 148, 151, 184, 185, 228, 231, 261, 263
DMEM, 42, 44, 45, 94, 95, 99, 100, 101, 107, 109, 116–119, 123, 146, 148, 157, 161, 166, 181, 183, 184, 194, 195, 197–199, 205–207, 212, 230, 231, 234, 237–240, 246, 264, 270, 271, 274, 275
DSIP, 7, 11, 12, 14, 16–19, 29, 32, 35, 51, 52, 54–57, 69, 70, 72, 73, 79, 97, 99, 101, 106–109, 121–123, 126–128, 131, 137, 145, 148, 159, 165, 166, 175, 203, 206, 208, 210, 212, 214, 219, 220, 228, 238, 241, 244–249, 251, 254–259, 261, 262, 265, 272, 275, 276, 279, 280, 282, 284
DSP, 6, 9, 11, 28, 29, 32, 33, 51, 56, 70, 164, 203, 254, 275, 281

E

Energy efficiency, 1, 3, 4, 6, 7, 10, 11, 12, 14, 16, 29, 43, 56, 57, 70, 71, 74, 79, 90, 100, 105, 114, 121, 123, 124, 138, 151, 167, 172, 187, 197, 199, 200, 206, 220, 222, 223, 241, 244, 245, 253, 279, 282
Engine, 17, 92, 93, 100–102, 115, 121, 148, 149, 154, 164, 183–185, 194, 195, 201, 203, 230–232, 234, 236, 237, 242, 247, 248, 257, 269–271, 274, 281

F

FFT, 18, 27, 32, 54, 57, 82, 85, 107, 114, 116–119, 145, 174, 180, 205, 207, 219–224, 226–228, 230, 232, 233, 236–239, 241–243, 245–249
Filter, 18, 19, 25, 32, 69, 84, 85, 124, 131, 172, 173, 177, 180, 194, 195, 199–203, 205–207, 210, 272
FIR filter, 18, 19, 51, 173–175, 199, 203, 206, 210
Flexibiliy, 1–3, 6, 11, 12, 16, 29, 32, 52, 56, 74, 100, 111, 112, 121, 139, 154, 172, 188, 219, 220, 223, 232
Front-end, 13, 14, 16, 17, 27, 283
Functional block, 28–32, 34, 35, 39, 69–71, 107, 118, 139, 222, 239, 256
Functionality design, 33, 37
Functional unit, 31, 42

H

Hardware accelerator, 31, 42
Hardware implementation, 33, 35, 36, 47, 57, 155, 156, 236, 238
High performance, 12, 18, 74, 172, 220
HLE, 255

I

ILP, 42–44, 82, 83, 86, 92, 95, 100, 101, 115, 184, 231, 263
IMEM, 42, 45, 95, 99, 119, 120, 146, 166, 181, 185, 193, 228, 230, 265, 269, 271, 275
Implementation, 232, 236
Inner modem, 27, 207
Instance, 4, 8, 13, 26, 32, 35, 36, 53, 71, 72, 75, 77, 82, 84, 92, 107–109, 114, 116, 118, 121, 123, 128, 131, 141, 144, 148, 157, 159, 162, 179, 181, 187, 194, 200, 201, 204, 208, 210, 223, 236, 239, 241, 248, 255, 256
Interconnect, 6, 43, 93, 94, 103, 114, 117, 123, 157, 166, 206, 270

L

Layout, 9, 10, 13, 14, 18, 19, 36, 49, 51, 76, 77, 108, 118, 126–128, 131, 197, 209–212, 282
Loop buffer, 17, 45, 56, 96, 99, 184, 238, 271
Low energy, 11, 54, 55, 70
Low power, 221, 284
LTE, 18, 74, 118, 124, 137, 141
LUTU, 103, 228, 230, 232, 269, 270

M

Mapping, 25, 33, 36, 46, 276, 282
Matched filter, 173, 174
MCO, 80, 82, 86, 91, 114, 144, 227
MIMO detector, 57, 114, 137, 138, 140, 143, 157, 167, 256, 276, 282
MU, 82, 93, 103, 113, 115
MVON, 114, 144, 151–154, 163, 179, 227, 266

N

NRE cost, 5, 12, 52, 110, 138, 172, 181, 220, 245

O

OFDM, 2, 27, 172, 219, 222, 223, 241
Outer modem, 27, 106, 172, 207

P
Performance, 1, 4–6, 8, 12, 15
Physical layer, 25, 27, 28, 33, 56
Platform, 69, 70, 72, 74, 83, 109, 120, 122, 256
Portability, 3, 4
Processor, 4, 6–8, 12, 13, 16
Productivity gap, 9, 10, 12
Programmability, 52

Q
Quantization, 38, 80, 89, 178, 180, 201, 202, 223, 226

R
RCU, 260
Research challenges, 6, 283
Reusability, 1, 12, 16, 29, 32, 43, 52, 56, 57, 70, 71, 74, 109, 123, 127, 153, 208, 220, 279
Run-time, 14–16, 37, 74, 85, 86, 107, 115, 119, 121, 141, 157, 241, 266, 280

S
Scalability, 9, 18, 37, 71, 75, 76, 107, 109, 112, 121, 124, 200, 240, 280
Schedule, 151, 176, 197, 246, 262
SDR, 1, 6, 10, 12, 15, 53, 163, 204, 280
Semi-custom, 13, 47, 49, 51, 57, 124, 125, 131
Shuffler, 18, 90, 101, 102, 105, 123, 153, 192, 232, 233, 240
SIMD, 12, 17, 241
Slice, 77, 93, 100–103, 149, 151, 185
Software mapping, 36, 46, 47, 155, 194, 219
Software SIMD, 19, 177, 279, 282
Software SIMD as alternative to Hardware SIMD:, 123
Spectrum usage, 2, 27, 56
Sphere detection, 140
SSFE, 101–103, 105, 141, 142
SSU, 18, 90, 94, 101–103, 105, 113, 117
Stride, 226, 232, 234, 236
System scenario, 15, 16, 248, 283

T
Target compiler, 283
Technology scaling, 1, 4–6, 9, 11, 16, 39, 281
Technology scaling friendly, 74, 75, 121, 173
Template, 12–14, 16, 18, 19, 31, 32, 53, 56, 69–72, 74, 75, 79, 80, 83, 87, 94, 109, 111, 116, 122, 128, 131, 137, 206, 212
Trends, 1, 107, 207
TSO, 81, 92, 227, 266
Twiddle factor, 224, 230, 234

V
VLIW, 42, 43, 53, 164
VWR, 18, 93–95, 97, 99–102, 107, 116, 117, 119, 123, 146, 183–185, 195, 230, 247, 264, 269, 270

W
Wire, 8, 9, 13, 19, 43, 48
Wireless communication, 2, 14, 25–27, 29, 69, 123
WLAN, 2, 18, 73, 124, 171, 219, 220, 280
WPAN, 2
WSU, 93, 101–103, 105, 113, 117, 118, 231, 233, 234, 245, 270

Printed by Printforce, the Netherlands